Print, Web & App

编辑设计人该会的基本功一次到位

EDITORIAL DESIGN: DIGITAL AND PRINT

[英] 凯丝·柯德威尔 [英] 约兰达·萨帕特拉 著 徐淑欣 译
Cath Caldwell Yolanda Zappaterra

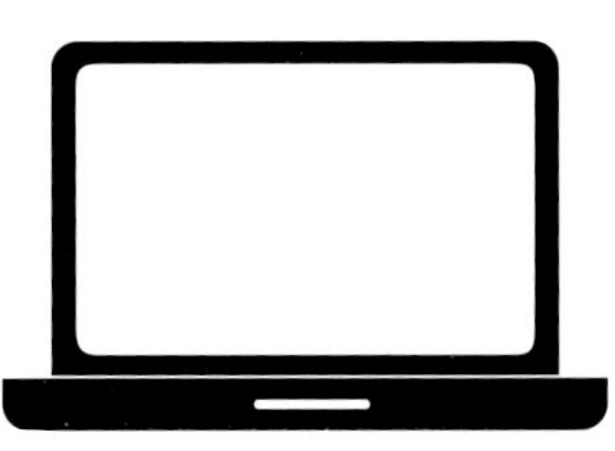

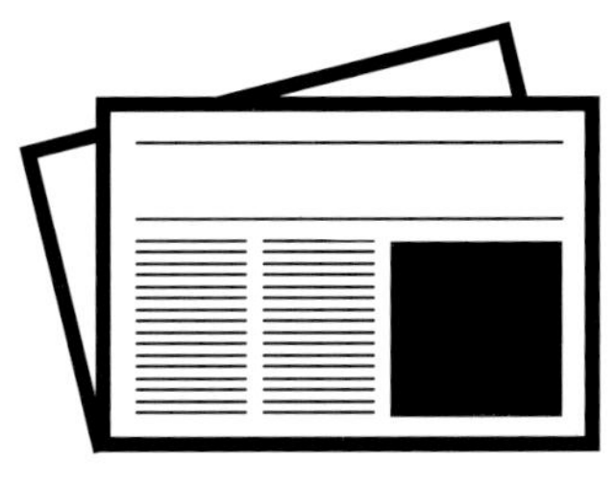

華中科技大學出版社
http://www.hustp.com
中国·武汉

图书在版编目(CIP)数据

Print,Web&App：编辑设计人该会的基本功一次到位/(英) 凯丝 · 柯德威尔，(英) 约兰达 · 萨帕特拉著；徐淑欣译. - 武汉 ：华中科技大学出版社，2020.9
ISBN 978-7-5680-6400-2

Ⅰ. ①P… Ⅱ. ①凯… ②约… ③徐… Ⅲ. ①移动终端-应用程序-程序设计 Ⅳ. ①TN929.53

中国版本图书馆CIP数据核字(2020)第131833号

Print, Web&App：编辑设计人该会的基本功一次到位

PRINT, WEB&APP : BIANJI SHEJIREN GAIHUI DE JIBENGONG YICI DAOWEI

[英] 凯丝 · 柯德威尔
[英] 约兰达 · 萨帕特拉 著
徐淑欣 译

出版发行：华中科技大学出版社（中国 · 武汉） 电话：（027）81321913
武汉市东湖新技术开发区华工科技园 邮编：430223
出 版 人：阮海洪

责任编辑：易彩萍 责任监印：朱 玢
责任校对：周怡露 美术编辑：张 靖

印 刷：武汉精一佳印刷有限公司
开 本：889 mm × 1194 mm 1/16
印 张：14.75
字 数：388千字
版 次：2020年9月第1版第1次印刷
定 价：88.00元

前言

本书旨在通过各种现代化纸质杂志与电子杂志等案例，指导读者如何将“文字编排”与“图片制作”相结合，展现新闻类版面设计的魅力。本书将帮助读者在已有的设计实践基础上，进一步打下坚实的知识基础。书中提供了丰富的参考案例以及实践指导，以启发读者。简言之，本书要告诉读者在纸质与电子版面中如何安排图片与文字，使之相映生辉，碰撞出不一样的火花。

2010年，平板电脑问世之后，人们对纸媒何去何从争论不休。然而事实证明，平面媒体的创意工作并未受到影响，并且关注的重心逐步向创意编辑设计回归。如今，有关纸质阅读与数字阅读的争论已经告一段落，编辑设计的黄金时代正在来临。纸质媒体与社交媒体、宣传活动、移动媒体产品等相互整合，形成了一个完整的生态体系，创造了更多发扬视觉传达设计的好机会。设计师必须熟悉字体、艺术设计以及版面编排的法则，才能好好把握住这些新生的设计需求。虽然要掌握新的设计需求，但先别急着把设计史的书籍都扔掉，设计发展的每一步都很重要，作为编辑设计师需要能够平衡过去、现在和未来。

本书第一章讲述编辑设计的概念与历史，第二章到第四章对不同的编辑设计形式进行概述；第五章、第六章侧重培养读者的设计能力；第七章为读者展现一系列历久弥新的经典之作。第二章至第六章结尾设有“工作坊”版块，为读者提供设计系学生实际操作的设计案例，读者可以从中获得启发，并设计自己的作品。

感谢珍妮特·佛罗里克（Janet Froelich）、杰里米·莱斯利（Jeremy Leslie）、马克·波特（Mark Porter）以及西蒙·艾斯特森（Simon Esterson）对我的启发，他们一直关注编辑设计领域的最新发展，不断为业界注入新的活力。我们需要在过去的兴趣与经验的基础上，学习利用新的技术与平台。编辑设计已经进入了新纪元，请在关注技术发展的同时，仔细阅读本书，它可以让你做好准备，迎接未来。

——凯斯·柯德威尔（Cath Caldwell）

目 录 Contents

7 第一章 编辑设计

23 第二章 编辑设计格式

41 第三章 封面

77 第四章 刊物内容

109 第五章 制作版面

153 第六章 编辑设计师的必备技能

205 第七章 回顾过去，展望未来

234 附录 纸质印刷演变史

236 作者致谢

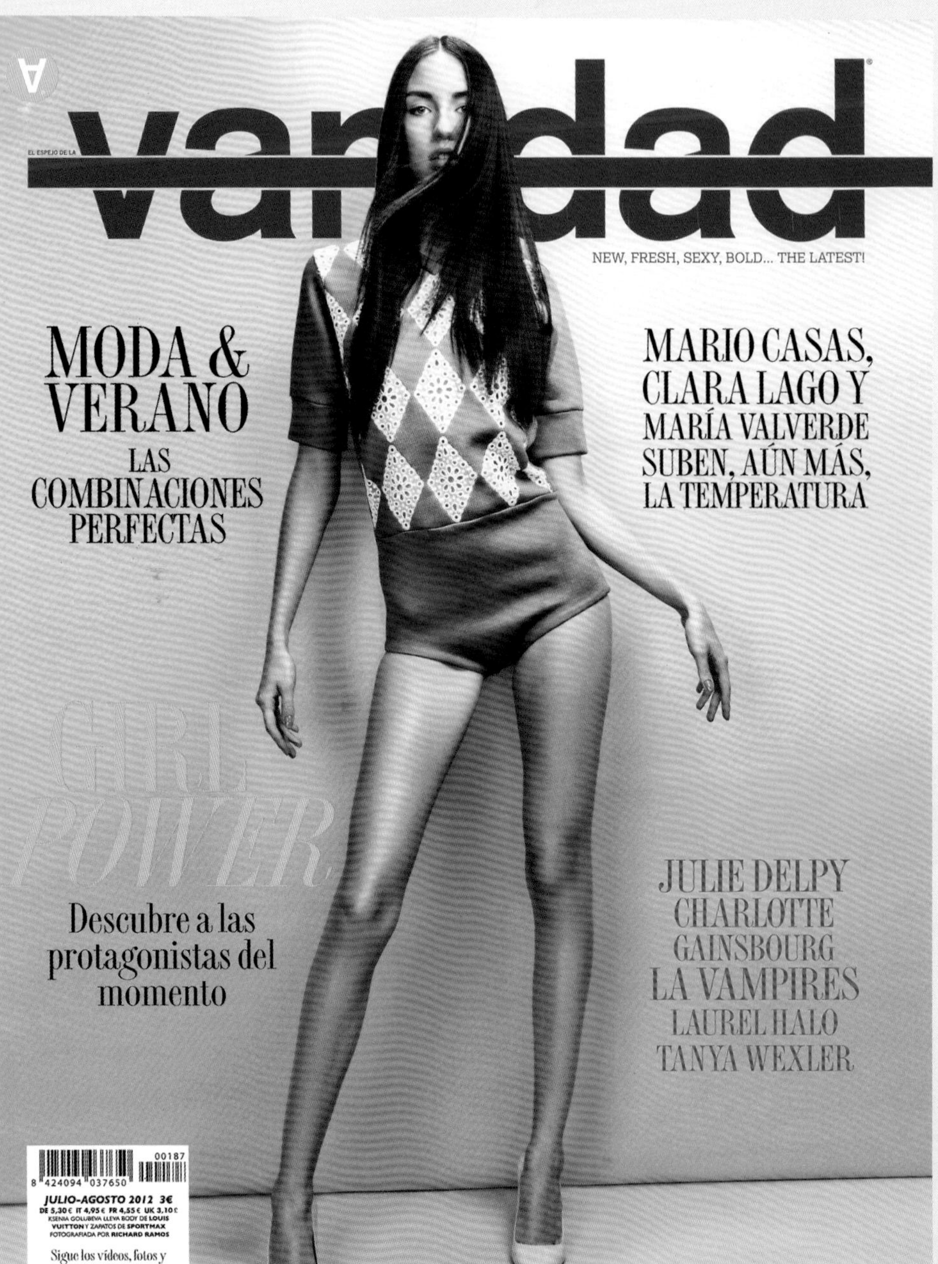
EL ESPEJO DE LA
vanidad
NEW, FRESH, SEXY, BOLD... THE LATEST!
MODA & VERANO
LAS COMBINACIONES PERFECTAS
MARIO CASAS, CLARA LAGO Y MARÍA VALVERDE SUBEN, AÚN MÁS, LA TEMPERATURA
GIRL POWER
Descubre a las protagonistas del momento
JULIE DELPY
CHARLOTTE GAINSBOURG
LA VAMPIRES
LAUREL HALO
TANYA WEXLER
8 424094 037650 00187
JULIO-AGOSTO 2012 3€
DE 5,30€ IT 4,95€ FR 4,55€ UK 3,10£
KSENIA GOLUBEVA LLEVA BODY DE LOUIS VUITTON Y ZAPATOS DE SPORTMAX
FOTOGRAFIADA POR RICHARD RAMOS
Sigue los vídeos, fotos y entrevistas en
www.vanidad.es

第一章　编辑设计

本书是一部面向纸质出版物与电子出版物的编辑设计指南。本书将编辑设计历史与当下实践贯通，逐一解读设计基本原则，以启发读者。“编辑设计”的英文为“editorial design”,其中“editorial”一词原指“社论”，即在特定时间就特定主题发表特定观点的文章。编辑设计，则指为要讲述的故事进行全面设计与策划，并通过设计安排好的出版物来分享观点、传递信息，甚至进行品牌传播。如今，编辑设计不再局限于纸张，越来越多的移动媒体也需要编辑设计。新老设计师都一致认同，无论传播介质如何改变，良好的沟通技巧与讲故事的热情依然是编辑设计最重要的能力。

让我们先从两个最基本的问题开始：什么是编辑设计？设计师在编辑设计中会扮演哪些不同角色？

什么是编辑设计？

想要学好编辑设计，首先要明确何为编辑设计，它与其他类型的设计有何不同？简单来说，编辑设计是具有新闻性或主题性的视觉传达设计，与其他类型的设计相比较，更侧重通过视觉策划与设计来传递新闻性、主题性的故事与内容。编辑设计的出版物对象，往往具有娱乐、传递信息、指导、沟通、教育等功能中的一项甚至多项，而且，往往需要在一个大的主题思想方向下，传递多种多样的观点与意见，报纸便是如此。如今，随着数字技术的发展，出版物可以实现交互，直接与读者沟通，接收读者的反馈。随着GPS（全球定位系统）等技术的加入，编辑和广告商可以有更多机会直接与读者互动。但万变不离其宗，从纸质出版物、网页出版到移动设备，无论介质如何发生改变，编辑设计的基本原则依然相同。接下来，本书将为你一一解读这些编辑设计领域的“金科玉律”。

!

编辑设计就是出版物的设计，例如怎么设计长期持续发行的纸质杂志——通常拥有独特的外观与质感。

——文斯·佛洛斯特，*Zembla* 杂志艺术总监

编辑设计的目标与要素

大多数期刊编辑的核心要义，是通过对文字与图片的组织呈现来传递思想、讲述故事。文字与图片在编辑设计中发挥着不同作用。在纸质杂志中，标题通常用来吸引读者注意，而图片则用来说明或者佐证正文观点与内容。在数字杂志中，标题、图像等元素则充当导航链接，吸引读者点击并阅读。

这些版面均以单一摄影作品作为主视觉元素，经过简单的裁剪，排满版面。设计师有意为之，展示了如何通过不同的方式向读者传递内容，与读者对话。

36
COUTURE
raphique

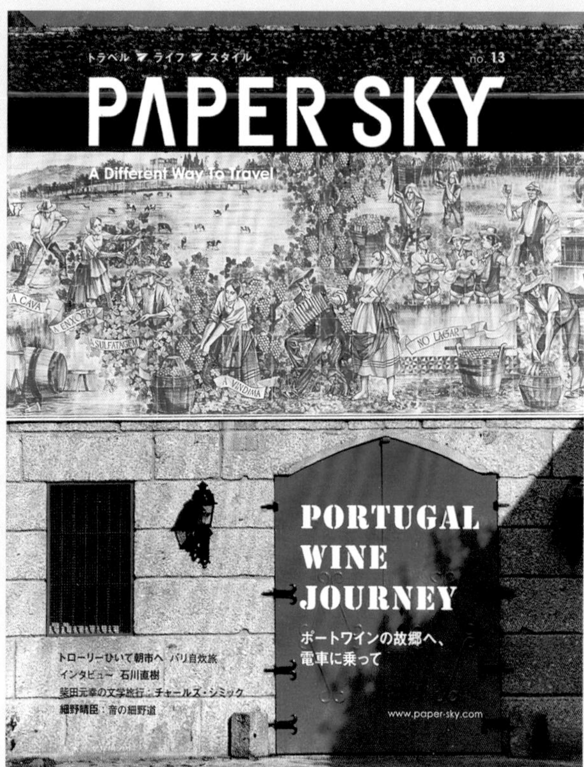
トラベル ライフ スタイル
no. 13
PAPER SKY
A Different Way to Travel
PORTUGAL
WINE
JOURNEY
ポートワインの故郷へ、
電車に乗って
トローリーひいて朝市へ パリ自炊旅
インタビュー 石川直樹
柴田元幸の文学旅行 チャールズ・シミック
細野晴臣：音の細野道
www.paper-sky.com

Time Out
London's weekly guide
August 11-18 1993 No.1199 £1.40
Essential Edinburgh
Late nights, northern lights
and wild sights in festival city
mr don and mr george
'Absolutely' hilarious
Steve Reich
Music from 'The Cave'
Complete
8-day
TV
guide
SOME LIKE IT
COOL
Why is Björk so hot?

Weekend FT. 26/02/00
the business
Who owns
Nazi art loot?
The British art world
confronts the past
Big brother
Noel Gallagher,
£20m Oasis star
City chic
Sir Terence Conran's
Great Eastern Hotel
FRAGILE
RETURN TO
RIGHTFUL OWNER
VALUABLE PAINTING

与图书和其他纸质出版物不同，杂志可以不断演化，每一期都可以不同，这也正是其迷人之处。

——杰瑞米·莱斯利（Jeremy Leslie） *magCulture*创意总监

编辑设计的功能

编辑设计有诸多功能，例如赋予内容更多的表现力与个性；吸引读者的注意并维系读者；使内容信息条理分明等。当然，这些功能要同时存在并相互协作，才能做出有趣、实用的优秀刊物。无论是为纸质期刊还是数字出版物做设计，都是一件令人兴奋的事情，因为可以不断尝试与创新，并带动其他视觉传达领域的设计效仿。

编辑设计是一个整体框架，在这个框架中，报道文章被阅读与诠释。编辑设计包含出版物的整体架构（指逻辑结构），以及对个别报道的特殊处理（因为个别报道可能曲解甚至违背出版物的整体调性）。

——马丁·威尼茨基 （Martin Venezky）*Speak*杂志艺术总监

此外，编辑设计也有其他功能，即生动形象、简明扼要地反映出所处时代的文化。例如，20世纪60年代的*Oz*与*Nova*杂志，不但拥有那个时代的活泼生气，还反映了注重实验、创新探索新方向的时代精神。

*Wallpaper**杂志的幕后制作

一家全球性出版品牌的纸质与数字出版探索之路。

“在图片拍摄时期，就要考虑到内容在不同载体上的组织与呈现方式。在纸质杂志上如何组织？在平板电脑上如何呈现？网页版又会有何不同？内容在不同载体上不能一模一样，而是要根据载体属性差异，相辅相成。”

——莎拉·道格拉斯（Sarah Douglas），*Wallpaper**设计总监

1996年，*Wallpaper*杂志由泰勒·布鲁雷（Tyler Br.lé）创刊，2004年，杂志推出网页版，每月的独立访问用户达500万。2007年，在其第100期杂志刊出时，其Logo上增加了一个星号（*），暗示其未来的数字化趋势。2010年，杂志推出平板电脑版，之后与纸质版一样按月发行。*Wallpaper**杂志内容涉及工业设计、室内设计、艺术、建筑、旅行、时尚与科技等领域。

*Wallpaper**常将合作关系拓展到时尚、建筑与设计领域。现今的编辑团队由托尼·钱伯斯（Tony Chambers）带领，他是为数不多能够从艺术总监转型成为编辑的奇才。他的团队与出版商关系密切，对出版保持开放态度，同时也会与广告商进行创意项目的合作。

在接下来的访谈中，莎拉·道格拉斯（设计总监）与麦里昂·普利查德（Meirion Pritchard）（前艺术总监），将向我们娓娓道来，*Wallpaper**是如何将品牌拓展至数字产品、设计活动、策展甚至房地产领域的。

问：在不同产品形式之间，您是如何控制安排各种设计元素的？

麦里昂·普利查德：最初引入网页版时，设计师与编辑们确实有一段时间有点儿措手不及。我们被告知这也不能做，那也不能做。但是在轮到平板电脑和手机版时，我们又重新掌握了主动权。现在一切尽在掌控之中，数字屏幕的改善带来了更好的阅读体验。

问：你们都使用哪些数字工具呢？

莎拉·道格拉斯：一切数字化形式都是为内容服务的，我们不会为了形式而形式。例如，在纸质杂志的建筑专题里，由于版面限制，可能只能展示几张图片。但是在平板电脑版上，我们已经可以展示更多图片，还可以加入建筑平面图。我们可以告诉读者，照片是从哪个角度拍摄的。这些可以帮助读者更好地理解建筑。同时，通过网站，读者可以在建筑中徜徉。

问：请问您的灵感从何而来？

麦里昂·普利查德：不要坐在办公桌前，而是要多出去走走。即使是在骑自行车或者散步时，也可以从不同视角看待事物，获得灵感。

莎拉·道格拉斯：我们需要其他文化产品的启发。在灵感方面，我很喜欢看安德鲁·迪普罗斯设计，担任编辑的《骑士》（*The Ride*）杂志。他在《连线》（*Wired*）杂志做得也很好，看得出他非常热爱杂志，所以有了这么棒的一份刊物。

BEIJING
The Chinese capital, past and present, page 181
SINGAPORE
Conversations and collages on the city-state, page 186
BANGKOK
Thailand's perpetual state of reinvention, page 191
ARCHIPELAGO CINEMA
The floating cinema heads to Venice, page 196
FUTURE
Douglas Coupland's snapshots from the future, page 198

OLE SCHEEREN

EDITORIAL TEAM Ole Scheeren, Dan Cheong, Eric Chang, Jasmin Dalle, Ken Burke, Kevin Ou

CHINA SPEED

Two weeks, 20 pages... *the only way to look at this was to simply declare it yet another exercise in making things happen at China speed.* Instead of filling the pages with views and statements of our own, I wanted to bring a group of people together to share their disparate views on the context we live in and on the work we do in these contexts. A group loosely centered around *three places, three cities*: Beijing – Singapore – Bangkok, and a series of individual (and unedited) perspectives as diverse as the conditions we find ourselves confronted with in these environments.

With his scenarios of the grotesque and absurd, **Wang Qingsong** seemed the perfect reflection on the state of architecture (and the physical and mental space of our office). **Cao Fei** has long intrigued me with her depiction of the new China through scenes of fantastic perpetual construction and a virtual world in which everything seems possible (albeit with the avatars of architecture threatening to erase the planet).

A judge in the *China's Got Talent* TV show and with a following of more than 5.5 million on his Weibo account (the Chinese equivalent to Twitter), **Gao Xiaosong** broadcasts a witty perspective on Beijing's CBD and the demons within (no account is given of his recent six-month detention which more than doubled his following). **Liu Heung Shing** offers his wealth of historical knowledge and long-term perspective on Beijing, having been the first accredited photojournalist for *Time* magazine in Beijing in the 1980s and awarded the Pulitzer Prize.

A conversation with Singapore's eminence grise of late modern architecture **William Lim** portrays a character that at the age of 80 is still one of Asia's most outspoken and socially motivated architectural protagonists, while **Valerie Stahl von Stromberg**'s photo-collages merge tropical reality with architectural ambition and fantasy.

Bangkok in flames... *no, it's a dance party that declares a revolution for fun* (**Surasi Kusolwong** says his friend suggested to just introduce him simply *as a fucking good artist and not say anything else*). To me, he remains one of the most political and subversive artists of his generation, with a particular psychology and sensibility. **Ann Phatarawadee Phataranawik** has been a journalist for *The Nation* for 20 years and was a complicit (media) member of the 'Cities on the Move' project in the late 1990s.

And finally, a global view from one of the writers I have admired from the beginning of my twenties as someone who so effortlessly characterised the whole notion of my generation: **Douglas Coupland**. Master of apocalyptic narratives, he has written *eight letters from the near future* that offer snapshots of our buildings as pieces of collective memory and as refuge for disaster struck victims of a global meltdown.

Before letting loose this group of creative minds on the following pages, I felt compelled to give the first word not to any specific creative individual, but to the guys without which our architectural musings would all remain bare theory. Written on a concrete column of the CCTV building, we discovered this poem (pictured right) one morning on our regular site visits...

Leaning tower, leaning tower,
You are famous in the world,
you are the greatness of China,
our migrant workers
come together to you,
from all over China,
you are the pride of ours.
No matter how high you are,
we will fight with you.
Our migrant workers are great,
we are everywhere.
It's us to build this country
a beautiful stage.

Photography: Dai Chang

OLE SCHEEREN
The German-born, China-based architect with a model of his Angkasa Raya skyscraper, now under construction in Kuala Lumpur, Malaysia
Portrait: Wai Yuan Ou
Writer: Jonathan Bell

Building the future

After a decade spent working on a variety of major projects across Asia – many of which will reach completion in the next couple of years – German architect Ole Scheeren is ready for a new challenge: Europe

When we last caught up with Ole Scheeren, (W*123), the architect was on the cusp of leaving the cradle of OMA – not that anyone suspected this at the time. The Dutch studio's landmark Chinese project, the CCTV Tower, was substantially complete and Scheeren's name was closely associated with some of OMA's most eagerly awaited projects, including the MahaNakhon tower in Bangkok and the Taipei Performing Arts Centre. As creative divorces go, the split didn't make headlines with outrageous claims for alimony or allegations of stylistic infidelities. Instead, Scheeren quietly slipped its moorings, to continue to work on some of these well-advanced megastructures and quickly acquire a few other substantial projects.

Two and a half years later, and Scheeren has effortlessly assumed the mantle of studio leader, with a new letterhead, an endless supply of crisp white shirts and a globe-spanning portfolio (and itinerary) that threatens to eclipse the schedule of his mentor. But where OMA is explicitly a globalised firm, deploying rhetoric and research in the name of new architectural paradigms, Büro Ole Scheeren has emerged out of Asia, primed to make the leap from East to West. This initial focus is unsurprising, given Scheeren's extensive experience of living and working in the region, but it has shaped the line-up of projects that formally launch the studio. These range from the racked and stacked blocks of The Interlace apartments and the side-hung Scotts Tower in Singapore, the greebled façade of the MahaNakhon in Bangkok, to the Angkasa Raya tower in Kuala Lumpur, sited up against the Moorish detail of Cesar Pelli's Petronas Towers, but a thousand miles apart in conception and delivery.

Scheeren's work resists categorisation. For the architect, variety is a crucial part of his studio's identity. 'It was important to me to open the practice with a range of different projects – it's the range that makes things so interesting,' he says, from a site visit in Bangkok, en route to his Beijing studio, via an earlier stopover in Singapore. Like many exponents of the new avant-garde, Scheeren is unafraid of the visual language of high modernism; indeed, many of the buildings developed by Büro-OS resemble the International Style on steroids, with their fragmented façades, casual structural gymnastics and juxtaposition of glassy geometric simplicity with rifts, ripples and shears.

Büro-OS has had to navigate the valley of expectations between rendered vision and built reality with more delicacy than most contemporary studios. CCTV has been on the design radar for a decade, with the first images emerging in late 2002. Although the building appeared to be complete by the 2008 Beijing Olympics, in truth the tenants, China's state TV station, have only just moved in after a complex fit-out. Scheeren is playing a long game and it's only now, two years after the studio was set up, that he is sharing projects as they approach completion. The transition from OMA to Büro-OS has resulted in a more explicit definition of authorship for many of the projects in his portfolio. Among the first to be won under his »

STUDIO SX
Model for a studio and gallery space for a painter and antiques collector in Beijing
Ole Scheeren © Büro-OS

THE INTERLACE
A large-scale housing development nearing completion in Singapore, The Interlace will feature 31 apartment blocks, landscaped roof terraces and pools
Ole Scheeren © OMA
Photography: Darren Soh

ANGKASA RAYA
Left, located next to Kuala Lumpur's Petronas Towers, this 268m-high skyscraper will consist of a stack of balancing volumes and should be completed by 2016
Ole Scheeren © Büro-OS

MAHANAKHON
Below, this 77-storey multi-use complex is due to become Bangkok's tallest skyscraper by 2014
Ole Scheeren © OMA

own name was Angkasa Raya. According to Scheeren, 'it marked a relatively significant step' in the evolution of the studio.

With three major projects on site, the studio also has smaller projects in the works, most notably a studio and gallery space for a leading, but publicity-shy, Chinese artist. 'It's a very personal project,' Scheeren explains. 'It's his very specific world meeting my own world – a house that places artistic space alongside a context of literal Chinese-ness.' The artist is a compulsive collector, not just of art. The entire context of the new structure will be assembled from his personal stash of traditional Chinese structures and landscape elements, creating a highly concentrated experience of the country's long tradition of art, design and landscape.

On a much larger scale, plans for a new Beijing auction house are nearing completion. Designed for the country's leading home-grown auctioneers, the project occupies a prime site off the commercial and cultural axes that run close to the Forbidden City. Both of these projects are in the final design stage, while the Bangkok tower is now taking shape after a spell of inaction, and The Interlace is rising fast above the low hills in the west of Singapore. 'Many of these buildings have an explicitly tropical context,' says Scheeren. 'The question is how you build in a place where nature is so dominant and beautiful, and where people live very differently to other climate zones.'

On the page, this scattering of projects, from high to low rise, small to large, appears rather disparate and disjointed, but what unites them is Scheeren's concerns about architecture's broader social role. Although Asia has been his focus for many years – Scheeren has lived around the region for nearly a decade, ever since relocating with OMA – the typologies that emerge from Büro-OS are stylistically disparate but focused, laser-like, on their immediate context. 'How can a building be relevant beyond its occupants?' the architect asks, and it's clear that this social concern is the key bridge that links his work to the European modernism of three quarters of a century ago, rather than any stylistic concerns. 'There's a sense of psychology at work: how do people live, how do they behave in these buildings?' he asks. 'A number of people in the studio also have degrees in history, sociology, psychology. We try to really bring local knowledge and perspectives to our projects.'

There is an overarching sense of anthropological exploration at work in the Büro-OS portfolio. As Scheeren himself acknowledges, 'We're at the stage where I still have the luxury of being personally engaged with projects. I enjoy spending time in locations and getting a real feel for them. It's really a goal to keep the practice at a scale where I can continue to do that.' Furthermore, he is all too aware of the pitfalls – both in terms of presentation and in approach – of digital architecture. 'There's a relatively big rift between my generation and what came after. I grew up in an analogue world. I learned to draw, to build models and to build houses by hand, I worked on a construction site for a long time... There is something very real about architecture, it's not all digital. But China is more manual; the West has turned very digital in the absence of built realities,' he says. Nowhere is this more evident than in the construction of CCTV. 'The way it was designed clearly took advantage of local conditions, labour and material costs. We also designed it entirely out of locally sourced Chinese materials and steel grades, which all had an impact, technologically and aesthetically. But the building clearly exceeded these qualifications.'

Maintaining a portfolio that runs from skyscrapers to studios has its benefits. 'You can engage in very specific issues of craft, how things come together. They're also about individual psychologies, not communal psychology. You can react very precisely to an individual, their tastes and fantasies,' Scheeren says. The Büro-OS Archipelago Cinema is another modest but high-profile work. First seen in a cove on the Thai island of Koh Yao Noi, it's currently being reconstructed for the 13th Venice Architecture Biennale, where it will host the premiere of *Against All Rules*, Horst Brandenburg's documentary about Scheeren's past six years in practice. 'The cinema is less of an architectural project, more an intervention in the landscape. It is a narrative about creativity,' says Scheeren.

The architect frequently returns to this idea of an architectural narrative, implying his buildings are about facilitating the way people live, rather than shaping them. 'It would be foolish to claim not to still be influenced by European modernists in some way,' he says, 'but I don't want to sound nostalgic or retro when I talk about these things. It's about ambition, about the impact on our society, rather than a formal concern.' In the next few years, the seeds currently being sown in the fertile urban landscape of Thailand, Singapore, Malaysia and China will grow into a more global approach. 'After basically a decade in Asia as an onlooker, I am now acutely complicit with this part of the world,' Scheeren says, 'but it feels like the right moment to orient myself back to Europe and re-engage with the West.'

CCTV
Below, Beijing's landmark 54-storey CCTV HQ tower was completed in May 2012
Rem Koolhaas and Ole Scheeren © OMA

CCTV
Top right, the main lobby. Above, the forum and event space at the top of the building
Photography: Shu He

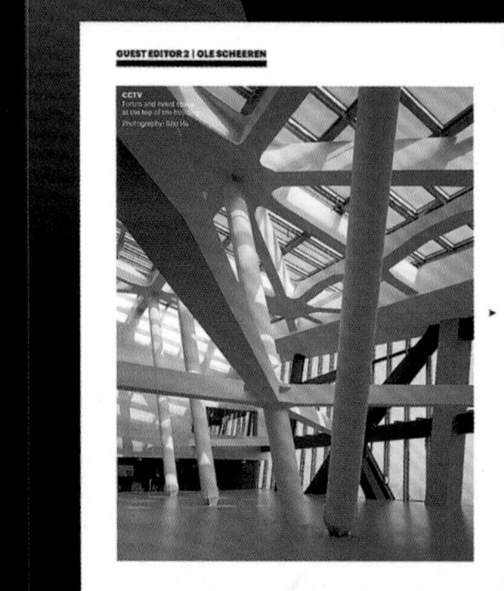

问：作为设计师，您在设计平板电脑版时，会着重考虑哪些因素？

莎拉·道格拉斯：需要重新构思。要结合平板电脑的特点，考虑读者在电子屏幕上的阅读差异。我们需要从读者阅读的视角来思考，结合读者的使用方式，重新进行设计。

问：请问杂志会有相关的数据反馈吗？

麦里昂·普利查德：通过推特粉丝的反馈，我们会收到大量反馈数据，可以知道哪些内容受欢迎。我们杂志的推特粉丝有50万，对于杂志来说很不错了。网页版也很受欢迎，并且最近有读者要求我们提供中文版。之前，已经在一个系列中，发行过一期《中国制造》。这很重要，因为虽然我们的总部在伦敦，但是通过互联网，*Wallpaper** 可以走得更远，现在杂志的发行量已经达到了每期15万册。

问：如何与广告商合作？

莎拉·道格拉斯：我们与*Wallpaper**的专门团队一起和广告商会谈讨论，编辑会沟通各自的需求与职责。之后，按照讨论内容进行合作，合作产品包括网站、活动或者拍摄等。

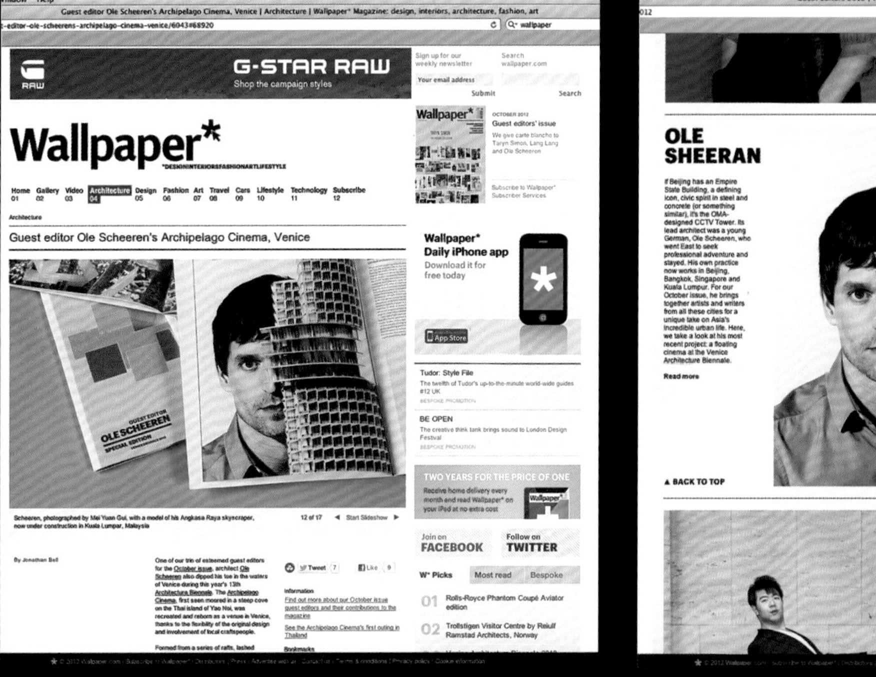

印刷版和数字版的大标题都有下划线。设计者注意使用最右手边的“蛞蝓”或标题来表示读者当前正在阅读的部分。

问：接下来有何安排？杂志会开拓更多的平台吗？

莎拉·道格拉斯：可开发之处颇多。*Wallpaper** 每年8月份有一期手工刊，对当代设计与工艺进行集中展示。艺术家们会拿自己的作品投稿，用于展示。借此，我们已经将品牌拓展至展会活动。Wallpaper* Composed项目应运而生。The Apartment（公寓）是另外一项品牌拓展。公寓可以用于举办活动，拍摄照片并向客户展示。这是一项风险投资，不过开发商已经在考虑如何与我们进行创意合作了。此外，它还可以成为我们的摄影师进行拍摄、休息之地。

问：你们要拍很多照片。在不同数字平台上发布，例如网站和平板电脑，会影响摄影师的拍摄方式吗？

莎拉·道格拉斯：还在适应中。就像从胶片时代转向数字摄影需要一个缓慢的过程一样，我们现在也处于这样的转型过程中。

问：未来静态摄影作品会加入动态图像元素吗？

莎拉·道格拉斯：会的，可能80%都会。

问：*Wallpaper 还有什么其他品牌拓展吗？**

*Wallpaper**旅行指南已经转为数字App应用，并且由另一家出版商运营。“Wallpaper*精选”（Wallpaper* Selects）这一品牌，则与艺术零售商Eyestorm合作，用于销售精选的当代摄影作品。“Wallpaper*设计大奖”则侧重发掘新兴设计才俊，并保持在国际当代设计领域的品牌定位。

这幅图片展示了伦敦中心圣马丁学院设计专业的学生们正在就运用本书设计原则设计的作品进行讨论。杂志设计需要协同合作，学生们正在以读者视角反观自己设计的图文作品，并听取他人的建议。

设计师在编辑团队中的不同角色

设计师与编辑的合作是编辑设计成功的关键，不过设计师与其他团队人员的配合也非常重要。设计师是除了编辑之外，每天与其他团队人员打交道最多的人。

编辑团队的主要构成

因出版物类型、团队规模、组织架构不同，团队成员的角色也各有差异。杂志编辑可能负责内容组稿，艺术总监、设计总监或总设计师则负责如何将内容进行设计呈现，以体现杂志的风格。

本书将对纸质版与数字版报纸、杂志编辑设计中设计师的不同角色与关系进行阐释。这些因媒介形式、规模、发行周期而异，独立双月刊杂志的人员需求肯定与数字日报的需求大有不同。下面是设计师需要与之密切打交道的人员简介。

主编：对出版物的内容负责，与艺术总监合作紧密。其下属编辑分别为专题编辑、图片编辑以及生产编辑。

艺术总监/美术编辑：负责安排所有文稿内容以及图像的架构和次序。他们必须掌控进度，按照制作经理或制作编辑指定的时间表完成工作，根据需求组织或制作相应图表、图片或插画。艺术总监、美术编辑与创意和制作人员、数字编辑等互动密切。

制作经理/网站编辑：负责制定时间进度表，整合全部材料。制定时间进度表时，需要从发刊日往回推算，来决定什么时候必须收到图文内容、什么时候进行编辑与设计，以及必须送印的日期。制作经理要负责制作落版单，更新落版单变化。他们通常与美编部门以及印刷厂合作密切，不但要负责到印厂看样，还要监督特殊印刷需求。

审稿编辑：负责校对、审阅文稿，确保文稿风格一致，字、句和标点符号没有错误等。需要撰写文案，对不理想的文章内容进行删改甚至重写。与编辑、美术团队、专题编辑或撰稿人合作密切。

图片编辑：主要负责寻找图片并厘清图片版权，必须与艺术总监以及主编合作，确保整个刊物的图片品质，也要和图片代理商、图片库公司以及图片版权公司合作。

设计师：根据艺术总监给出的方针或指示，设计编排刊物版面。设计师与艺术总监的合作模式及其自主程度，取决于多方因素。例如设计师资深程度，艺术总监的工作风格（有人事必躬亲，喜欢把控每一项细节；有人乐于放手，由设计师自由发挥），团队人数，刊物页数，以及出版的前置时间。一般而言，前置时间越短，设计师要负责的事情就越多。

编辑室经理：并非所有刊物都有这一职位，编辑室经理负责项目管理，与制作经理有职能上的重叠。但编辑室经理是很好的协助者，不但能统筹事务，也能处理日常沟通。编辑室经理须保证刊物进度计划按时进行，不同素材能符合页面编排的需求，并摆放正确。

来吸引读者。黑白报纸是20世纪的产物，新世纪的读者不会买账的”。

当时，读者与日报的关系发生变化，英国各大报为迎合读者新的需求，纷纷改为采用小报版式。但《卫报》却另辟蹊径，效仿法国《世界报》（*Le Monde*）的“柏林版式”，走出了自己的路。波特表示这种版式“能方便读者阅读严肃新闻，符合当代的设计手法，同时顾及广告主的需求，没有其他版式能与之相比”。《卫报》的设计一直都很有智慧，且有先见之明。1988年，五角星设计公司（Pentagram）的大卫·希尔曼（David Hillman）开创先河，将报纸版面一分为二，同时推出新的报头，最重要的是将杂志式的留白版面，应用到报纸设计上。

《卫报》经过希尔曼的重新设计，完全改头换面，不仅焕然一新，更拥有了完整的设计理念，或许是第一份拥有具体设计理念的报纸。波特说：“迈克·唛可奈（Mike McNay）、西门·艾斯特森（Simon Esterson）与我，都遵循这些设计理念原则行事。”希尔曼表示，理想的报纸设计“不该仅仅是沿用新闻界的习惯与惯性，应该有完整的设计原则”。2005年的改变也遵循了这些原则，沿着希尔曼的理想前进。

《卫报》除了报纸刊物，还有即时更新的网站与App，并随时发布新动态。波特在2010年离开了《卫报》，他和团队

How dismal England crashed to earth
Mike Selvey in the 28-page sport tabloid

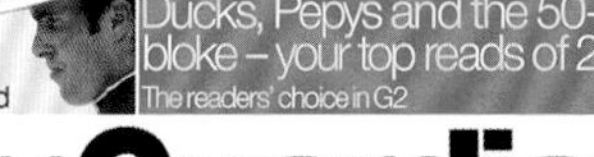

Ducks, Pepys and the 50-quid bloke – your top reads of 2004
The readers' choice in G2

60p
Monday
December 27 2004
Published in London and Manchester
guardian.co.uk

The Guardian

Giant waves kill thousands after calls for warning system ignored

Inside

'It tossed boats around like toys'

Ukraine opposition in big exit poll lead

Hell hath no fury as a dad scorned

Quick Index

的App等纳入考虑。只设计纸质版的时代已经过去，新闻媒体必须同步整合不同平台的需求，并保持品牌视觉识别的一致性。”

《卫报》的网站团队秉持波特勾勒出的设计核心价值，坚信无论何种呈现形式，都需要有均衡的文字编排与强有力的新闻照片，只有这样，才能设计出好的充满智慧的刊物。《卫报》的平板电脑版除了坚守这些原则，也善于运用平板电脑的多媒体交互功能，使其比报纸更有阅读价值。《卫报》平板电脑版不但可以顺畅地滑动浏览，而且具备移动属性，可以通过精密的数据收集软件与GPS定位系统为编辑和广告商提供更多与读者互动的机会。波特表示：“平板电脑带来更多契机，使我们能够将在传统报纸领域积累的知识与技能，再次应用到数字出版领域。平板电脑发展日新月异，市场不断扩大。未来我们可以创作出更多比以前更为优秀的编辑设计作品。电脑端浏览新闻的时代已一去不复返，移动阅读才是未来主流的阅读方式，从而取代家用电脑在阅读新闻方面的地位。纸质书依然会存在，但不再会是主导。”

£0.60
Monday 12.09.05
Published in London and Manchester
guardian.co.uk

the guardian

Backlash over Blair's school revolution

City academy plans condemned by ex-education secretary Morris

Patrick Wintour and Rebecca Smithers

UK link to terror snatches

Ian Cobain and Richard Norton-Taylor

Bad'day mate Aussies lose their grip

Shane Warne at the Oval yesterday. Sport 9 Photograph: Kieran Doherty/Reuters

Column five
The shape of things to come

Alan Rusbridger

Continued on page 2

National
Police chief blames Orangemen for riots

Law
Judges may block deportations

International
Israeli troops leave Gaza after 38 years

Financial
Sky's Premiership rights under threat

Bigger isn't always better...

文字编排的层级，能够显示出报道的重要程度。设计简洁利落的标题，可以使图文在视觉上达到平衡。只要充分利用图片的视觉影响力，没有无谓的噱头或者哗众取宠的图片，只采用简单的导航与设计，就可以为读者提供丰富的阅读体验，并能让读者轻松找到想看的内容。

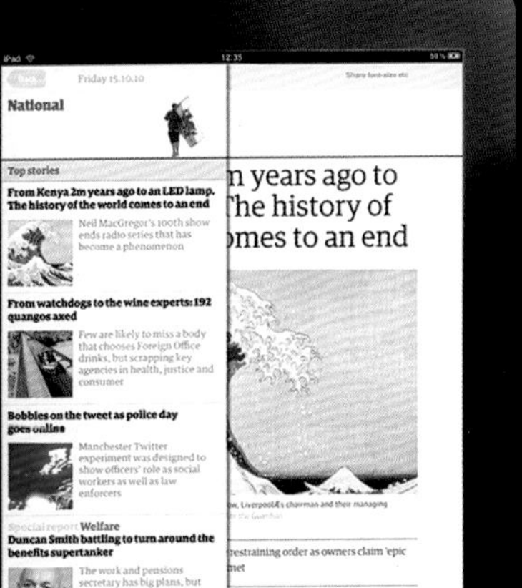

Friday 15.10.10
National
Top stories
From Kenya 2m years ago to an LED lamp. The history of the world comes to an end
Neil MacGregor's 100th show ends radio series that has become a phenomenon
From watchdogs to the wine experts: 192 quangos axed
Few are likely to miss a body that chooses Foreign Office drinks, but scrapping key agencies in health, justice and consumer
Bobbies on the tweet as police day goes online
Manchester Twitter experiment was designed to show officers' role as social workers as well as law enforcers
Welfare
Duncan Smith battling to turn around the benefits supertanker
The work and pensions secretary has big plans, but little room to manoeuvre
n years ago to
The history of
omes to an end
theguardian
12.10.10
Home
News
National
International
Financial
Comment & debate
Reply
Obituaries
Reviews
Weather & Crossword
Sport
theguardian
12.10.10
National
Families 'devastated' after Red Cap trial
Blair wanted to sack Sir Richard Dannatt, says aide
General said deployment in Iraq boosted militants
North Korea's Young General meets his public
Seconds from safety, then the bomb went off: how hostage rescue ended in tragedy
theguardian
Friday 15.10.10 2:35pm
film
music
some text goes here
iPad
9:41 AM
Front page
National
International
Financial
theguardian
Issues
Sections
Settings
International
Murder and torture 'carried out by both sides' of uprising against Libyan regime
David Smith in Tripoli
Greeks born with €35,000 debt
Cyberbullying 'caused boy's gun rampage'
Sichuan quake evacuees relax and draw the crowds at Beijing zoo
Ed Pilkington in Honolulu
Maldivian atolls sinking at rate of three per year
'Blasphemous' Potter film banned in Iran

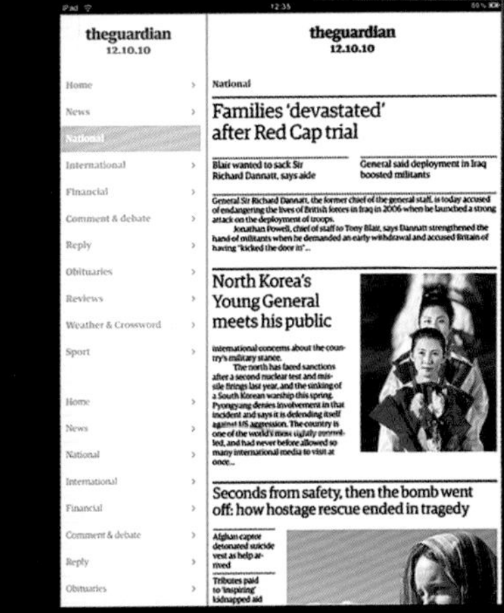

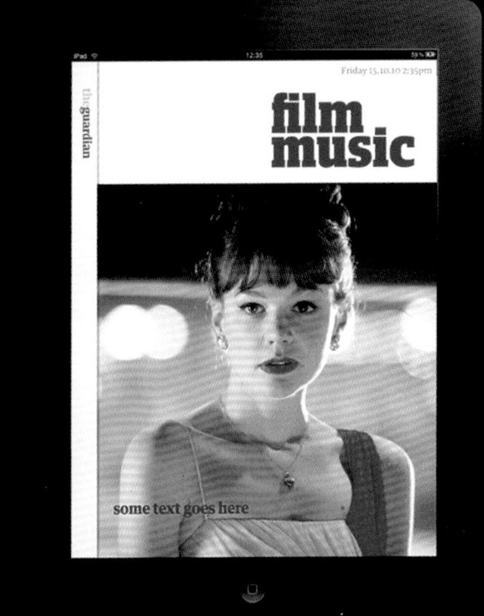

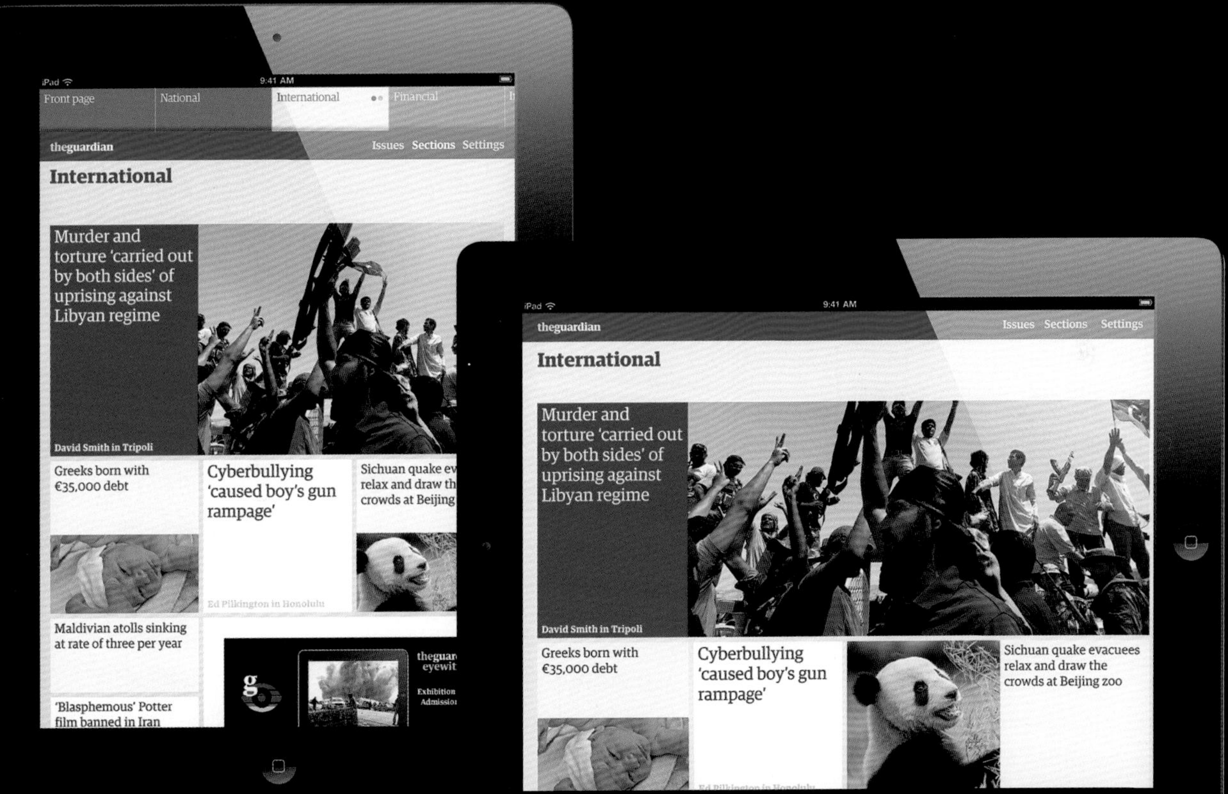

杂志的内容基本上是建立在编辑打破广告的理念之上，对于很多杂志来说，这就是它的全部意义：销售广告。

——文斯·弗罗斯特，美术指导，赞巴拉

编辑设计师需要具备哪些特质？

美国平面设计大师提波·卡尔曼（Tibor Kalman）有句名言："如果设计师发现编辑的工作没做好，就有责任把编辑炒掉。"他的意思是说，编辑设计师应该和编辑一样重视刊物内容，因为编辑设计是在编辑工作基础上进行的。编辑设计师与编辑都必须发挥创意，而他们如何妥善地分工合作，几乎决定了整本刊物的成败。

正如编辑设计师需要懂编辑知识一样，编辑也应该懂一些编辑设计，至少双方需要了解彼此的态度、角色、擅长领域以及工作内容，这样才能相互建立信任，合力打造一流出版物。所有的编辑设计师以及编辑都认同这一点，有些人甚至还有其他技能与背景为工作加分。在担任《卫报》创意总监之前，马克·波特曾担任《色彩》（*Colors*）以及《连线》（*Wired*）杂志的设计师。然而，他却并非科班出身，而是毕业于牛津大学语言学系。波特曾表示，大学教育对自己的设计方式影响深远：

"我会从读者的视角来设计刊物。优秀的编辑设计首先要能够通过编排引起读者的阅读兴趣，其次是讲述故事的技巧与方式。"平面设计在刊物中扮演的角色一般不易被读者所察觉，他们在阅读时看到的只是文字所提供的信息，例如想法、人物、地点等。因为编辑们大多有大学教育背景，所以我的大学教育背景有助于和编辑沟通。报社有很多非常聪明、敏捷的记者，与他们打交道同样也是对我个人智慧的挑战。如果我不能为设计方案提出清楚、有说服力的理由，方案就会被打回重做。除非在海外设计项目，不然语言学的专业在我的工作中并没有很大作用。但是我坚信设计也是一种语言，设计和所有语言一样，本身没有价值，当通过设计很成功地讲述值得诉说的内容时，才真正发挥了自己的价值。"

戴伦·琼斯（Dylan Jones）在担任*GQ*杂志的编辑前，曾是*i-D m*，*The Face*，*Arena and Arena Homme Plus*等诸多杂志的编辑，但其实他是平面设计师出身。威利·弗雷克豪斯（Willy Fleckhaus）是20世纪60年代德国重要杂志*Twen*的艺术总监，在此之前，他是名记者。五角星设计公司合伙人希尔曼曾是*New Statesman and Society*（《新政治任务与社会》）杂志与《卫报》的设计师，此前也担任过*Nova*的艺术总监与副主编。他曾说："艺术总监的主要作用不是简单地进行格线设计，制作精美的报头，甚至不是如何让图文排版更加美观。优秀的艺术总监，需要深刻且全面地理解杂志的宗旨，并且通过设计来影响与引导内容的传达与讲述方式。"

也许你曾听过这些设计大师的事迹，读过他们的作品，并好奇他们的成就从何而来。接下来的三篇访谈，访问了在业界不同层级的设计师，他们将为大家分享自身的工作内容与入行故事。

初级设计师

爱莎·马丁斯娃（Esa Martinesva），*Port* 电子杂志

请问电子杂志的初级设计师或者实习生，他们的工作内容是什么？

平板电脑是新兴产物，大家都在摸索中，因此即使是资深设计师也未必能完全掌握所有细节。身为初级/实习设计师，我主要负责研究读者的阅读方式以及与杂志的互动模式，同时也会辅助排版，思考屏幕上的文字编排与呈现方式。

你的日常工作与纸质版刊物的设计师有何不同？

虽然媒介是新的，但是沟通与互动的基本原则都一样。因此我必须彻底理解文字、图片（动态或静态）与音效的整合方式，也要明白如何避免过多使用非必要特效。

为满足在不同的平台做设计，是不是需要额外学习更多的内容？

许多App都想要尽可能同时具备多种功能，尽管平板电脑的多媒体呈现能力很棒，但是读者最终会喜欢的仍然是简洁细腻的设计。

我不认为设计师有必要彻底了解每种媒介的技术细节，但是必须理解不同媒介的差异以及如何呈现和传播内容。像Adobe Creative Suite这样的设计软件已经可以让设计师不用管代码的事情，而是把重点放在安排按钮与超链接等互动元素上。其实最耗费心思的，是需要了解文字与图像如何在屏幕上呈现，以及这种呈现方式与纸质刊物有何差异。

如何跟上最近技术的发展脚步？

勇于尝试，别无他法。只能在新工具与新的媒介推出时立刻尝试与研究。

你是如何进入杂志设计这一行的？

我的作品集中有很多编辑设计的作品。有位老师发现了我对编辑设计的兴趣，便推荐我入行。于是我进入《Port》杂志，协助设计杂志的平板电脑版。参与这个项目后，公司希望我能继续留下来工作。目前我主要负责设计纸质杂志，同时也负责数字版的部分工作。

在实习时是没有薪水的，是吗？

最初确实没有薪资，因为当时*Port*是新创刊的杂志，主要靠志愿者来运行。这是新创刊的独立杂志的常态，等预算来源稳定后，成员才有钱拿。后来我就有薪水了，现在的实习生也有少许薪水。

PICTURE PERFECT
“You can translate an emotion into something that other people can then read again. And that’s what art should always be about: communicating certain ideas and emotions through work without having to explain them”
Interview Phil Rhys Thomas
Photography Eva Vermandel
Back
View
Next

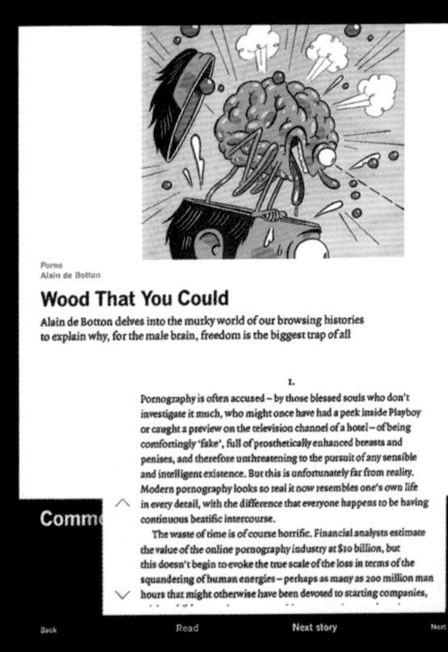
Porno
Alain de Botton

Wood That You Could

Alain de Botton delves into the murky world of our browsing histories to explain why, for the male brain, freedom is the biggest trap of all

1.

Pornography is often accused – by those blessed souls who don’t investigate it much, who might once have had a peek inside Playboy or caught a preview on the television channel of a hotel – of being comfortingly ‘fake’, full of prosthetically enhanced breasts and penises, and therefore unthreatening to the pursuit of any sensible and intelligent existence. But this is unfortunately far from reality. Modern pornography looks so real it now resembles one’s own life in every detail, with the difference that everyone happens to be having continuous beatific intercourse.

The waste of time is of course horrific. Financial analysts estimate the value of the online pornography industry at $10 billion, but this doesn’t begin to evoke the true scale of the loss in terms of the squandering of human energies – perhaps as many as 200 million man hours that might otherwise have been devoted to starting companies,

Comme

Back Read Next story Next

“身为实习生，你很难指着一个成品说这是自己的作品，因为刊物通常是在艺术总监的带领下，由团队成员共同完成的。但即使是只参与了简单的版面设计，也能收获宝贵的经验。”

资深设计师

嘉玛·斯塔克（Gemma Stark），*Net-a-Porter* 数字杂志

身为数字杂志的资深设计师，具体的工作内容是什么呢？

某种程度上，数字杂志设计师的工作与传统杂志的设计师差别并不大。都要指导时尚照片拍摄，设计专题，参加选题策划会。我通常同时设计几篇不同的报道，还要寻找摄影师并与他一起讨论拍摄理念。有一点不同的是，要能掌握最终成品的数字特性，所以必须要和技术人员密切合作，一起探讨页面如何在屏幕上呈现。

你的日常工作和纸质刊物的设计师有何差异呢？节奏是否会更快？

因为我们的杂志是周刊，因此节奏非常快。星期二，大部分内容就会进入设计阶段，到了星期四午餐时间，编辑会对整本杂志进行审阅。星期一，我们完成设计的版面就会交给技术团队，星期三就会放到网站上。数字杂志的校对过程与纸质杂志大不相同，我们利用公司内网页预览来校对图片和文字，而不是看纸质打样。发现修改的地方后，交给技术团队直接更新修正。

是否需要在不同的平台工作？

我们的杂志会以Flash格式进行设计，之后会以HTML5形式做成平板电脑上的App，每周还会选一篇杂志报道，做成迷你的PDF版本供手机用户阅读，偶尔也会推出纸质版。我们用InDesign排版，但网页与电子邮件广告也会用到Photoshop。确实，我们会用到各种各样的设计软件，但是*Net-a-Porter* 的所有设计都是从内容角度出发，服务于内容的，就连横幅广告区域都会放入刊物信息。电子杂志的每一页通常不会放太多东西，但是可以用动态画面来传递丰富的内容与图像信息。在开始设计任何版面时，就要考虑到是否适合动态效果这些因素，因为这往往会大大改变整个版面的风格。

如何跟上最新的技术脚步？

我会通过不断上网搜索，浏览网页和博客以寻找灵感。同时会通过团队合作，定期讨论并交换心得。如果有人看到一些很棒的东西就会和整个团队分享，虽然不一定和每个人的工作都相关，但是总有人会用得上。技术部门也很负责，他们总是会提醒我们有哪些新工具可以使用。

你是如何进入杂志设计这一行的呢？

我还在圣马丁学院就读时，一位老师推荐我到*Elle* 杂志做暑期实习，我在那里工作了几个月，发现这份工作结合了我最喜欢的两个方向，平面设计与时尚，所以我很喜欢。于是我尽可能多地积累人脉与工作经验，并在那里认识了我现在的上司。在她邀请我加入*Net-a-Porter*工作之后，我就开始跃跃欲试。*Net-a-Porter* 是一家线上精品商店，非常重视编辑设计，正符合我的需求。毕业后我就立刻入职公司，当年9月成为初级设计师。

约翰·贝克耐普（John Belknap），《犹太纪事报》

报纸设计总监的工作内容是什么？

报纸设计总监的工作主要分为两部分。

第一部分，是为报纸打造出具有一致性的视觉形象，从报头到分类广告都要在符合整体调性的前提下独具一格，并且不断关注、维护与提升。此外，编辑有时候要做新的专题，那么我负责监督或设计这些专题的呈现方式。

第二部分，就是参与制作日报或者周报。这需要与编辑密切沟通，并协助准备版面需要的照片、插图或者信息图表。大型报纸的设计室通常只负责设计最为复杂的几块版面，剩下的页面则由审稿编辑（在美国称为文字编辑）按照报纸已订立的排版指南来完成。设计总监会雇用其他的艺术总监来设计不同专题区域的页面，例如商业版、运动版、评论版等。设计总监与图片编辑、各版面编辑、平面设计师合作紧密，在后期环节还要和总是急于将刊物送印的制作经理合作。

你们是怎么设计出可以适用于网络版本的刊物内容的？

大多数报道内容会先存进资料库，再由系统从资料库选取并发布到网页上。这项工作可以通过一些安装在电脑中的排版软件如Quark或InDesign上的外挂系统（plug-in）来实现。虽然把报道内容放到网站上的方法有很多，但是我们基本上都是从资料库中提取的，因为资料库同时也是历史文章的档案库。以前我们的报道是先印成纸质版，再在网站上刊出，现在则是反过来。资料库的管理多半交给制作或者信息技术系统，系统设计也会影响到我们的工作方式。例如，大标题与副标题一定要贴在不同的栏位，或者至少使用不同的样式表，这样发布到网站上时，系统才能够正确识别并排列。

如何跟上最新科技的脚步？

IT技术部门的经理会告诉我要安装哪些最新的软件，而我会尽力学习使用它们。

你是如何进入报纸这一行的呢？

我在读高中时，曾参加制作校报。从那时期就对报纸这种一开始节奏慢悠悠，后来节奏快到让人肾上腺激素飙升。大学时期我开始在报社兼职，毕业后，留下来继续在那里工作了十年。

▶ Orthodox to hold a mixed minyan P6 ▶ Sparks fly over gay marriage P7

THE JC

LIFE MAGAZINE INSIDE

MEHTA AND BETTER P29

REVITALISING ISRAEL'S ORCHESTRA

7 ways to make your holiday an adventure

THE JEWISH CHRONICLE

Together

CLOSER TO ISRAEL PARADE — SEE P2-5

▶ Why we should forgive Ken P27 ▶ How City traders make their money P17

THE JC

EXCLUSIVE INTERVIEW: LEONARD COHEN

LOOTED ART COMES TO LONDON

THE JEWISH CHRONICLE

ON PARADE...AND COMING TO A STREET NEAR YOU

Boycott battle to cost £1m a year

Businessman pays £300,000 up-front

Israel disappears from tour map

THE JC LAUNCHES A CHARITY APPEAL

右页：约翰与编辑密切合作，一起制作了大胆抢眼的头版，以确保每一部分内容都能够抓住读者的目光。左边的图片由约翰·里夫金（John Rifkin）拍摄。

左页：从嘉玛设计的作品可以看出，她已经很了解时尚版读者喜欢干净清爽的页面。同时，读者通过滑动页面，就可以进一步阅读不同的标题与内容。

T
The New York Times Style Magazine
WOMEN'S FASHION FALL 2005
Tilda Swinton

第二章　编辑设计格式

2010年平板电脑问世之前，编辑设计界地位最高的工作是为全国性报纸、消费与生活类杂志以及制作精良的增刊（ supplements）做设计。

然而，随着数字出版物的快速发展，编辑设计师、出版社以及广告商们获得了全新的机遇与舞台。新的数字出版物发布平台包括网站、手机、平板电脑，还有各种各样的App。这些内容为设计报纸和期刊提供了更多的发挥空间，设计师可以添加图片与互动效果等。在几乎人手一部手机的时代，出版商获得了比纸质杂志时代更多的受众，这是过去纸质版刊物所无法比拟的优势。

本章将介绍常用的编辑设计格式，适用于定期出刊的纸质版与电子版杂志或报纸。同时，也会和读者一起欣赏引领国际潮流趋势的刊物设计，以供学习。

数字出版物简史

早期的数字出版物以PDF格式来构成网站呈现形式，读者通过逐页翻阅来进行阅读，这种方式与传统的报纸和杂志如出一辙，仅仅是对纸质版的简单数字化复制。但因PDF文件通常很大，能用的字体有限，所以发挥有限。后来，美国出版商康泰纳仕（Condé Nast）自主开发了一套软件，制作了*Wired*（《连线》）、*GQ*（《智族》）以及*Vanity Fair*（《名利场》）等杂志的电子版。20世纪90年代HTML格式出现后，设计师可以运用这种程序语言将动态内容嵌入网站之中。网页浏览器也能够阅读这些标记语言，将程序代码转换为图文。

随着多媒体互动技术的不断进步，App中已经可以加入动态图像以及互动内容，深受广告商喜爱。2010年，平板电脑问世后，读者拥有了更为便携的数字出版物阅读体验。平板电脑提供了更多的工具和可能性，让编辑设计可以将生活中各种好玩的事情集于一体，包括邮件、照片、购物、上网和阅读在内的各种功能，都可以用在编辑设计的数字刊物上。不过，自2011年底开始，很多平板电脑的App开始化繁为简，返璞归真，那些同时拥有一大堆功能的App反而不再受到青睐。数字出版物的设计方向再次回到以优质的平面设计与精良的编辑为重心。虽然，最初数字出版物曾一窝蜂地过度使用交互设计等复杂的互动功能，但是在2012年后，这种趋势已经逐渐回归正常。

信息分享的陷阱

未经允许擅自使用其他公司制作的内容，势必会引起原创者的不满。以英国和美国为例，全国新闻报道的制作成本几乎全落在了《纽约时报》与BBC之类的大型新闻媒体身上。此外，2011年维基解密（WikiLeaks）的丑闻发生后，新闻版权的道德问题引发关注，人们也开始思考与质疑原创内容的归属权问题，相关法案需要进行更新，相关概念需要进行界定。2012年，英国爆出新媒体非法监听个人手机的丑闻。此案件暴露出相关机构根本无法完全掌控媒体，因此，英国政府召开勒弗森听证调查委员会（Leveson Inquiry），调查新媒体的文化、业务运作与伦理。听证会的调查结果呼吁政府成立新的主管机构，为著作权设立相应原则，厘清合法性、诚实性与隐私的议题。

“在设计杂志之初，就要考虑到内容在平板电脑上的呈现效果。平板电脑使我们重新思考，如何运用纸质杂志与网站上的内容。平板电脑具有优秀的互动功能，虽然阅读时很像纸质杂志，但是却拥有网站那种互动能力。”

——梅隆·普利查特，*Wallpaper 前艺术总监**

Real Simple 杂志的艺术总监珍妮特·佛罗里克（Janet Froelich）选择了这篇专题作为刊头，在平板电脑上，这张由格雷·科特勒（Craig Culter）拍摄的照片非常吸引读者的目光，而且图片的饱和度也明显提升。很多出版物数字版只是纸质版的PDF形式，但是*Real Simple* 杂志却选择尽量利用好平板电脑的优势，并考虑到读者以厨房为主的阅读场景，突出设计。

oops, I did it again

Don't let holiday spills mess with your spirit. Here are solutions (and soaps and sprays) for every **seasonal stain.**

REPORTING BY **NATALIE ERMANN RUSSELL**
PHOTOGRAPHS BY **CRAIG CUTLER**

Inside The Inauguration 20-page souvenir

THE TIMES

Wednesday January 21 2009 timesonline.co.uk

90p

Obama's promise

'We must begin again the work of remaking America'

Inside The Times today

'We have chosen hope over fear, unity of purpose over conflict and discord'

President Obama's inauguration speech is translated into what it really means by **Bronwen Maddox**, chief foreign commentator, and assessed for technique and craft by **Phillip Collins**, former speechwriter to Tony Blair

Bronwen Maddox

Phillip Collins

THE TIMES Wednesday January 21 2009

The speech in full Obama supplement

Barack Obama News

News Barack Obama

THE TIMES Wednesday January 21 2009

Americans of all colours get the real feel for national celebration

Huge crowds on the streets and in the bars of Washington gave an emotional welcome to their new President, writes **Robert Crampton**

Thunder-struck Obama lost for words

Fluffing their lines

《泰晤士报》（*The Times*）运用简明扼要的头条，搭配生动形象的照片，营造出很好的视觉效果。美国总统就职这样极为特别的历史时刻，《泰晤士报》除了用头版刊登该则新闻外，还将国外特派记者的报道整合起来，并搭配丰富的图像信息资讯，精心编排，为读者提供一份充实的新闻作品。

报纸

哈罗德·伊万斯（Harold Evans）在1967年到1981年担任《周日泰晤士报》（*The Sunday Times*）的编辑。他撰写过一系列有关报纸编辑、排版与文字编排的著作，至今仍然被高校新闻系采用。他在《第五册：报纸设计》（*Book Five: Newspaper Design*）一书中指出：

> “报纸是传播新闻与观点的载体，版面设计也是这种传播中的重要一环，最初，我们只有一张空白的新闻纸和大脑中想要传播的内容，这些内容就像马赛克一样东一块西一块，报纸的功能就是通过版面设计，将这些零散无序的内容进行组织呈现，使之更为有序，易于理解。为了达成这一目标，报纸设计师会使用内文字体、特排字体、照片、线条、留白以及页面顺序等元素，进行最佳组合展示。”

这段话言简意赅地描述了报纸版面的功能。但是在哈罗德·伊万斯写作该书时，网络与移动媒体尚未出现。随着新兴媒体的问世，报纸必须为读者提供与传统纸质媒体不同的服务，这就要求设计师也要做出相应的改变。《卫报》的前创意总监马克·波特表示：

> “现今，绝大多数报纸都不仅仅单纯报道新闻事件，而是更加重视提供新闻背景、观点与论述。报纸不能只告诉读者发生了什么事件，而且要帮助读者理解事件的重要意义，鼓励读者思考。设计必须以多种方式来满足这一目的。随着新闻报道越来越长，越来越复杂，逻辑合理，可读性强的版面设计也越来越重要。巧妙地使用照片、信息图表以及版式，是可视化新闻制作过程中不可或缺的编辑手段。

葡萄牙彩色日报《公共报》（*Publico*）的头版，拥有非常显眼的Logo。下图中头版图片选用红花的照片，既抢眼又具有象征意义，下方放置正式的大头照，在视觉上达到平衡。这份小报版式的头版运用大小比例展现出张力，在文字编排上也非常有现代感。

EDIÇÃO LISBOA QUA 25 ABR 2012 | HOJE Colecção Thorgal 4.º Vol. - A Guardiã das Chaves | Por + 7,90€

publico.pt

A fragilidade da democracia preocupa-nos

Observadores analisam o estado da nossa democracia | Históricos dos quatro maiores partidos escrevem textos inéditos, 38 anos depois do 25 de Abril | Na Primeira Pessoa: O meu pai estava na coluna de Salgueiro Maia
Págs. 2 a 7 e Editorial

A austeridade conduz-nos de mal a pior. Mas o povo não vai deixar de reagir Mário Soares

Temos de aproveitar [o 25 de Abril] como um motivo para novas mudanças Francisco Pinto Balsemão

Temos [na UE] uma nefasta 'ditadura sobre democracias'. Até quando? Diogo Freitas do Amaral

Somos nós [do PCP] os mais convictos defensores das conquistas alcançadas Margarida Tengarrinha

Situação familiar vai travar mobilidade na função pública
Governo disponível na mobilidade temporária para analisar a situação familiar de quem seja escolhido p16

Estrangeiros estão de volta mas só têm 6% da dívida nacional
Em 2007, os estrangeiros detinham 86% da dívida nacional. Hoje esse valor está no nível mais baixo (6%) p15

1958-2012 MIGUEL PORTAS, UMA VIDA CONTRA A CORRENTE
Portugal, 8/9 e textos de Rubén de Carvalho e Rui Tavares

Menos 4 mil cirurgias nos hospitais em Janeiro e Fevereiro
Há menos idas às urgências e as cirurgias diminuíram. Este indicador preocupa o bastonário dos médicos p14

Ano XXIII | n.º 8052 | 1€ | Directora: Bárbara Reis | Directores adjuntos: Nuno Pacheco, Manuel Carvalho, Miguel Gaspar | Directora executiva Online: Simone Duarte | Directora de Arte: Sónia Matos

6 | PORTUGAL | PÚBLICO, SÁB 28 ABR 2012

PSD e CDS esperam que Assunção Esteves chumbe Conde Rodrigues

O PS garante que o seu candidato pode concorrer ao Constitucional, mas não exclui apresentar nome alternativo, já que o processo regressou à estaca zero. O dossier está nas mãos da presidente do Parlamento

"Esta lista é letra morta. Ainda vamos ter de verificar por que razão o PSD interrompeu este processo e se há condições para fazer outro acordo com o PSD"

PÚBLICO, SÁB 28 ABR 2012 | PORTUGAL | 7

Cavaco Silva avisa que a retórica não resolve desemprego

Breves

MAS começou recolha de assinaturas

PCP manifesta-se contra "pacto de agressão"

Ministério Público investiga "campeão" dos ajustes directos e obras públicas

《公共报》的内页设计颇具杂志风格，将密集的内容结构化，在同一页面上以四篇报道的形式进行集中展示。这四篇报道都有副标题，帮助读者快速了解每篇报道的重点。通过背景底色将栏与栏之间的内容进行区分，同时垂直栏平衡了水平走向的导语和标题。重要引语旁边放置去掉背景的人物图片，又为版面增添了一丝生气。

编辑设计工作的美妙之处，在于可以与一群有着聪明才智的人一起工作。但同时，无法掌控细节时也更糟糕。许多报纸的版面都不是由专业平面设计师设计的，这使得很多杂志出身的艺术总监很难适应。”

数字报纸

报纸刚开始数字化的时候，只是将纸质版报纸直接转换成PDF放到网站上。后来随着智能移动设备的普及，数字版报纸开始加入更多的互动手段，并将GPS的定位功能应用到广告中。这样新闻服务和广告可以进一步结合起来，根据读者的兴趣或者地理位置来推送广告。这样一来，新闻机构就可以获得广告收益，恢复在纸质刊物时代广告收入的无限风光。

现在编辑设计师需要与各个数字平台的开发者合作，因此需要了解一些编程的基本知识。设计师需要了解程序员的术语，以保证沟通顺畅，构建完整的内容管理系统，保证经过其处理的内容拥有合适的视觉呈现。因为读者在不同的平台有着不同的阅读体验，编辑设计师要知道不同平台和设备所能展示的页面参数，才能保证内容在每种数字设备和平台上得到完整呈现。

报纸尺寸

报纸的版式大小多种多样，欧洲就有很多报纸介于大报和柏林版式之间。然而全球大多数报纸所采用的版式，依然以下列三种为主。

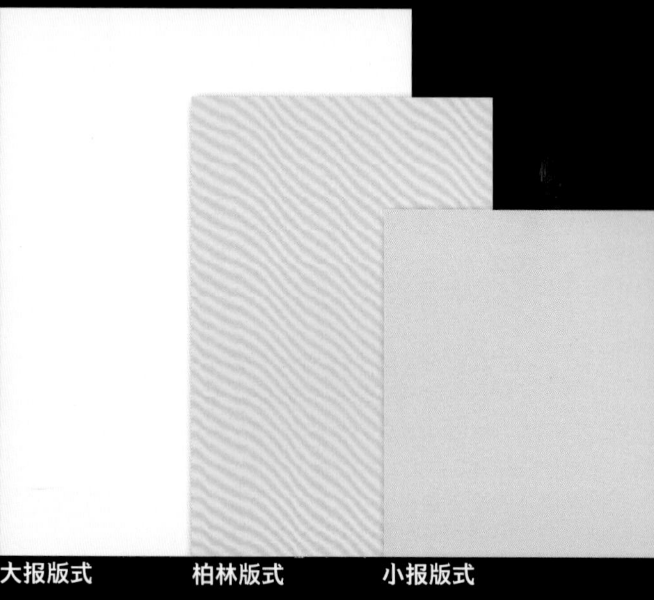

大报版式 **柏林版式** **小报版式**

大报版式（也称北欧版式）：
约56×43.2厘米（22×17寸）

柏林版式（也称Midi版式）：
约47×31.5厘米（18.5×12.5寸）

小报版式（也称半北欧或Compact版式）：
约35.5×25.5-30.5厘米（14×10~12寸）

数字报纸的常见迷思

平板电脑的问世标志着纸质报纸时代的结束？

错！ 2010年平板电脑的问世，并未改变一切，原来的杂志和报纸的品牌依然活得好好的。平板电脑是很棒的设备，提供了更多的互动可能性。

读者不会用数字设备阅读长篇文章？

错！ 如果认为移动设备阅读只是人们在路上时的随意浏览与消遣，那就大错特错了。现在的屏幕经过改良后更适合长时间阅读了，读者会利用它们阅读更长篇幅的文章，还能通过快速浏览及时获取信息，或者将文章存储，稍后阅读（例如Instapaper这样的App）。人们的阅读习惯发生了改变，2012年的统计数据显示，读者线上阅读的平均时间已增长至17~31分钟。

读者更喜欢在电脑端阅读数字报刊？

错！ 本书出版之时，移动端的报刊下载量已远超电脑端，显然可随身携带的移动设备更受读者青睐。根据2012年的数据统计，人们不仅在外出场合中携带使用移动设备，有84%的用户在家中也喜欢使用移动设备。移动设备已经改变了人们的行为，我们越来越习惯一边阅读报纸杂志，一边看电视或者使用其他媒体。任何内容都无法再独占读者的注意力。

小发行量的纸质出版物能够生存。

对！ 很多独立出版商仍然大量投入纸质刊物的制作，只要能负担得起纸张费用，纸质刊物就会一直存在。编辑设计界会继续出版各种各样的纸质刊物，读者也乐见其成。数字印刷技术降低了印刷成本，因此小发行量的出版物也更好做了。例如资助出版网站新闻俱乐部（Newspaper Club），就可以帮助读者将上传到网站的内容按需打印或者生成在线版。

发行量大的出版商将无法生存。

错！ 因为传统印刷与运营费用高昂，因此大型媒体公司在纸质出版物方面不如独立出版商灵活。然而知名的大型刊物懂得变通，他们把内容从纸质刊物转移至付费网站，并利用响应式设计（responsive design）让读者可以在移动端方便地阅读。许多刊物活了下来，还有些在2010年获得了国际新闻设计协会（SND）的最佳新闻网站设计奖。就算《新闻周刊》（*Newsweek*）在2012年曾一度终止过纸质版业务，但大型出版商的生存之路未来可期。

数字出版物术语

无缝体验（Frictionless experience）

用户在浏览网站或者使用App时，拥有相当流畅的体验，能够顺畅地浏览不同区域，无须切换到首页，或者面临重新登录等恼人的层层阻碍。

地理定位（Geo-targeting）

这是营销术语，意思是能够依据读者的实际位置，为读者提供定制化内容。出版社可以利用这一点，为读者定向提供专有的刊物与广告内容，例如附近的餐厅名单等。当然，这需要用户的移动设备开启了GPS功能授权才能实现。

全球定位系统（GPS）

利用全球卫星导航系统，为车载装置、智能手机、平板电脑等提供位置与时间信息。

流式布局（Liquid layout ）

运用设计软件，为不用设备的屏幕尺寸微调页面的比例，流式布局很有效，但是编辑和摄影师往往不喜欢自己的作品被调整变形。

付费墙（Paywall ）

付费墙是指网络用户必须先付费才能获取内容的系统。一些报纸在纸质版销量与广告收益下降的情况下，会在网站设立付费墙，需要付费才能阅读，以增加收益。

平板电脑手机（Phablet ）

具有手机功能的平板电脑。

杂志

主流消费类杂志与新闻杂志

无论是在纽约与伦敦的书店或巴塞罗那的报摊，还是在中国的杂志专卖店，无数消费类杂志都在使出浑身解数，以封面图像、搭售赠品、封面文案、品牌知名度、呼唤读者忠诚度等方式，争相吸引顾客的注意力。在网络不断发展之际，曾有人预言杂志将死，然而放眼全球，杂志依然蓬勃发展。不过，纸质刊物现在只是杂志产品线中的一员。以时尚与生活风格类杂志来说，因为页面印刷精美，纸质版触感很好，所以纸质杂志依然有自己的优势。但是对新闻与观点类杂志来说，纸质杂志可能开始退居二线。

消费类杂志包括女性、男性、商业、休闲、新闻、时尚与特别关注等类型，每一种又可以进一步细分为不同领域、关注话题以及类别，分别拥有各自的目标读者，也有不同国家的专属版。

独立出版物（自助出版杂志）

世界各地的杂志不仅阅读人数多，连投身杂志出版的人也很多。如独立刊物（发行量小，专注于服务少数兴趣爱好者的杂志）与专业杂志的种类和数量激增，也证明了这一点；全球各地都有服务于小众市场读者的刊物。这些出版物有别于那些追求发行量的主流刊物，能够提供这些主流刊物无法提供的内容。独立杂志不介意刊载长篇的网络文章，可以说是一种新的趋势，内容常常涵盖艺术、建筑、摄影、时尚与音乐等领域。

《广告克星》（*Adbuster*）是独立杂志中的翘楚。这份杂志1989年在加拿大创刊，主张反消费主义，去提倡环保，持续以国际活动来挑战体制，例如“无消费日”（Buy Nothing Day）和“无数字阅读周”（Digital Detox Week）。

THE SAME SEX SCENE

BY CHRISTOPHER HEREDIA

A SEROSORTING STORY

Dating within the HIV positive or negative population has reduced the HIV infection rate in San Francisco. It also allows for an intimacy previously missing

San Francisco lovers James Nykolay and Brian Basinger met in 2002 at the testoster-one clinic at San Francisco General Hospital Medical Center's Ward 86. Both long-term HIV survivors, the men had appointments every Friday for treatment for AIDS-related wasting.

It took several times seeing each other before Basinger got up the nerve to ask Nykolay for his name and number.

Basinger, director of the AIDS Housing Alliance, wasn't feeling up to par that January day. His weight was down to 135 pounds, from his usual 170. He had just gotten over a bad fever and a fungal infection in his mouth, all signs his medicine was failing to fight the virus. On top of it all, he was still grieving the loss of a partner to AIDS the previous summer.

"My head was so thick," Basinger, 39, recalled in a recent interview, while seated on a couch at his condominium in South Beach. "I was faking it. It hurt. It was an effort to really get the words together. I fell flat on my face."

Nykolay, who is also 39 and co-founder of the Alliance, remembered their first exchange differently.

"Looking back, it was amazing," Nykolay said, sitting in Basinger's living room on a separate couch. "He was his ebullient self. He had been through a hard place, but he was free flowing."

Nykolay was also going through a difficult time, battling spinal myelopathy. His virus was concentrated in his spine, which made it difficult for him to walk. He also had to wear diapers because the condition made him incontinent.

The pair hit it off almost immediately and became fast boyfriends.

"We could lean on each other," Nykolay said. Both men had struggled with symptoms of HIV. Each could see that in the other. It might seem that would be a barrier to getting involved, but it was the opposite. It was part of the attraction. "When I was sick, people would stare," Basinger recalled. "I was 'the AIDS guy.' That's an isolating feeling. I was underweight, gaunt and anemic. My color was all off. I was visibly ill. We could share that and understand that."

This made the two men part of what health experts in San Francisco believe is a significant trend. They say that HIV positive couples — as well as their negative counterparts — are helping reduce HIV infections when they choose to limit sexual encounters to partners of like HIV status. In scientific terms, it's being called serosorting.

In San Francisco, the estimated HIV incidence rate among gay and bisexual men is down to 1.2 percent, from the city's previous estimate of 2.2 percent a year (first reported in 2001), according to a Centers for Disease Control and Prevention study published in June.

"Serosorting is happening," said Dr. Mitch Katz, San Francisco's public health director. "It's working. It's one of the explanations we have for the flattening of the (HIV) seroincidence curve.

"As an HIV prevention strategy, serosorting is quite effective for positive and negative couples," Katz said. "It enables men to have the sex they want without worrying about contracting HIV."

Katz said another probable cause for the leveling off of infections is that more HIV-positive people have access to medications, which when adhered to make them less infectious to sexual partners by reducing their viral loads.

"For positive men, they don't have to worry about HIV transmission to somebody who is negative," Katz said. "You do still have to worry about sexually transmitted diseases. But for many HIV-positive men, serosorting brings with it a tremendous psychological relief."

Serosorting is a more successful prevention strategy for HIV-positive men, Katz said, because with HIV-negative men there is the issue of trust. "Can you trust what your partner is telling you?" Katz said. "How recently was he tested? People worry about the window period. Even if the person is telling you the truth as they know it, how certain can someone be about the truth?"

The answer is that health experts are more certain today about diagnosis of HIV than ever before, Katz said. The HIV antibody test is more sensitive than when it was first introduced in 1986. Katz said 90 percent of people who contract the AIDS virus will test HIV positive within six weeks of being exposed. "Someone with a (recent) negative test is very likely to be negative," Katz said.

For Basinger, serosorting is like choosing a partner because he shares your values or beliefs.

"Seroselecting is like choosing somebody who is in the same church, like being Episcopalian or Methodist and wanting to marry an Episcopalian or a Methodist," he said. "They just get it. You don't have to explain."

Lee Jewell, a 44-year-old parttime usher who lives in Hayes Valley, prefers to only date HIV-positive men because he would rather have sex without condoms.

4 San Francisco Chronicle Magazine, February 12, 2006

ILLUSTRATION BY MARK TODD

增刊会沿用主刊的元素，以确保品牌的一致性。然而，享誉国际的设计顾问马里奥·加西亚（Mario Garcia）认为："除了在增刊某个位置放置Logo外，增刊应该有属于自己的识别特性与生命力。读者很聪明，他们可以分辨出母刊和增刊，因此增刊不妨在设计上冒些险，使用更为灵活的文字编排，不要沿用报纸的做法。而应该将照片放大，使用更丰富的色彩，用更高质量的纸张。"《旧金山纪事报杂志》（*The San Francisco Chronicle Magazine*）便运用这种理念，利用出血图片和加宽栏位，和母刊呈现不同的调性与风格。

The Economist

Fear and loathing at the BBC

Sex and the Pentagon

Japan's new video-game champions

Gay marriage goes global

NOVEMBER 17TH–23RD 2012

Economist.com

The time-bomb at the heart of Europe

A 14-PAGE SPECIAL REPORT ON FRANCE

《经济学人》（*The Economist*）除了有纸质刊物外，也有很受欢迎的免费App。订阅用户可以通过简单清晰的设计轻松阅读新闻。这个App没有太多花哨的互动内容，顶多加上一些文章与网站的链接，不会抢了主刊的风头。

理查德·特里，《彭博商业周刊》

理查德·特里（Richard Turley）曾与波特一起在《卫报》共事，合作过各种项目，包括专刊、图书与报纸改版。特里目前在纽约担任《彭博商业周刊》（*Bloomberg Businessweek*）的艺术总监。2009年，特里在接手这一职位后，带领团队为这份财经杂志改头换面，效果惊人。他以引人注目的图片和犀利的语言，制作出拥有海报般品质、令人过目难忘的精彩封面。如此现代感的视觉风格，在财经杂志中可以说是独树一帜，他的作风快准狠，不禁让人想起20世纪60年代雄心勃勃、蠢蠢欲动的图片风格。由于杂志的封面设计富于变化，常常出人意料，因此读者总是在期待下一期封面会有什么样的精彩创意。这份周刊的编辑设计灵活独特，观点犀利鲜明（有时会将文字当作图片，也很善于利用信息图表），还配以丰富的分析性内容，深受读者喜爱，非常成功。

经特里重新设计的《彭博商业周刊》屡获好评，曾获得设计与艺术总监协会（D&AD）及出版设计协会（SPD）等机构颁发的奖项。出版设计协会曾称赞道：

> “《彭博商业周刊》以纽约设计界擅长的版式，展现《纽约时报》（*The New York Times Magazine*）”美观的视觉，有能结合英国和欧洲一流刊物的层次架构（如《卫报》）。《彭博商业周刊》有的设计非常有智慧，能够打破读者预期，突破传统新闻杂志与商业杂志的设计界限，令人佩服。”
>
> ——鲍勃·纽曼（Bob Newman），*Grids* 杂志

在接下来的访谈中，特里谈到以下关于数字设计的想法。

可以从哪些地方获得关于未来杂志想法的启发？

小发行量的独立杂志会给我以启发。除了偶尔翻阅*Vanity Fair*之外，我很少细看大型商业杂志，更不会主动去买它们。我对数字杂志也没什么兴趣，不太愿意看。我知道自己很守旧，认为杂志就应该是纸质版。真不好意思，其实我是会看一些数字杂志的：《伦敦标准晚报》（*London Evening Standard*）的*ES Magazine*，虽然我是看PDF版本，但是对我来说够了。我也喜欢他们的网站，尤其喜欢用平板电脑看。

在处理非常密集的信息内容时，数字版能不能处理得和纸质版一样好？

我做的数字杂志设计不多，不过对自己的成果还算满意。数字设计最令人兴奋的是，“设计”其实是最不重要的一环。数字设计讲究的是强烈、单纯的刊物理念。一旦确定了精确的架构，就能避免数字杂志为了形式而形式，加上一些花里胡哨的功能和特效，再说这些功能多半都很蠢（我是过来人才这样说的）。

作为编辑设计师，除了要具备传统平面设计师的能力之外，还需要哪些数字方面的技能呢？

我其实算不上数字编辑设计师，所以不确定我是否能回答这个问题。我认为编辑设计师最重要的特质，在于极强的好奇心以及沟通与分享能力。

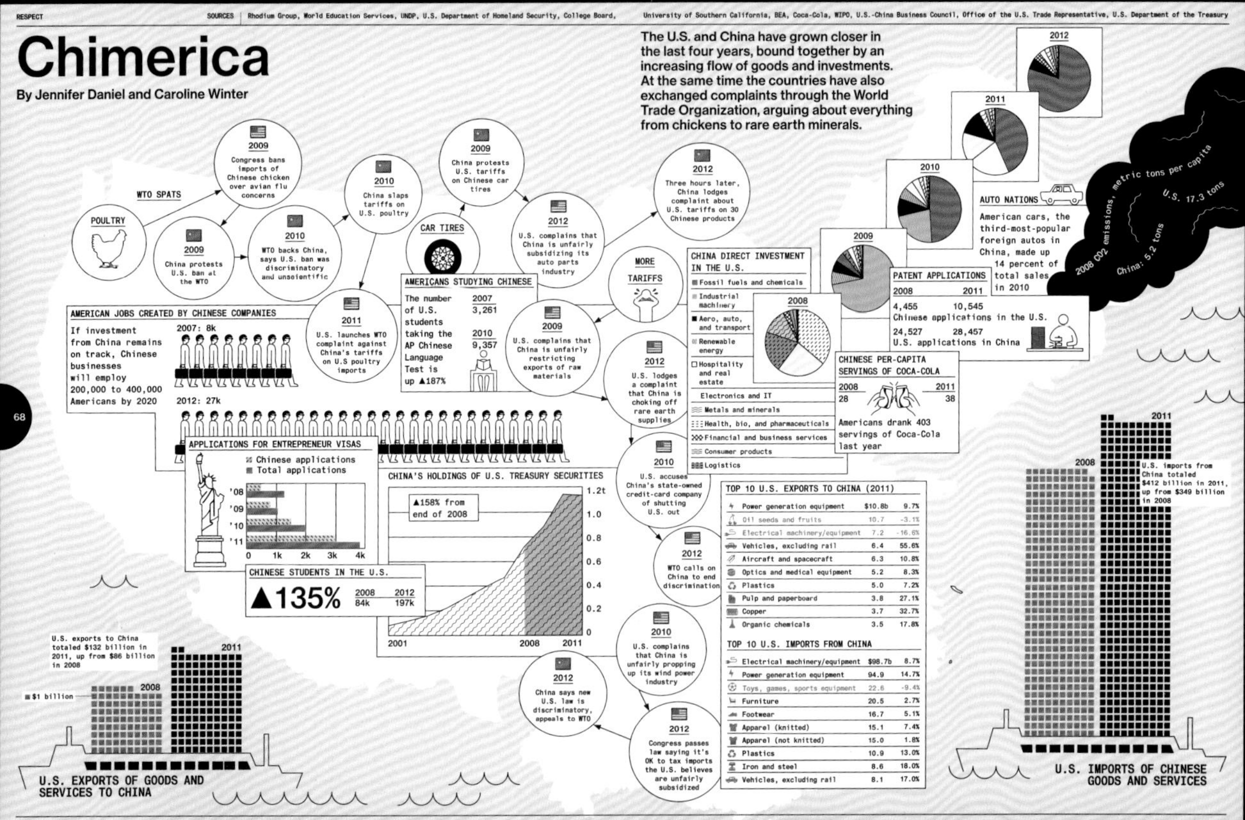

RESPECT

SOURCES | Rhodium Group, World Education Services, UNDP, U.S. Department of Homeland Security, College Board, University of Southern California, BEA, Coca-Cola, WIPO, U.S.-China Business Council, Office of the U.S. Trade Representative, U.S. Department of the Treasury

Chimerica

By Jennifer Daniel and Caroline Winter

The U.S. and China have grown closer in the last four years, bound together by an increasing flow of goods and investments. At the same time the countries have also exchanged complaints through the World Trade Organization, arguing about everything from chickens to rare earth minerals.

WTO SPATS

POULTRY

2009 Congress bans imports of Chinese chicken over avian flu concerns

2009 China protests U.S. ban at the WTO

2010 WTO backs China, says U.S. ban was discriminatory and unscientific

2010 China slaps tariffs on U.S. poultry

2011 U.S. launches WTO complaint against China's tariffs on U.S poultry imports

CAR TIRES

2009 China protests U.S. tariffs on Chinese car tires

2012 U.S. complains that China is unfairly subsidizing its auto parts industry

2012 Three hours later, China lodges complaint about U.S. tariffs on 30 Chinese products

MORE TARIFFS

2009 U.S. complains that China is unfairly restricting exports of raw materials

2012 U.S. lodges a complaint that China is choking off rare earth supplies

2010 U.S. accuses China's state-owned credit-card company of shutting U.S. out

2012 WTO calls on China to end discrimination

2010 U.S. complains that China is unfairly propping up its wind power industry

2012 Congress passes law saying it's OK to tax imports the U.S. believes are unfairly subsidized

2012 China says new U.S. law is discriminatory, appeals to WTO

AMERICAN JOBS CREATED BY CHINESE COMPANIES

If investment from China remains on track, Chinese businesses will employ 200,000 to 400,000 Americans by 2020

2007: 8k

2012: 27k

AMERICANS STUDYING CHINESE

The number of U.S. students taking the AP Chinese Language Test is up ▲187%

2007 3,261

2010 9,357

APPLICATIONS FOR ENTREPRENEUR VISAS

Chinese applications

Total applications

'08 '09 '10 '11

0 1k 2k 3k 4k

CHINESE STUDENTS IN THE U.S.

▲135%

2008 84k

2012 197k

CHINA'S HOLDINGS OF U.S. TREASURY SECURITIES

▲158% from end of 2008

1.2t 1.0 0.8 0.6 0.4 0.2 0

2001 2008 2011

CHINA DIRECT INVESTMENT IN THE U.S.

2008

Fossil fuels and chemicals

Industrial machinery

Aero, auto, and transport

Renewable energy

Hospitality and real estate

Electronics and IT

Metals and minerals

Health, bio, and pharmaceuticals

Financial and business services

Consumer products

Logistics

2009 2010 2011 2012

PATENT APPLICATIONS

	2008	2011
Chinese applications in the U.S.	4,455	10,545
U.S. applications in China	24,527	28,457

CHINESE PER-CAPITA SERVINGS OF COCA-COLA

2008 28

2011 38

Americans drank 403 servings of Coca-Cola last year

AUTO NATIONS

American cars, the third-most-popular foreign autos in China, made up 14 percent of total sales in 2010

2008 CO_2 emissions, metric tons per capita

U.S. 17.3 tons

China: 5.2 tons

TOP 10 U.S. EXPORTS TO CHINA (2011)

Power generation equipment	$10.8b	9.7%
Oil seeds and fruits	10.7	-3.1%
Electrical machinery/equipment	7.2	-16.6%
Vehicles, excluding rail	6.4	55.6%
Aircraft and spacecraft	6.3	10.8%
Optics and medical equipment	5.2	8.3%
Plastics	5.0	7.2%
Pulp and paperboard	3.8	27.1%
Copper	3.7	32.7%
Organic chemicals	3.5	17.8%

TOP 10 U.S. IMPORTS FROM CHINA

Electrical machinery/equipment	$98.7b	8.7%
Power generation equipment	94.9	14.7%
Toys, games, sports equipment	22.6	-9.4%
Furniture	20.5	2.7%
Footwear	16.7	5.1%
Apparel (knitted)	15.1	7.4%
Apparel (not knitted)	15.0	1.8%
Plastics	10.9	13.0%
Iron and steel	8.6	18.0%
Vehicles, excluding rail	8.1	17.0%

U.S. exports to China totaled $132 billion in 2011, up from $86 billion in 2008

$1 billion

2008 2011

U.S. EXPORTS OF GOODS AND SERVICES TO CHINA

U.S. imports from China totaled $412 billion in 2011, up from $349 billion in 2008

2008 2011

U.S. IMPORTS OF CHINESE GOODS AND SERVICES

68

《彭博商业周刊》的平板电脑版与纸质版各有不同的封面，皆隶属于知名的彭博公司网站，该网站提供丰富的商业数据。通常这类图片会由艺术总监德利制作，再由编辑决定如何使用。

德利为2011年9月29日的刊物设计的封面，看起来像被什么油污泼了一样，体现了数字媒体能够让纸质媒体如虎添翼，设计也很活泼有感染力。《彭博商业周刊》的文图有摇滚感，让人联想到《滚石杂志》（*Rolling Stone*）。像这样抢眼的封面，证明《彭博商业周刊》改版后不仅展现了刊物的自信，也展现了对读者领悟能力的信心。

这份《彭博商业周刊》的封面，用大胆、极简的方式来处理日本国旗（右图），在周刊市场过分装饰的封面中显得特别抢眼。

哥特字体“T”这个Logo表明这份杂志来自《纽约时报》。然而吸引读者阅读这份秋季时尚特刊的元素其实是抢眼的封面人物，摄影师以绝美的灯光风格营造出这种版面的魅力。

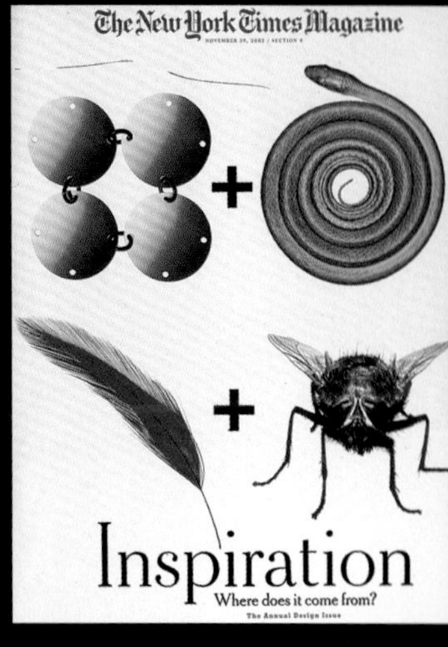

增刊

1962年《周日泰晤士报》最先推出了用铜版纸印刷的全彩杂志，开创了这种新型杂志类型。其实早在19世纪末，美国就出现过增刊，然而因为《周日泰晤士报》的制作水准颇高，设计精美，形式活泼，重新定义了增刊这种刊物。增刊很快在设计圈声名鹊起，与优质杂志不相上下，在长达十五年的发行期间，《每周泰晤士报》吸引了很多全球一流的设计师加入，推出过很多在全球范围内首屈一指的增刊设计。对设计师来说，增刊最大的挑战在于一方面要传达其所属母刊的主旨，另一方面也要兼顾自身的调性、立场和读者群体。当然，设计师可以尝试不同于母刊的字体、版面与版式，为增刊赋予独特的视觉识别，这是多数正刊没有的自由空间。增刊可以说是报社的宣传，因为报社老板知道，如果读者喜欢这份杂志或者增刊，就会爱屋及乌购买这份报纸，因此设计报纸增刊变成一份令人艳羡的工作。

“编辑设计师处在培养刊物与读者养成良好关系的前沿阵地。设计风格会透露出这份刊物的内容、精神、思想等特质，让读者马上领会杂志的核心价值，与杂志精神贯通。”

——珍妮特·佛罗里克，*Real Simple* 杂志创意指导

上面《纽约时报》杂志的例子，可以清楚感受到图片所传达的纯粹力量，艺术总监佛罗里克与摄影师密切合作，根据设计规则，拍出概念性很强的照片。文字编排又进一步凸显出摄影的细腻之处，夸张的首字放大能把读者的目光带到服装的特殊造型上。

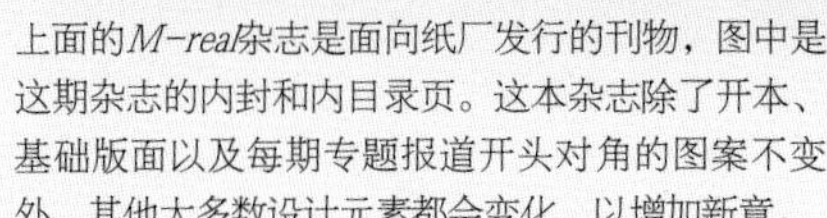

上面的*M-real*杂志是面向纸厂发行的刊物，图中是这期杂志的内封和内目录页。这本杂志除了开本、基础版面以及每期专题报道开头对角的图案不变外，其他大多数设计元素都会变化，以增加新意。

客户杂志与B2B（企业对企业）杂志

客户杂志能够将品牌的视觉形象和植入式广告以及市场营销相结合，成为品牌营销的重要阵地。客户杂志最初只提供给特定消费产品服务的用户，但如今已经出现许多创新，跨越多种平台，为特定产品和服务的用户创建自己的社群，这些用户喜欢在社交媒体上讨论产品或服务。和一般消费类杂志不同，客户杂志由企业赞助，是进行品牌宣传、产品推广的重要渠道。品牌营销者明白，要想让客户杂志有很好的效果，需要有丰富有趣的内容，而且品牌推广元素和手法则必须细腻低调，把握好分寸。*magCulture*创意总监杰里米·莱斯利（Jeremy Leslie）表示，因为客户杂志有服务读者和品牌营销的双重需求，因此对设计师的要求也比较多：

> “客户杂志和一般杂志在设计原则和技巧方面差异不大，但是需更注重策略、思维与更广的创意。消费类杂志既要在展架上够突出以吸引新读者，但也不能冒着疏远现有读者的风险来吸引新的读者。客户杂志也讲究创新，但需要符合读者们对品牌或服务的期待。设计师在设计客户杂志时，更需要进行概念性的思考，探讨杂志更多的可能性。”

B2B杂志的涉及领域更广，这是一般消费性杂志不具备的，这个优势也容易被忽视。一般国有企业和私营企业的品牌传播刊物就属于B2B刊物。通常这些刊物、新闻通讯和博客会优先向所属用户提供，而不是面向大众销售，因此没那么重视设计，但也有例外。*The Lawyers* 杂志与网站都是由艾斯特森设计事务所进行设计，就是很好的案例。另外，期刊出版商服务协会（Periodical Publishers Association Services）也有专门的奖项来鼓励优秀的B2B杂志设计。

数字杂志出版

杂志的市场类型会影响杂志的内容与发行方式。以时尚杂志为例，纸质杂志会刊登精美的广告，杂志品质精良，深受读者喜爱，其网站设计也用动态效果来强化内容，还搭配线上专题的读者特供优惠。在生活方式类杂志中，例如美国的杂志*Real Simple*，则将纸质版内容放在杂志网站，并大篇幅介绍纸质杂志，以吸引读者订阅。

左方是*Carlos*杂志的封面与内页。*Carlos*是英国维珍大西洋航空公司（Virgin Atlantic）头等舱的乘客杂志，视觉识别非常突出。有趣的是，它的优势也是它的弱点，没有摄影照片和彩色版面。

社群杂志

一些极具创意的创新公司会根据读者浏览与点赞记录来收集分析读者的需求，将高度受欢迎的内容聚合，并以杂志形式出版。在美国，这种杂志被称为“社群杂志（social magazines）”，由Flipboard与美国在线（AOL）的Editions等App首创。社群杂志是以推特和Facebook等社交媒体为基础发展起来的，例如Editions就能把美国线上新闻提供者发布的新闻资讯，通过GPS与本地内容相结合，整合出属于读者所在地的内容，提供给读者。社群杂志是经过严格编辑的品牌产品与聚合内容的集合，同时鼓励读者积极分享。这种做法有点像以前人们会把喜欢的刊物页面撕下来，和朋友分享传阅。

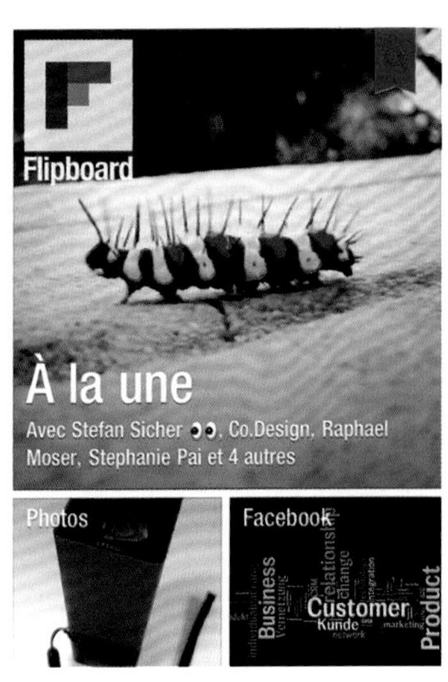

Flipboard 成功利用社交网站的力量，根据读者的资料分析，并结合其他网站取得内容，很受欢迎。刊物内容是自动生成的，用户就像是在逛一站式商店，只要愿意，一次就可以看完所有感兴趣的文章。

Real Simple 杂志以简洁的外观，凸显出刊物的品牌价值。它采用的精选配色，不仅印刷后呈现精美质感，在平板电脑上也赏心悦目。并通过交互视频形式展示那些需要一步一步操作的内容，可以说是增强内容呈现的范例。

平板电脑能以简洁的方式呈现复杂的刊物内容，对读者与设计师来说，都是一种乐趣。在左图中，只要在屏幕上用手指轻轻一点，食谱就会从中跳出。看似简单的页面，其实隐藏丰富的内容。

这是学生萨洛米·德赛（Salomi Desai）构思的杂志，以日常生活中的迷信为主题。德赛用这个概念以不同角度拍摄日常生活中的事物，让它们显得十分特别，让摄影和图片变得趣味盎然。

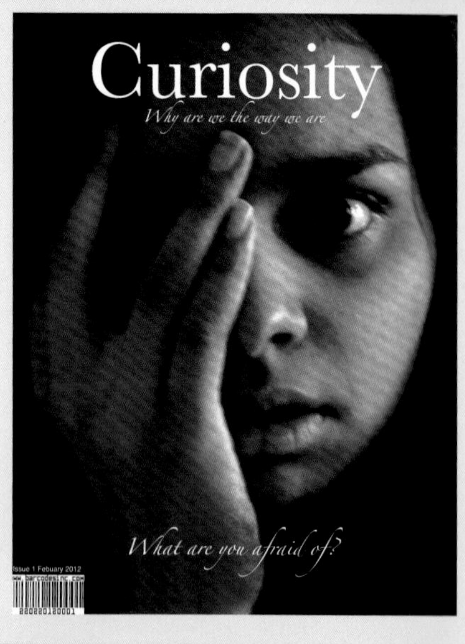

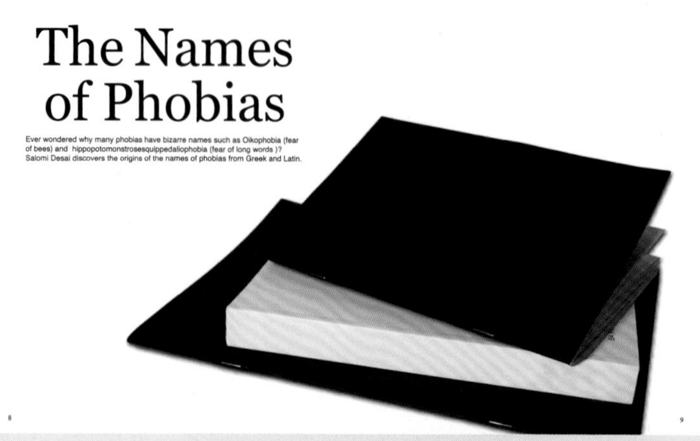

工作坊1概念

如果你是老师，可以以此作为教学计划的基础，然后根据需要进行延伸，并加上与其他领域相关的部分。这份纲要的主要适用对象是艺术或美术专业一年级学生，进行调整后，也可以供高年级学生使用。

如果你是学生或者想要自学，可以根据本书中的工作坊单元建立起自学的框架。请自行拍摄照片，让朋友评论你的作品，并接受他们的意见反馈。

目标

以某个当代主题为出发点，进行发散，策划一本杂志，来捕捉时代的脉搏。

操作步骤

发挥你的想象力，构思一个属于当代的主题，以此为基础进行深入研究，并发展为杂志中的专题报道。

首先，进行头脑风暴，提出至少五种不同的杂志专题概念。在纸上把每种概念写下来，只要写出刊头、Logo和轮廓草图就够了。然后把你的概念整合起来。

选择其中一个概念制作杂志，并收集适合这份杂志内容的视觉材料。研究其他有相似内容的杂志，以及关注类似主题的俱乐部或者社群。它们是怎么进行内容编辑的?到图书馆或独立书店找资料，也可以参考线上杂志。从报纸杂志上剪下适合的照片，或把你认为适合的内容复印下来。可以记下任何可以让故事更精彩的字体或插图。持续收集资料，把杂志的概念精简成连贯的十几个字，例如我的杂志叫作什么？它讨论的主题是什么？

制作“情绪版（mood board）”，用以说明杂志的概念，再把你所制作与研究的图像汇集修订，贴到一张A3或A2的卡纸上，这是为了找出这本杂志的视觉风格。为杂志写几条封面文案，正式替你的杂志取名，作为刊头。

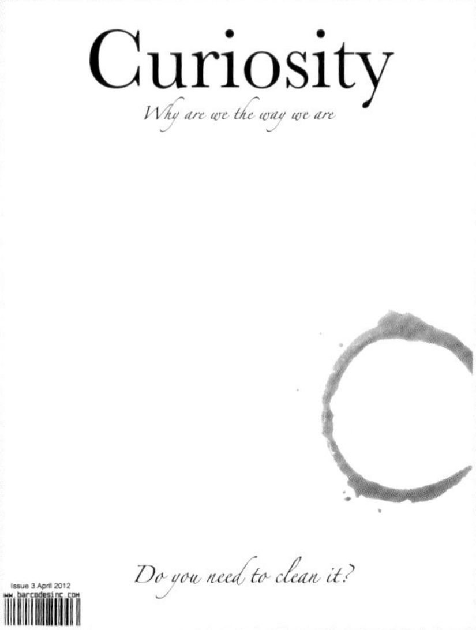

Curiosity

Why are we the way we are

Do you need to clean it?

Issue 3 April 2012

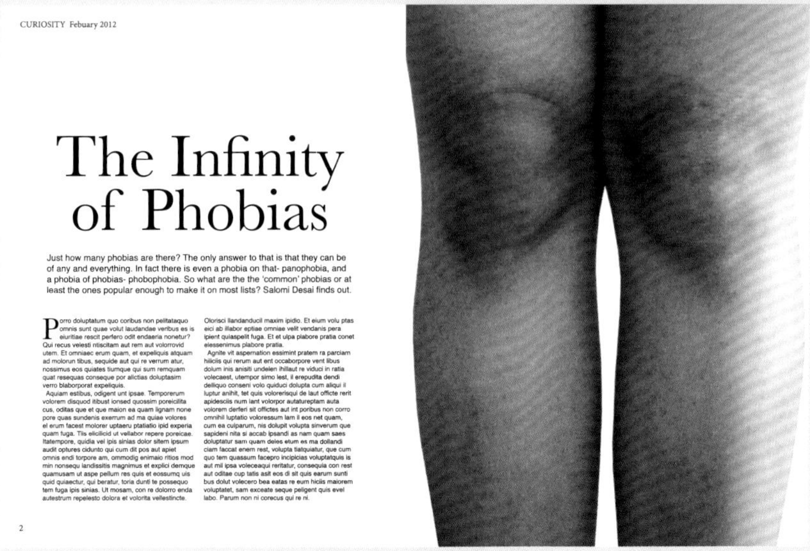

CURIOSITY Febuary 2012

The Infinity of Phobias

Just how many phobias are there? The only answer to that is that they can be of any and everything. In fact there is even a phobia on that- panophobia, and a phobia of phobias- phobophobia. So what are the the 'common' phobias or at least the ones popular enough to make it on most lists? Salomi Desai finds out.

2

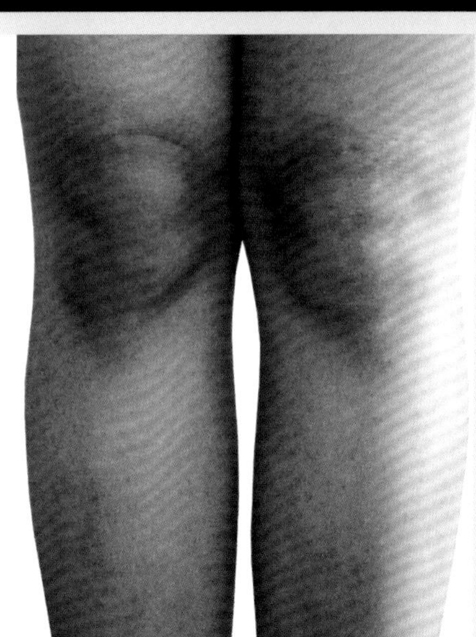

Curiosity

Why are we the way we are

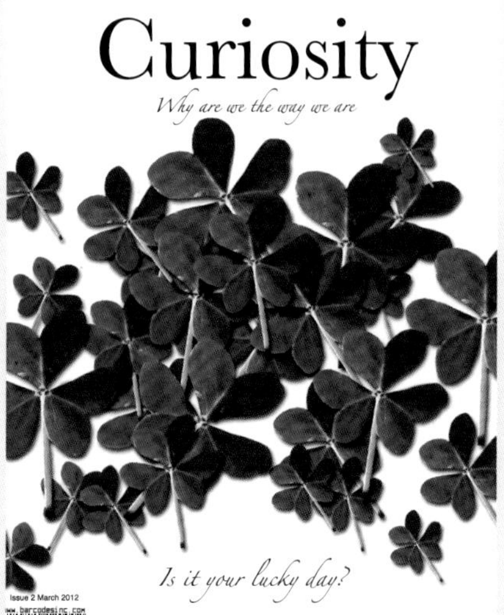

Is it your lucky day?

Issue 2 March 2012

Wisdom or Folly?
Are superstitions really the result of feebie minds (As Edmund Burke suggested), or have we learnt to keep them for more practical and commonsensical reasons? Salomi Desai finds out more.

一开始，德赛只是用铅笔勾勒出概念草图，然后做成实际大小。她利用简单的图像来展开想法，逐步发展成报道内容与标题。参考维多利亚时期古玩店稀奇古怪的收藏，将杂志命名为*Curiosity*。

CURIOSITY March 2012

Bizarre Superstions?

6

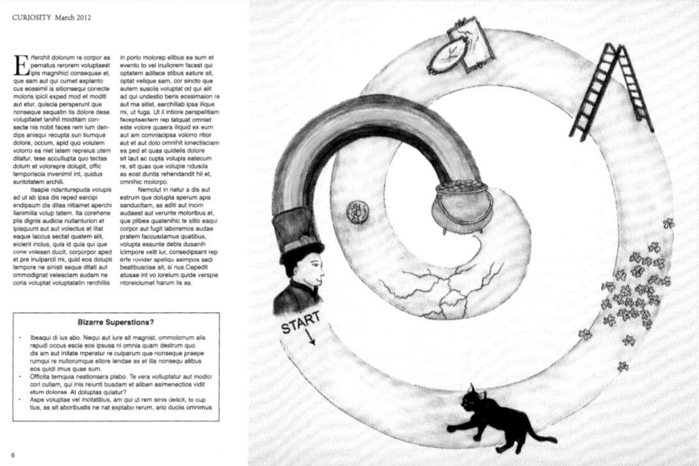

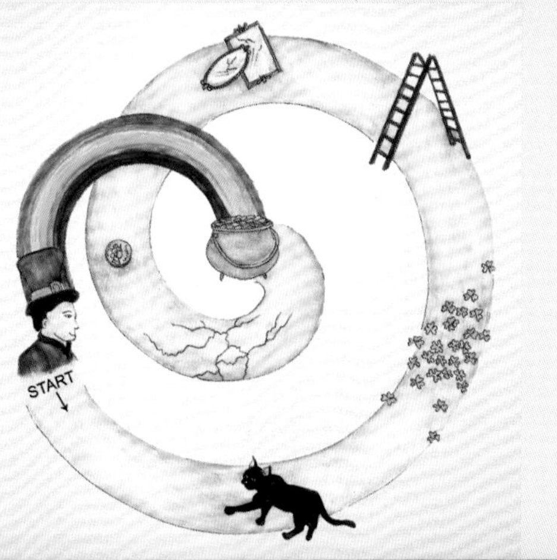
START

Wallpaper*
APRIL 2012
*DESIGNINTERIORSFASHIONARTLIFESTYLE
UK £4.99
US $10.00
AUS $ 10.50
CDN $ 10.00
DKK 75.00
F € 8.50
D € 10.50
NL € 8.50
I € 9.00
J ¥ 1740
SGP $ 18.20
E € 8.50
SEK 75.00
CHF 16.00
AED 45.00

THE GREAT INDOORS

Your guided tour to new global design and tomorrow's national treasures

+
FRANCE
GERMANY
ITALY
JAPAN
SCANDINAVIA
SPAIN
USA

Boarding now for

BELGIUM

第三章　封面

看到引人瞩目的杂志封面和刊头，谁能不拿起来翻阅一下呢？如果能在展架上抓住来往人群的目光，也就是抓住了潜在的读者。数字刊物的封面更是吸引读者阅读的入口，读者只要轻触封面，就能进入编辑好的内容、动态和交互元素中。封面是传达品牌信息的关键，本章首先要探讨的就是刊物的封面以及如何通过封面打造品牌，这对任何新刊物来说都是重中之重。

品牌识别

对于新刊物来说，最重要的是建立品牌信息，以及这份刊物的标识、表现与质感。要成功诠释这些元素，需要靠编辑与设计师的共同努力，构建一座坚固的桥梁，让出版社（或自费出版者）能把品牌与品牌价值传达给客户及读者。完成品牌建立的工作后，才能开始制作刊物内容，也就是本书第四章、第五章和第六章所述的内容。品牌识别的设计内容包含了Logo、配色、字体、照片与插图等。同时要建立一套规则进行使用管理，让普通人接手该刊物的设计时也能够按照规则来使用。这些原则结合起来，就会形成品牌的视觉标识。每一期刊物的识别，不是保持原貌，而是要有变化，以保证刊物新颖有活力，绝不能为了遵循设计规则而设计，牺牲核心品牌的价值，本末倒置。每次标识延伸都需要与设定好的格式相互契合，不能只是在不同产品上复制相同外观。要达到这个目标，就要刊物保持有辨识度的风格，但每期也要有足够的差异，让读者与潜在读者看到喜爱的刊物在不断变换与创新。

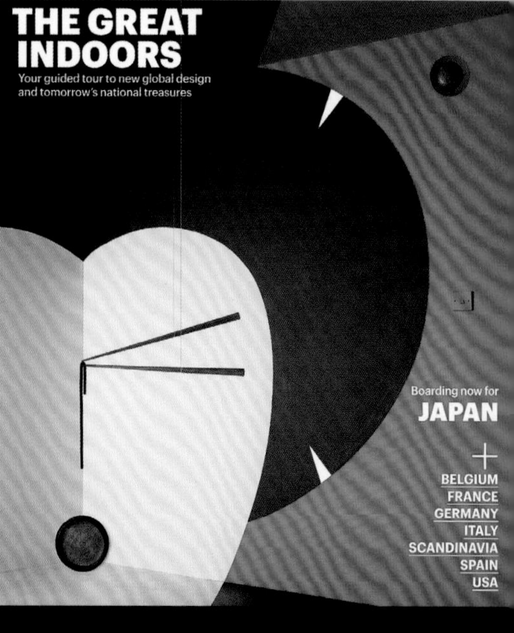

这些奇怪的图画是*Wallpaper**委托艺术家Noma Barwere设计的8个全球版本，所以无论你在世界的哪里，一个“本地”的封面都会出现在你的面前。一开始，它们看起来都像是以传统的方式绘制，但仔细观察就会发现，这些画面实际上是3D的房间布景，全尺寸的绘制，并结合了道具，比如家具和灯具。最后，一系列图片呈现出精美的艺术魅力，创造了一个可变的形象空间，与比例和负空间相互配合。它们标志性的自然元素推动了这个品牌的发展，其隐含的思想为其使用者提供了在不同国家旅行和工作的便利。

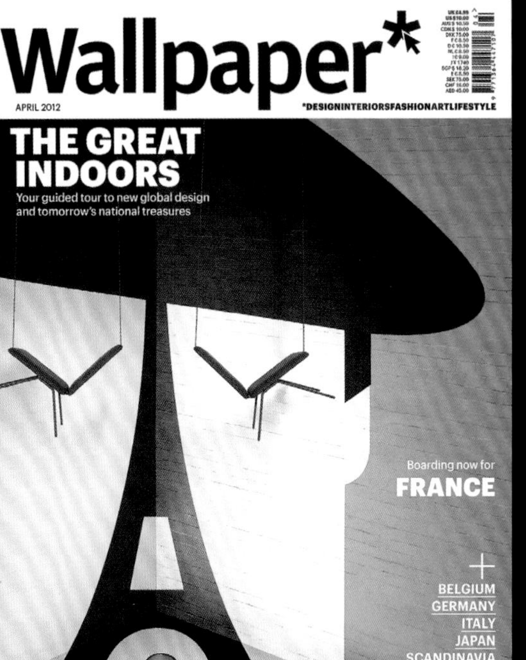
Wallpaper*
APRIL 2012
*DESIGNINTERIORSFASHIONARTLIFESTYLE
THE GREAT INDOORS
Your guided tour to new global design and tomorrow's national treasures
Boarding now for
FRANCE
+
BELGIUM
GERMANY
ITALY
JAPAN
SCANDINAVIA
SPAIN
USA

Wallpaper*
APRIL 2012
*DESIGNINTERIORSFASHIONARTLIFESTYLE
THE GREAT INDOORS
Your guided tour to new global design and tomorrow's national treasures
Boarding now for
GERMANY
84 pages of Deutsches design in our special German supplement!
+
BELGIUM
FRANCE
JAPAN
SCANDINAVIA
SPAIN
ITALY
USA

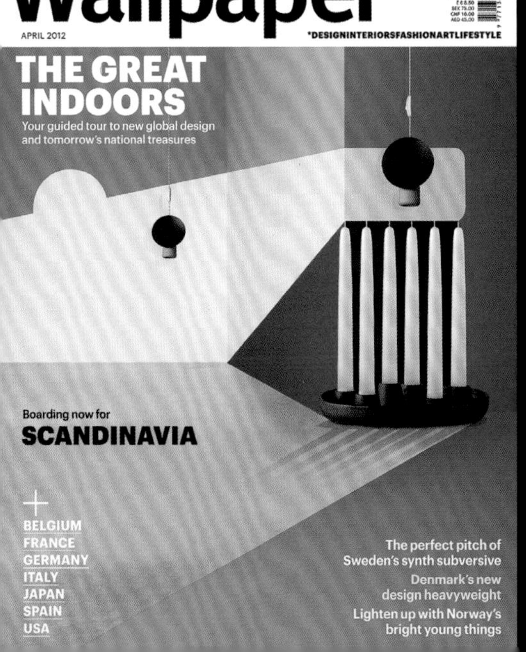
Wallpaper*
APRIL 2012
*DESIGNINTERIORSFASHIONARTLIFESTYLE
THE GREAT INDOORS
Your guided tour to new global design and tomorrow's national treasures
Boarding now for
SCANDINAVIA
+
BELGIUM
FRANCE
GERMANY
ITALY
JAPAN
SPAIN
USA
The perfect pitch of Sweden's synth subversive
Denmark's new design heavyweight
Lighten up with Norway's bright young things

Wallpaper*
APRIL 2012
*DESIGNINTERIORSFASHIONARTLIFESTYLE
THE GREAT INDOORS
Your guided tour to new global design and tomorrow's national treasures
Boarding now for
SPAIN
+
BELGIUM
FRANCE
GERMANY
ITALY
JAPAN
SCANDINAVIA
USA

封面

无论是哪一种出版物，封面都是清楚传达品牌特色与价值的首要因素。纸质杂志在设计封面时总是绞尽脑汁，希望杂志在展架上能够传达信息，脱颖而出。希望读者购买杂志后，通过封面影响读者和其他看到刊物的人，从而进一步宣传刊物杂志的品牌价值与理念。对电子版杂志而言，封面不仅能强化品牌，也是浏览工具，读者会从封面点进去阅读刊物内容，同一张封面图可能出现在纸质图书、网站、平板电脑以及App应用上。因此负责设计封面的艺术总监，必须要考虑各个版本之间呈现效果的差异。封面承担着多重任务，既要呈现丰富含义，让不同读者有不同解读，还要能促进销售，封面要足够抢眼，才能脱颖而出，吸引读者目光，避免读者被竞争者抢走。定期发行的刊物封面既要让固定的读者觉得熟悉，又必须每期都要有差异与变化，以便读者一眼就能看出这是最新的一期。刊物封面要传达刊物的特性与内容，还要获得潜在读者的青睐，吸引读者进一步阅读，但同时也不能让原有读者感到陌生。这样看，许多出版社和设计师在设计一页封面上所耗的时间、金钱和精力几乎与设计剩下的所有内容一样多，也就不足为奇了。数字刊物的封面可能出现在Facebook或博客上，所以要足够醒目，即使被缩小也能看得清楚，每期封面都要有吸引读者的亮点。

EL PAÍS

www.elpais.com

EL PERIÓDICO GLOBAL EN ESPAÑOL

JUEVES 29 DE AGOSTO DE 2013 | Año XXXVIII | Número 13.210 | EDICIÓN EUROPA

revistaverano

El cine más guapo se sube a 'vaporetto'

Clooney y Bullock abren la Mostra de Venecia PÁGINAS 33 A 35

El regreso del pecado original

'El diario de Adán y Eva' vuelve a los escenarios PÁGINA 37

Aquí hay tomate, pero menos

La 'privatización' de la Tomatina desluce la fiesta PÁGINA 39

JEWEL SAMAD (AFP)

Obama llama a que el sueño de King se cumpla para todos

Barack Obama aprovechó ayer su discurso en recuerdo del 50º aniversario del *I have a dream* de Martin Luther King por los derechos civiles para advertir de que, mientras no se avance en la justicia social en EE UU y se acabe con lo que llamó "discriminación legalizada", el sueño de verdadera igualdad de oportunidades del predicador no se habrá cumplido aún. PÁGINA 8

EE UU y Reino Unido agotan las vías legales para el ataque a Siria

▶ Londres acude a la ONU, que pide tiempo para sus inspectores
▶ Obama se dispone a presentar pruebas del bombardeo químico

ANTONIO CAÑO, Washington

Los tres países que encabezan la inminente operación contra el régimen sirio —EE UU, Francia y Reino Unido— pretenden agotar las vías diplomáticas antes de atacar y se disponen a presentar en breve pruebas de la responsabilidad de Bachar el Asad. No quieren repetir errores del pasado en Irak y aspiran a que la operación esté lo más justificada posible desde el punto de vista político. El secretario general de la ONU, Ban Ki-moon, pidió ayer tiempo para que los inspectores de armas terminen su trabajo sobre el terreno. PÁGINAS 2 A 7

EDITORIAL EN LA PÁGINA 26

Tarde y mal

El ex secretario general de la OTAN avisa de los riesgos de una acción que se prevé rápida, pero podría no serlo

JAVIER SOLANA

Lo que más sorprende es que hace varios meses no hubiéramos escuchado al presidente de EE UU y al Consejo Europeo un anuncio de esta naturaleza: "Hemos decidido enviar un equipo con capacidad negociadora a los países implicados y a la Liga Árabe para que con la mayor brevedad se pueda encontrar una solución al conflicto de Siria". Los resultados de la reunión de Ginebra del 30 de junio de 2012 podrían haber sido la base. PASA A LA PÁGINA 3

Es hora de actuar en Oriente Próximo

Por Tony Blair PÁGINA 27

El presidente del Constitucional deja el puesto en la OIT que ejerció siendo magistrado

Cobos aseguró que tenía autorización verbal del Tribunal

MARÍA FABRA, Madrid

Francisco Pérez de los Cobos, presidente del Constitucional, simultaneó su cargo de magistrado con el de asesor en una comisión de expertos de la OIT, por lo que cobraba dietas. La ley impide a los magistrados actividades de asesoramiento durante su mandato, pero Pérez de los Cobos adujo que tuvo autorización verbal del entonces presidente del Constitucional, Pascual Sala. Cobos decide ahora dejar el puesto en la OIT que ejerció 15 meses siendo magistrado. PÁGINA 12

Wert promete más dinero para becas, pero con menos beneficiarios

J. A. AUNIÓN, Madrid

El ministro de Educación, José Ignacio Wert, intentó ayer acallar en el Congreso las críticas a su reforma del sistema de becas universitarias con el anuncio de que esta partida crecerá este curso 250 millones (un 20%). En todo caso, será un dinero a repartir entre menos becarios, tras el endurecimiento de los requisitos para acceder a las ayudas. PÁGINA 32

Científicos crean microcerebros a partir de células madre humanas

JAVIER SAMPEDRO, Madrid

Científicos vieneses han traspasado una nueva barrera en la investigación con células madre humanas: han creado microcerebros del tamaño de un guisante que reproducen las estructuras y capas neuronales de la corteza cerebral. Es un modelo sin precedentes para estudiar el desarrollo de este órgano y las enfermedades neuronales. PÁGINA 3

Why women identify with Kylie
Libby Brooks in Comment Page 28

In search of Britain's worst commute G2 Page 10

How Massive Attack discovered politics G2 Page 23

£0.70
Tuesday 06.02.07
Published in London and Manchester
guardian.co.uk

the guardian

Newspaper of the year

Children of war: the generation traumatised by violence in Iraq

Growing up in a war zone takes its toll as young play games of murder and mayhem

Michael Howard Baghdad

Dangerous games
There are fears among parents, teachers and doctors across Iraq that the experience of growing up in a war zone is creating a generation prone to violence, panic attacks and nightmares

Continued on page 2

Playtime stories

National
Britain worst for crime, says EU poll

National
Fashion is taking a backseat to the growing debate over stick-thin models at New York fashion week

International
Bush seeks $100bn extra for war in Iraq

Financial
Ryanair's profits climb by 30%

Sport
Barton booked by England captain

报纸头版

如今报纸已经无法简单凭借新闻内容来吸引读者了。互联网和移动媒体提供丰富多样、及时快捷的新闻内容，报纸变得没那么重要，因此报纸需要重新寻找定位，以适应新的变化。传统定义认为新闻是指昨天不知道，今天才发现的事，担任全球多种报纸顾问的马里奥・加西亚表示：“现在我和广告客户们分享的一个新的新闻定义是，新闻是指我昨天就发现了但今天才明白的事。”因此，21世纪之后，许多报纸纷纷改版，最明显的就是报纸头版的变化，希望能够焕发青春。加西亚表示报纸必须给读者提供能让他们觉得惊讶的好故事，上面的照片也必须是过去24小时内，在电视与网络上找不到的。新闻媒体应当给予读者新奇，而不是重复已知的事情。

报纸的头版依然相当依赖抢眼的图片以及煽动性的标题，例如《卫报》。报纸会在头版上放新闻摘要，来吸引读者注意。例如左页的《西班牙国家报》，在头版用铺底色对横幅以及垂直对栏位，先告诉读者内页报道的提要。即使是叙利亚的头条新闻也仅占17行，以便挤进一篇广告。

同样，《卫报》也在头版下方提供四五篇精简的新闻摘要。设计师必须要找到一种设计平衡，毕竟，所有的优质报纸都想在头版多列出几篇报道，马克・波特（Mark Porter）曾表示：“用新闻摘要头版能够展现许多故事，对当天的新闻提供广泛平衡的报道，所以利用好新闻摘要很重要。”

Boston Sunday Globe

Ideas

APRIL 24, 2005

D

& Books

CHILDHOOD'S END Sue Miller's new novel, "Lost in the Forest" D6
MURDER AND MAJESTY Mysteries in LA and Elizabethan England, in "On Crime" D6
WAR STORY Reimagining a brother's experiences in the SS D7
REFLECTIONS Gail Caldwell on a novel about an octogenarian writer D7
UNREALITY SHOWS Trompe l'oeil, from ancient Greece to today D9
PARADOXICAL Works about Gödel and other thinkers, in "On Science" D9
STRING THEORY In "A Reading Life," bluegrass and fiddling pioneers D9
Thinking Big: The new insecurity D10-12

Filmmakers on the war path D2
By Thom Powers

Orgasmic science D4
By Christopher Shea

Kosher in Concord D5
By Ben Birnbaum

AND...
Loathsome Bostonians, Jane Fonda's latest goof, and more D2-3

Ideas online
In "The politics of pain," Drake Bennett discusses the long-running debate over the nature of pain and the proper medical use of powerful–and dangerous–opioid pain killers in treating it. Should such drugs be prescribed liberally to combat the chronic pain estimated to afflict as many as 50 million Americans? Or are the risks of addiction and abuse just too high? Record your thoughts on a message board by visiting

THE POLITICS OF PAIN

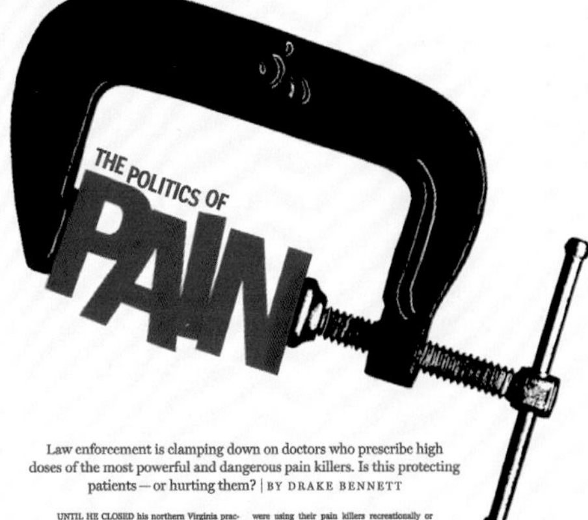

Law enforcement is clamping down on doctors who prescribe high doses of the most powerful and dangerous pain killers. Is this protecting patients – or hurting them? | BY DRAKE BENNETT

UNTIL HE CLOSED his northern Virginia practice in 2002, Dr. William E. Hurwitz was a nationally known pain specialist whose willingness to treat chronic pain with high doses of powerful narcotic pain killers like Oxycontin and Dilaudid had attracted patients from around the country. Many of them saw Hurwitz as a savior offering deliverance from years of agony that other doctors had been unwilling to treat. Hurwitz's liberal prescribing got him profiled on "60 Minutes." Twice, in 1991 and 1996, it also got his medical license suspended. And a week and a half ago, it got him sentenced to 25 years in a federal prison.

Hurwitz's was the most visible conviction in a three-year federal investigation of prescription drug abuse, a crackdown triggered by a widely reported rash of Oxycontin addictions in the late 1990s. According to prosecutors, Hurwitz's willful ignorance of the fact that some of his patients were using their pain killers recreationally or turning around and selling them was tantamount to running a drug ring out of his office. A few of his patients, the prosecution charged, suffered severe overdoses at his hands, one dying after Hurwitz prescribed her morphine at a dose 45 times higher than anything she had previously taken. As Drug Enforcement Administration chief Karen P. Tandy put it, "Dr. Hurwitz was no different than a cocaine or heroin dealer peddling poison on a street corner."

To Hurwitz's defenders, however, he was a victim of drug hysteria and of a cruel disregard for the destructive power of chronic pain. And while few pain doctors would defend all the particulars of Hurwitz's practice, many are deeply worried about the example that his conviction sets. According to Russell K. Portenoy, a neurologist and leading pain care specialist, PAIN, D5

Drake Bennett is the staff writer for Ideas. Email drbennett@globe.com.

Common ground

A group of historians want to reconsider the 1915 Armenian genocide – and prove that Turkish and Armenian scholars really can get along
By Meline Toumani

FIVE YEARS AGO, Ronald Grigor Suny, a professor of political science at the University of Chicago, sat in a tiny room on campus and waited nervously for a group of colleagues to arrive. "What have we done?" he asked his wife. "What if these people choke each other to death?"

The conflict that Suny feared was no arcane ivory tower dispute. It was the first meeting of the Workshop for Armenian-Turkish Scholarship, and most of the participants were of Armenian or Turkish descent. In other words, in addition to being historians, sociologists, and political scientists, they were members of ethnic groups that – particularly in the diaspora – view one another as sworn enemies.

Animosity between the groups stems from events in 1915 in Ottoman Turkey that Armenians – along with most prominent historians worldwide – call the "Armenian genocide," and that many Turks call the "so-called genocide" or the "Armenian allegations," if they don't use the phrase employed by Turkey's foreign minister, Abdullah Gul, at a press conference last month: "unacceptable claims by the [Armenian] diaspora to continue its existence." The Turkish government promulgates a view that the number of Armenians who died is much lower than Armenians claim – around 300,000 instead of 1.5 million – and that their deaths were the consequence of their collusion with Russian forces in World War I, not HISTORIANS, D6

Meline Toumani is a writer living in Brooklyn.

AP PHOTO/HERBERT BAGDASARIAN

A boy paused last week in front of a poster in Yerevan, Armenia, depicting survivors of the mass killings of Armenians that took place in eastern Turkey between 1915 and 1923. In recent years, a group of Armenian and Turkish historians have been working together to bridge the gap between the two sides' sharply polarized views of the events.

GL D1 20:20 FIRST RED BLUE YELLOW Black

Boston Sunday Globe

Ideas

AUGUST 14, 2005

E

& Books

CITY OF SHADOWS An anonymous diary of the fall of Berlin E6
LAUGHING MATTERS Romantic comedies, in "Pop Lit" E6
TEN FATHOMS DEEP Profiting from marine disaster E7
"LUNAR" LANDING From Bret Easton Ellis, a new novel E7
MISSISSIPPI MASTER An ambitious biography of Eudora Welty E8
UP AGAINST THE EVIL EMPIRE The Dodgers vs. the Yankees, 1955 E9
ICONOCLAST In "A Reading Life," remembering the British writer B.S. Johnson E9
Thinking big: Going to college in high school E12

Literary hoaxes E2
By Hua Hsu

Was the New Deal racist? E5
By Christopher Shea

Hiding at the movies E5
By Mark Pothier

AND...
Cap tricks, words to sail by, and more E2-3

IDEAS ONLINE
In "Tanked," Thomas C. Palmer Jr. discusses the rise and fall of Pioneer Institute, a local conservative think tank that helped bring market-oriented policy ideas to predominately liberal Massachusetts government in areas ranging from education to welfare reform to cutting red tape. Does the Bay State need a greater infusion of free-market thinking on the big policy issues facing it today? Record your thoughts on a message board by visiting www.boston.com/ideas.

Tanked

In the 1990s, Pioneer Institute helped introduce market-oriented ideas to Massachusetts government. So why did the upstart conservative think tank run out of gas? | BY THOMAS C. PALMER JR.

WHEN A COUPLE of hundred people got dressed up to go to Pioneer Institute's 14th annual Better Government Competition awards dinner on a splendid summer evening in late June, they put on a good face. They congratulated each other on saving state and local government an estimated $300 million through innovative ideas they have generated over the years, heard state administration and finance secretary Eric Kriss compare altering government to the evolutionary change in coloration of the upside down catfish, and listened politely as adviser-to-four-presidents David Gergen lamented the disappearance of the good old days in Washington, when ideological foes such as Ronald Reagan and Tip O'Neill would drop their differences at 5 o'clock and hoist glasses like old buddies.

But many of those present at the Boston Harbor Hotel were aware it was a time of crisis for a think tank that throughout the 1990s had made dramatic headway in rolling the rock of free-market ideas up the steep hill of Massachusetts' political culture. They didn't let on, but they knew Pioneer had lost its way. A glaring clue: The name of the person who had led the institute since 2001, president and chief executive Stephen J. Adams, was mentioned nowhere in the four-page program.

In the same way that nationally oriented conservative think tanks like the Cato Institute and the Heritage Foundation provided fresh ideas that helped bring the Republican Party out of the wilderness after Barry Goldwater's defeat in 1964 and fueled the subsequent Reagan Revolution, Pioneer Institute was instrumental in many of the victories won by lonely Republican governors of Massachusetts on overwhelmingly Democratic Beacon Hill, on issues ranging from education to PIONEER, E4

Thomas C. Palmer Jr. is a member of the Globe staff. He can be reached at tpalmer@globe.com.

Orange alert

As Israel prepares to withdraw from Gaza, pigment gets political | BY BORIS FISHMAN

THE ISRAELI WITHDRAWAL from the Gaza Strip, scheduled to begin tomorrow, has provoked ferocious opposition from those who view it as a concession to Palestinian terrorism. There have been periodic warnings of civil war, but, as of press time, the only war to break out in Israel has been between colors: anti-disengagement forces are orange, whereas the government's supporters are blue, after the Israeli flag.

These days, especially after the resounding success of the Orange Revolution in Ukraine, color-coding is de rigueur for fledgling political movements. Earlier this year, Kuwaiti advocates of women's suffrage marched in blue. Mongolian reformers are yellow. Azerbaijanis and Moldovans, not wishing to stray from the path of success paved by the Ukrainians, adopted orange.

"Orange is an excellent choice for a political movement," said Leatrice Eiseman, executive director of the Pantone Color Institute in New Jersey. "It's the marriage of red, which is exciting and dynamic, with yellow, which is friendly and convivial. It's the best of both worlds."

Indeed, in Israel, orange seems to have bested blue, prompting observers to wonder whether it was the anti-pullout campaign's bid to share in the magic of the Ukrainian moment. The reality is more mundane. Orange, meant to evoke the surrounding sun and sand, is one of the municipal colors of Gush Katif, the main group of Jewish settlements in Gaza. Earlier this year, the Israeli fashion establishment had anointed it as the "in" color of the summer. "If someone doesn't choose orange," ORANGE, E4

Boris Fishman is the editor of "Wild East: Stories from the Last Frontier." He lives in New York.

PHOTO: EITAN ABRAMOVICH/AFP/GETTY IMAGES

COLOR ME RESISTANT. An Israeli policeman arrests an opponent of Israel's disengagement plan from the Gaza Strip, Aug. 8. Anti-withdrawal activists have adopted orange as their color, while the government supporters embrace the blue of the Israeli flag.

GL E1 20:23 FIRST RED BLUE YELLOW Black

报纸的编辑设计师和杂志设计师不一样，他们没有办法指望像杂志一样用高清大图、醒目色彩或者光亮的纸张来吸引读者。所以必须另辟蹊径。《波士顿环球报》就是一例。“文字编排是体现报纸外面最初十秒钟的时间里影响读者判断的首要因素。通过不同的文字编排，可以传达出严肃、年轻、趣味等不同感受。其次是配色，读者会对文字和配色的搭配产生即时、直观的判断。最后就是留白对读者感

《骑行》（*The Ride Journal*）杂志的艺术总监安迪·迪普罗斯（Andy Diprose）将封面图片直接跨页铺到封底。作为一份独立的自出版杂志，它可以不受传统杂志封面文案、条形码等的限制。

数字刊物封面

出版商已经看到了封面的潜力，意识到这是一个编辑、出版商与读者进行互动的入口。数字屏幕的尺寸也会影响封面图片的特点：例如，智能手机屏幕上的封面图片就会比纸质版要小得多。因此设计数字刊物的封面时，有两个基本原则：一是要有强有力的图像和封面字体，能够吸引读者、勾起阅读兴趣；二是数字刊物的封面要和纸质版的封面有所关联。例如《纽约时报》（*The New York Times*）的数字版，虽然把Logo做得很小来增加趣味，但同时也暗示读者，数字版和纸质版内容一样丰富。在移动设备的屏幕上阅读刊物和阅读纸质刊物的体验很不同。读者阅读数字刊物的方式很多，可以是网站、App应用甚至是朋友发来的一个链接。因此，设计师必须要综合考虑、用心设计，封面和刊头要精彩，同时保持简洁、清晰的品牌辨识度。

在触摸屏上，封面就是一张图片。读者可以通过滑动或者触摸封面阅读数字刊物。一般的销售模式是付费订阅数字版杂志，或者在订阅纸质刊物时，获得免费的电子版或App版。

自出版杂志封面和小众杂志

如今小公司有很多机会自助出版刊物。通过按需印刷技术进行自助出版，可以绕过传统印刷与发行营销的门槛，降低成本，减少资源浪费。独立杂志界创意不断、欣欣向荣。不少有志之士投身其中，以满足新兴细分市场和受众的需求。独立出版商可以利用社交媒体进行刊物宣传，省去了广告费用。举例来说，史蒂夫·沃森（Steve Watson）独立创办的Stack Magazines自出版杂志网站，就是位于伦敦的独立杂志集团。在这个平台上可以买到很多有特色的刊物，例如*Oh Comely*、*Anorak*、*RiDE*、*Port*和*Huck*等。这些自出版杂志的封面和一般书店里的刊物不同，没有过多的文案，整体简洁、美观。

《火与刀》（*Fire & Knives*）杂志，是一家出版A5开本杂志的小众美食杂志。封面用插图替代图片，以凸显和其他美食杂志的差异。杂志广告不多，文艺气息浓厚，积极评论美食行业以及各种饮食现象。杂志主要由艺术总监罗伯·洛（Rob Lowe）手工制作，有些页面是画好之后直接扫描的，以降低使用设计软件的设计成本。加之其出版规模不大，所以绕开经销商自行发行。

这些小众杂志，是在悉尼通过丝网印刷工艺生产的，采用单色印刷。制作者是尼尔·爱德华兹（Neil Edwards），曾任职于*i-D*杂志。爱德华兹说：“它们就像我在圣马丁艺术与设计学院（Saint Martins College）时影印的小书一样，尽量保留了不完美的痕迹，这样每一本都独一无二。”

小众杂志也是独立出版杂志产业的一分子，发行量小，关注小众话题，通常使用影印机复印就可装订成刊。曾经，这种小众杂志多数是黑白的宣传手册，非主流意味很浓。内容也常常被视为不够得体或稍有争议。英国时尚杂志*Dazed & Confused*以及 *i-D* 都是从小众杂志起家，后来因为思想创意为公众所知，之后成长为大型主流杂志。20世纪90年代，纽约暴动女孩（New York Riot Girl）现象推动了很多自制的小众杂志出现，有的甚至挑战主流的政治观。这些杂志通过口耳相传为愤怒的女性发言，使该运动名声大振。

定制封面

2010年，*Wallpaper** 杂志邀请读者利用指定的一些元素通过线上App制作自己的刊物封面。这些元素由奈杰尔·罗宾逊、詹姆斯·乔伊斯、甘棠、安东尼·伯里尔等艺术家和设计师设计。读者利用模板和指定元素进行自由组合，创作出属于自己的定制封面，提交后会收到寄来的有专属封面的纸质杂志。艺术总监麦里昂·普利查德（Meirion Pritchard）解释说：

> “我们第一次尝试手工特刊。实际上是由印刷厂商、设计师和读者一起创造了这期刊物。首先，我们制作了一款软件，让读者能够自行组合设计师设计的元素。之后需要把读者的作品印刷出来。但是因为每种设计都各不相同，所以我们需要事先找到可以配合的印刷厂，才能确保刊物品质。最后，每位读者都可以收到自己设计封面的定制杂志。我们一共印刷了21000份不同封面的杂志。”

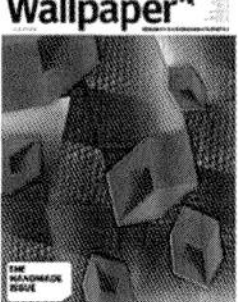
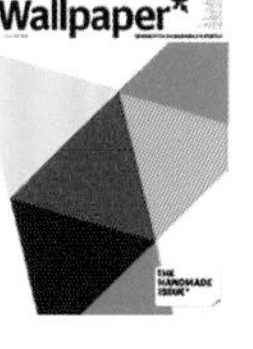

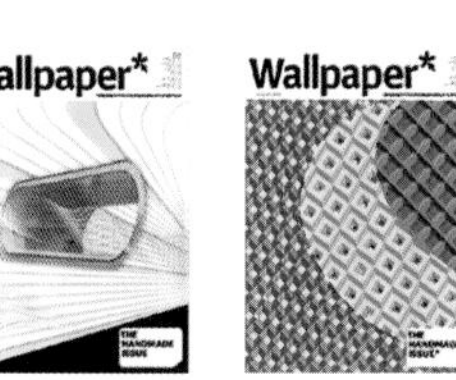

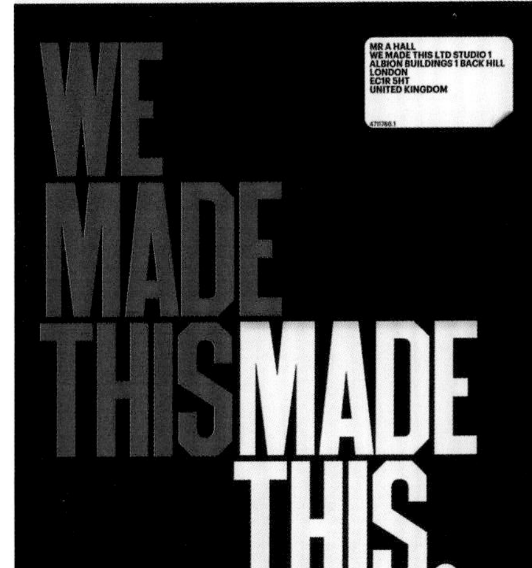

阿利斯泰尔·霍尔（Alistair Hall）设计了自己的封面，并将之发布在自己的博客上。

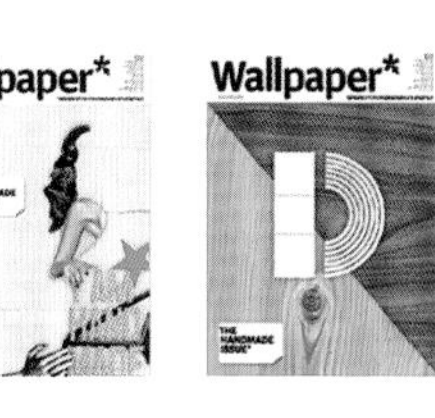

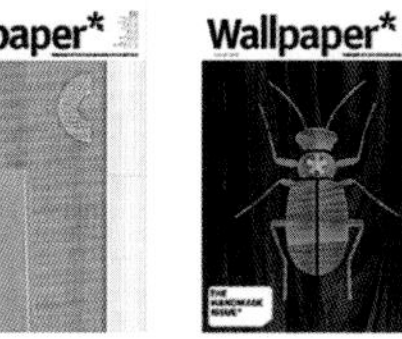
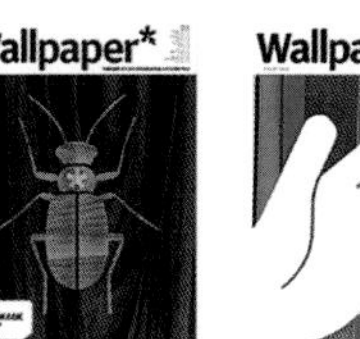

设计与艺术总监协会将2011年的黄铅笔奖（Yellow Pencil），颁发给了*Wallpaper**杂志这项邀请读者做封面的创举。

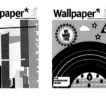

1940年迄今的封面发展

1940—1950

20世纪40年代的英国杂志多为黑白印刷，有些搭配彩色封面，有些杂志和20世纪30年代的娱乐资讯杂志一样，是靠排字工人用金属铅字排版，这样的制作过程限制了字体的使用以及广告的空间。电影类杂志可以说是最早的大众流行专业领域的杂志，与当时电影院日渐普遍的现象进行呼应。这些杂志很便宜，不过在第二次世界大战期间，英国出现纸张短缺现象，部分杂志不得不停刊。

美国《时尚先生》（*Esquire*）杂志在1933年创刊，邀请了知名作家弗朗西斯·斯科特·菲茨杰拉德（F. Scott Fitzgerald）、欧内斯特·海明威（Ernest Hemingway）等撰写专栏。这时，杂志开始使用彩色封面，以便能够在报刊亭里脱颖而出。战争使得当时很多躲避战火的欧洲英才流向美国，奥地利的亨利·沃尔夫（见211页）就是如此。战后很多才华横溢的设计师，如《时尚芭莎》（*Harper's Bazaar*）的阿列克谢·布罗多维奇（Alexey Brodovitch）（见208页）等，为美国大众引入包括超现实主义画家萨尔瓦多·达利（Salvador Dalí）、A.M. Cassandre等在内的欧洲艺术家。1943年，亚历山大·利伯曼（Alexander Liebermann）成为美国版时尚杂志*Vogue*的艺术总监，于是他发挥自己的特长，引进现代感的摄影艺术与印刷技术，给美国以及欧洲很多艺术总监带来了启发。

女性在战后回归家庭，*Good Housekeeping*与*Better Homes and Gardens*等女性杂志应运而生，这时更讲究封面图片设计，以吸引读者目光。

1950—1960

现代广告诞生于20世纪50年代，杂志开始刊登一些针对战后大批回归家庭的女性受众的产品与服务的广告。第二次世界大战后，一批欧洲设计师为了逃避战争余波，纷纷涌入美国，女性时尚杂志也从中

《时尚先生》（*Esquire*）的乔治·路易斯（George Lois）开始利用简洁的标志性照片以及拼贴画来讲述故事。图文相互衬托、相辅相成、颇具美感。一些寻找创意以吸引读者的艺术总监们，深受这种风格的启发。

视觉识别对品牌的影响可能被误解了。《时尚芭莎》和*Vogue*就是很好的例子。即使它们几乎没什么Logo，还是会被识别。而那些不知名的杂志则相反，Logo再显眼，也没什么影响。

HARPER'S
BAZAAR
ch 1959 60 cents
Eyes
on
Paris
and
America

第七期*Oz*杂志是其最出名的一期。因为这期封面是当时的标志性人物鲍勃·迪伦（Bob Dylan），是马丁·夏普（Martin Sharp）利用当时的新印刷技术创作的图像，用来阐释当时实验性的、带有迷幻情绪的音乐和文化风格。

受益。亚历山大·利伯曼（Alexander Leiberman）开始为康泰纳仕集团（Condé Nast）工作，为其带来了欧式美感。

曼·雷（Man Ray）和李·米勒（Lee Miller）是当时的知名搭档。《时尚先生》杂志那时开始采用彩色印刷，并很快吸引了广告商的目光。虽然英国战后的恢复速度比美国慢，但是伦敦和美国的杂志人才却在这期间交流频繁。这一时期，杂志封面开始推陈出新，常常采用新闻纪实摄影作品做封面，曝光了许多战争期间震撼人心的画面。

1960—1970

《周日泰晤士报》（*The Sunday Times Magazine*）的构图和跨页排版设计，奠定了英式刊物的风格基调。设计师大卫·金（David King）带领着一支年轻的设计师团队，交出了漂亮的成绩，还发掘了像大卫·贝利（David Bailey）这样的摄影师。20世纪60年代的杂志封面反映了当时社会的重大变革，关注越南战争、性别革命等主题。《生活》（*Life*）杂志在电视尚未普及之前，就已经把新闻照片带入了美国的百姓家中。《插画新闻杂志》在英国也扮演着同样的角色。杂志封面成为读者了解世界的一个重要窗口。越来越多的杂志采用彩色印刷，美国和英国的杂志上刊登的广告也反映着生活形态的变化。这一时期，大众偶像开始登上杂志的封面，像大卫·贝利（David Bailey）、理查德是·阿维顿（Richard Avedon）、诺曼·帕金森（Norman Parkinson）等著名摄影师都曾拍出过非常隽永的照片，至今依然令人赞叹。

在《访谈》（Interview）杂志上，安迪·沃霍尔（Andy Warhol）曾在封面和封底刊登全幅肖像画。从封面上看起来就像是传统的明星特写，然而，展开封底，读者会对拍摄视角和主题有更深的理解。

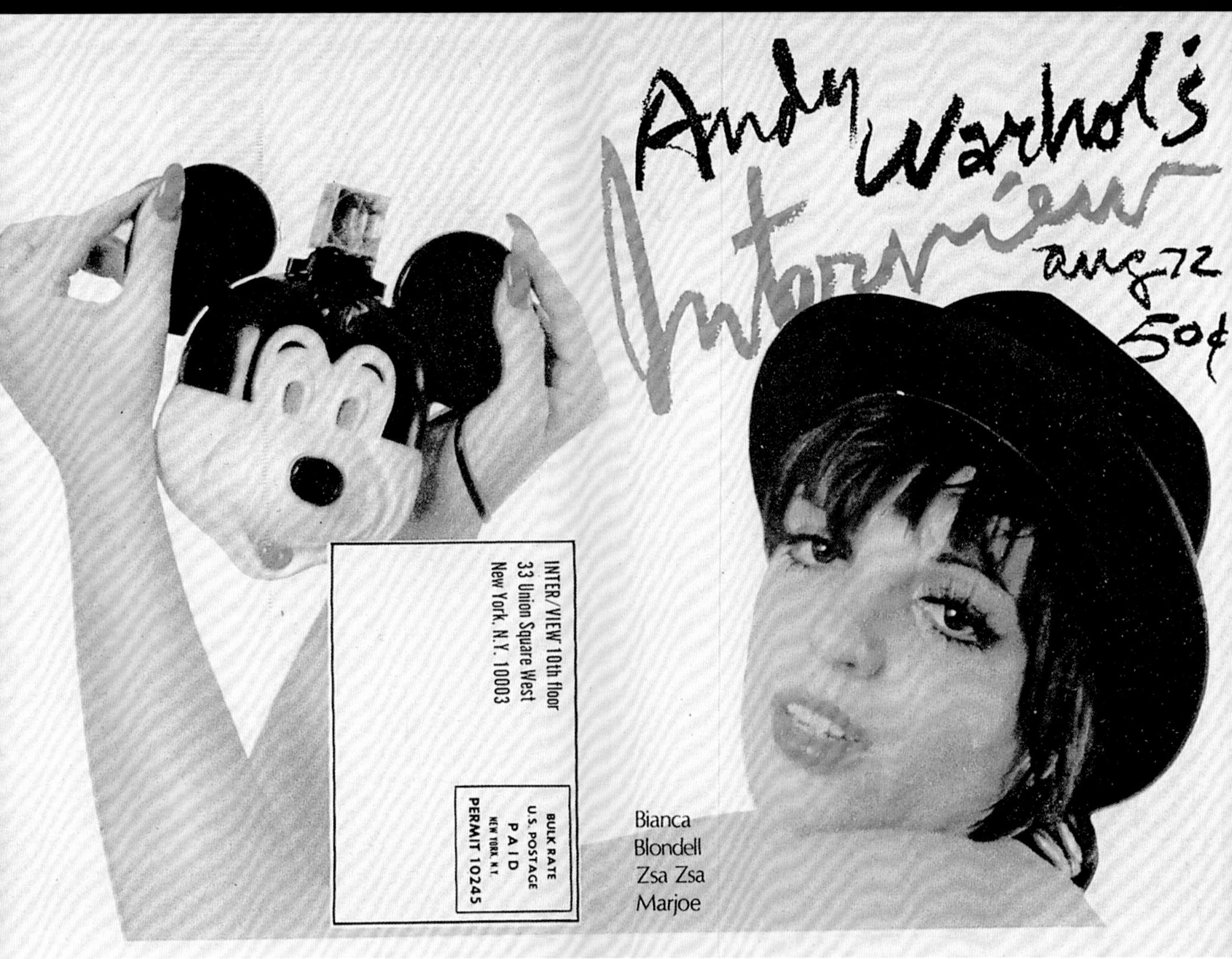

1970—1980

20世纪70年代是文化变革蓬勃发展的十年，这也反映在刊物设计的变化上。《滚石》（*Rolling Stone*）和*Nova*等杂志，建立了令人赞赏的新标准。美国的《滚石》杂志封面颇具政治色彩，内容则以音乐流行文化为主。英国的*Nova*是有自由派政治倾向的女性杂志。这些杂志就像十年前德国的*Twen*一样，用大胆的封面照片反映所处社会的背景，把目标读者定位为那些不安、精力旺盛、充满反派精神的年轻人。在欧洲，政治动荡成为新闻摄影崛起的绝佳机会，德国的《明镜周刊》（*Der Speigel*），利用强烈的图像封面来呈现政治主题。

这时期设计师开始大胆探索跨页排版。《世界时装之苑》（*Elle*）杂志另辟蹊径，大量使用倾斜的版式，这与20世纪60年代印刷精美的时尚杂志的垂直水平排版方式大相径庭。一时间杂志格式变得活泼。《访谈》杂志很懂得活用封面名人照片，意大利版的*Vogue*杂志则靠着法比恩·贝伦（Fabien Baron）的卓越设计而闻名。著名设计师内维尔·布罗迪（Neville Brody）在杂志上尽情发挥文字编排

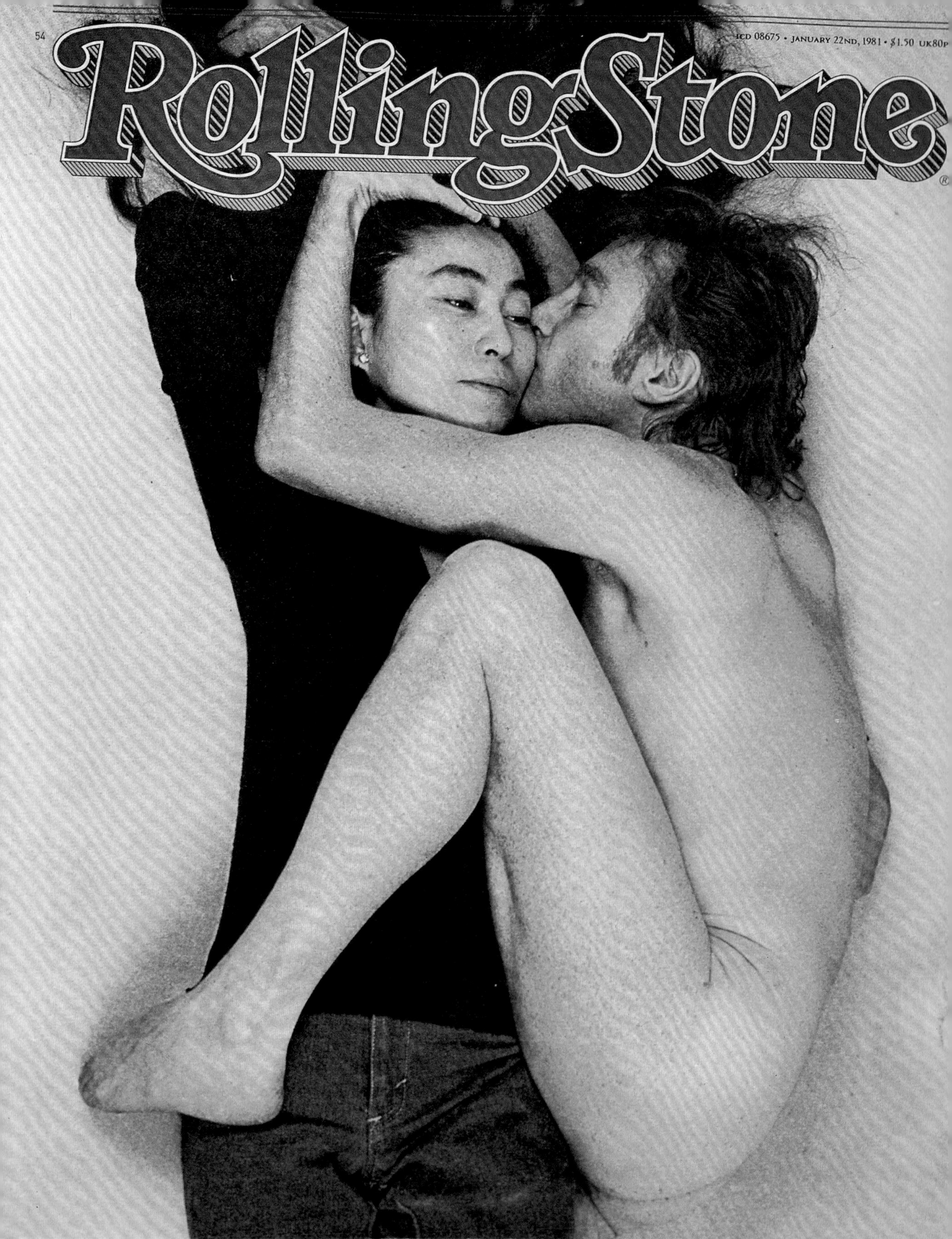
RollingStone
ICD 08675 • JANUARY 22ND, 1981 • $1.50 UK80P

的创意，带来了新的排版风格，其他设计师更有信心探索以格线为基础的排版形式，进一步挑战极限，排版工作原本是交给专业人士的，但是在20世纪80年代苹果电脑上市后，设计师也能自行制作封面了。

20世纪70年代初期，杂志还是黑白印刷，只有封面和专题报道的少部分内容是以彩色印刷的。但这十年来，随着印刷技术的飞速发展，四色印刷的价格降低，因此彩色印刷也更加普遍，刊物开始使用铜版纸，并且尝试不同的开本。

1980—1990

20世纪80年代初期，出现了许多优秀的时尚杂志与男性杂志。报社出版的报纸增刊类型杂志，是时尚类平面设计的实验场所，也可以说是纸质刊物纪实摄影报道的最后堡垒。放弃了密密麻麻的封面文案，这些杂志的封面凭借极具视觉张力的照片脱颖而出。在英国，像*i-D* 和*The Face*这样的杂志善于抓住时尚氛围。至于美国，民众开始购买电脑，对科技类杂志的兴趣随之增加，电脑类杂志应运而生，例如1984年上市的*MacUser*杂志。在杂志出版界，设计师开始玩起了数字设计，首次在电脑上将整本出版物设计完成。加州*Emigre*杂志的设计师开始试验把像素作为设计元素，并纳入排版设计。大卫·卡尔森通过*Beach Culture* 和*RayRun* 杂志，将解构主义带到平面设计中。英国设计师纳威·布罗迪（Neville Brody）、沃恩·奥利弗（Vaughan Oliver）和文斯·弗罗斯特（Vince Frost）也在媒体上展现出大胆的设计。

《滚石》把安妮·莱博维茨（Annie Leibovitz）为约翰·列侬（John Lennon）和小野洋子（Yoko Ono）拍摄的照片作为当期杂志的封面。也正是这张照片，帮助《滚石》一举成为当时炙手可热的杂志，这张照片很美，充满象征意义，完美地捕捉了时代精神，约翰·列侬在1980年12月拍完这张照片5小时后就在曼哈顿的达科他大楼外被枪杀，《滚石》杂志仍如期发表这张照片，向这位披头士成员致敬。

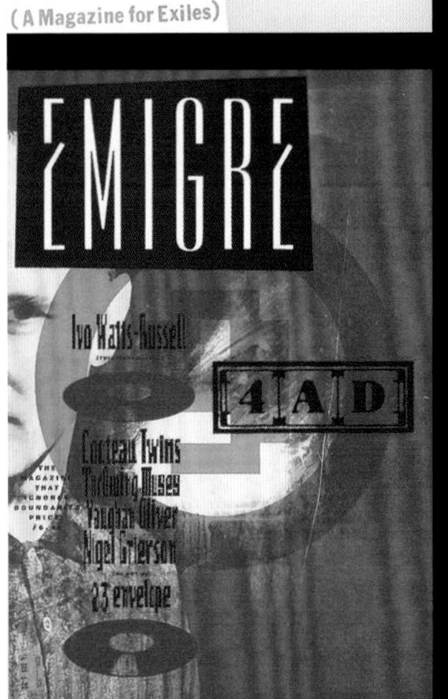

1984年，*Emigre*杂志反映出旧金山的氛围变动。这份杂志是由荷兰设计师鲁迪·范德兰茨（Rudy VanderLans）和他在捷克出生的妻子祖莎娜·利科（Zuzana Licko）创办的新潮杂志。利用可设计字体的麦金塔电脑，来试验数字化的字体和版式。回顾过去，会发现这是一个真正的转折点，桌面排版系统让设计师能够用像素来创造自己的字体，打破固有的规则。

DENMARK DKR37 GERMANY DM10 GREECE DRA470 HOLLAND DFL8.25 ITALY L4700 SPAIN PTAS420 SINGAPORE M$8.90 USA $4.50

JANUARY 1989 No. 73 £1.65

BLITZ

INTO THE FUTURE...

And so to bed...

review 88

preview 89

What to expect
in media,
film, music and print

plus
herb ritts
pj o'rourke
peter o'toole
siobhan fahey
mike leigh

LONDON'S POLITICS ★ ART ★ BOOKS ★ CINEMA
NEWS ★ SPORT ★ CLASSIFIEDS ★
CITY LIMITS
MARCH 12 TO 18 1982. NO.23 50P
16 PAGE PHOTO SPECIAL!
LIFE AND DEATH IN APARTHEID SOUTH AFRIC

上图是戴夫·金（Dave King）为伦敦娱乐资讯杂志《城市界限》（*City Limits*）设计的封面。他大胆采用了一些政治类海报的设计技巧，把群众集结的画面用三种不同颜色重复印刷三次，刊头文字则做了镂空处理。戴夫·金经常改变刊名“City Limits”的字号与位置，但是字体不变，这样既保持了设计的灵活性，也保持了强烈的刊物辨识度。

左图是1989年一期*Blitz*杂志的封面，巧妙的布局让歌手麦当娜（Madonna）看起来像正在深情地凝望杂志的刊头，在图像和文字之间营造了一种互动氛围。

下图的*Dazed & Confused* 杂志中，在封面增加刮刮卡涂层，营造一种趣味性。最下图的另一本杂志，封面文字编排很吸引人，用非常简单的方式，凸显出杂志讽刺性的品牌特质。

这一时期崛起不少明星设计师，他们促进了时尚出版物的蓬勃发展。春秋两季出版的刊物中，都是时装设计的大师之作。这两季期刊深受广告商的喜爱，反过来，他们也会雇用同一批摄影师来制作自家的广告页面。随着经济繁荣，大众渐渐开始消费奢侈品，时尚产业随之欣欣向荣。20世纪80年代末，英美杂志业迎来了黄金时代，人才济济，相互合作，迸发出精彩火花，编辑设计和新闻摄影专业变得很受尊重。

1990—2000

拍摄名人照片作为杂志封面几乎主导了整个20世纪90年代，有些公众人物、演员和音乐家在登上封面杂志后，事业也更上一层楼。这股风潮也帮助杂志保持新鲜和活力，杂志封面直接反映出读者的文化偏好。随着品牌的发展和廉价的数字印刷技术的创新，消费者杂志也加入了这个市场。设计师们不再需要在大型印刷机上印刷客户杂志，有更小的数字打印机，可以满足更小印量的需求。数字印刷采用了包括喷墨在内的更简单的技术，而且不再需要制版等印前工序，节省时间的同时还能降低成本。

随着杂志的广告收入日渐增多，报纸不得不重新认识自己曾经在新闻市场的霸主地位。新闻出版从业者倍增，竞争越来越激烈。要想维护住其霸主地位，就需要另辟蹊径。新闻网站最初是免费推出的，随后很快就引入付费墙机制。此外数字印刷技术的出现以及品质的日渐提升，也大幅改善了报纸的生产方式。旧的印刷机被废弃，大报缩小尺寸，数字文件可以直接送印，不再需要印版或制作彩色样张，成本降低。于是不仅是设计师，连企业也纷纷采用数字印刷。

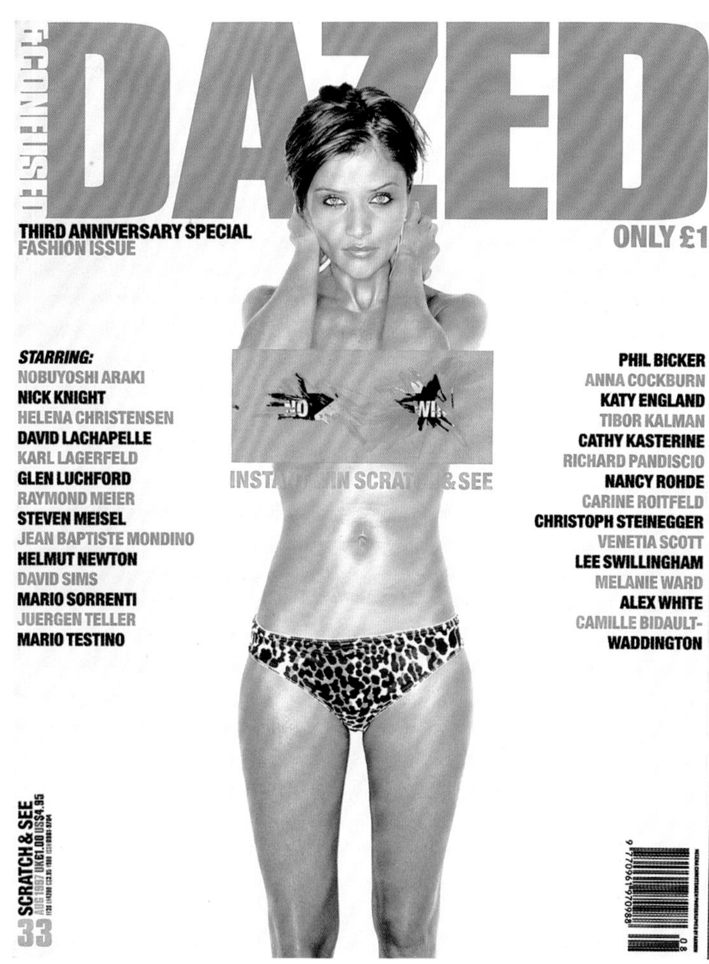

ABONNEZ-VOUS       

Rechercher *Kate Moss*

CONNECTEZ-VOUS

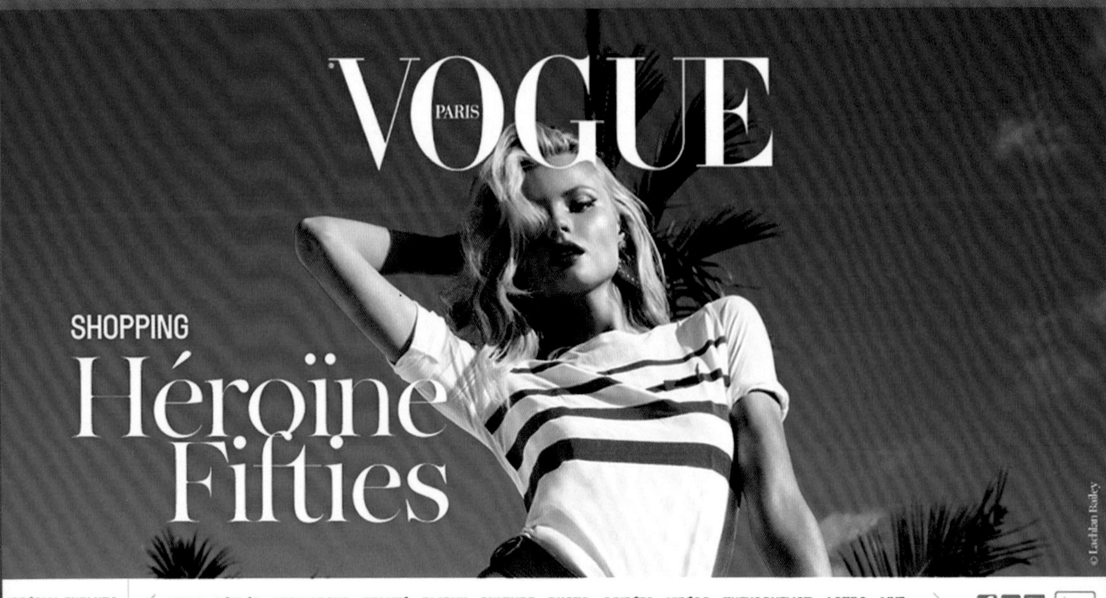

SPÉCIAL ENFANTS | MODE | DÉFILÉS | LOOK BOOKS | BEAUTÉ | BIJOUX | CULTURE | PHOTO | SOIRÉES | VIDÉOS | THEVOGUELIST | ASTRO | LIVE

26 Avr 2013

PARTAGEZ

SHOPPING

Vernis parade

A effet velours, thermoactifs ou endiamantés...
Les vernis font leur grand show pour les fêtes de fin d'année.

Lire la suite

2000—2010

21世纪前十年，平板电脑尚未问世，杂志一时独领风骚。大型出版公司把旗下的纸质杂志放到互联网上，例如2009年的《纽约客》（*The New Yorker*）和《连线》（*Wired*），以及2011年的《卫报》（*The Guardian*）等熟悉的报纸，很快就出现了网页版和移动应用程序。互联网成为杂志广告模式的最大威胁，新闻机构不得不重新思考自身定位。此外，这时期爆出了维基解密事件和英国电话窃听丑闻，媒体诚信受到质疑，知名媒体大亨的威望不断下滑。在美国，老牌报纸对员工进行重组，大量员工失业。例如，2009年《波士顿环球报》（*Boston Globe*）进行数字化转型。纸媒时代受过专业训练的设计师必须重新学习编程等技能，以适应数字交互的设计需求。年轻的“数字土著”设计师倒是很适应，在不同平台设计间的转换游刃有余。

文艺杂志*Zembla*，由文斯·弗罗斯特（Vince Frost）设计，图文编排非常有视觉冲击力，并采用大胆、清晰的版面结构。

Vogue 杂志是全面发展的一个范例，为不同平台设计不同版本。纸质杂志可以借助网站来扩大读者群，也为广告商接触读者提供了更多方式。

2010年至今

2010年苹果平板电脑上市后，出版商和设计师可以在设计中加入交互元素。凭借轻薄的外形以及可触摸的屏幕，苹果平板电脑吸引了苹果手机和其他苹果技术粉丝，也比其他正在开发的平板电脑更有档次，把其他公司的平板电脑远远甩在后面。后来，其他平板设备也纷纷推出了新的操作系统，例如谷歌的安卓操作系统，就是基于开源模式开发而来。设计师利用触摸屏的特点，开发出了新的视觉浏览模式，并强化了平板电脑及手机中的动态内容。消费类杂志的平板电脑版本，封面首次增加了交互功能，封面上的设计元素成为内容的入口，读者只要轻轻滑动手指，就可以浏览想看的内容。

这时，以读者的历史搜索记录和浏览偏好数据为基础的个性化推荐出现了，这些个性化的聚合内容可以被整合到读者的平板电脑和手机上。如此一来，编辑、艺术总监、出版商和开发人员的角色也随之发生改变。对于有些人而言，不能接受没有编辑人员处理的内容，但对另外一些人来说，摆脱编辑反而使得阅读变得更自由。如今，每个人都可以定制自己的个性化杂志。例如，Flipboard这款软件，就可以将特定内容聚合后供读者订阅。只要下载其应用程序，所订阅的社交媒体、网站和博客上的内容，就能被整合成精美的个性化数字杂志，供百万读者阅读。

浏览杂志不再是线性方式了。以《卫报》苹果平板电脑版为例，设计师的任务是利用视觉暗示引导读者阅读，并避免阅读过程中出现任何阻碍。

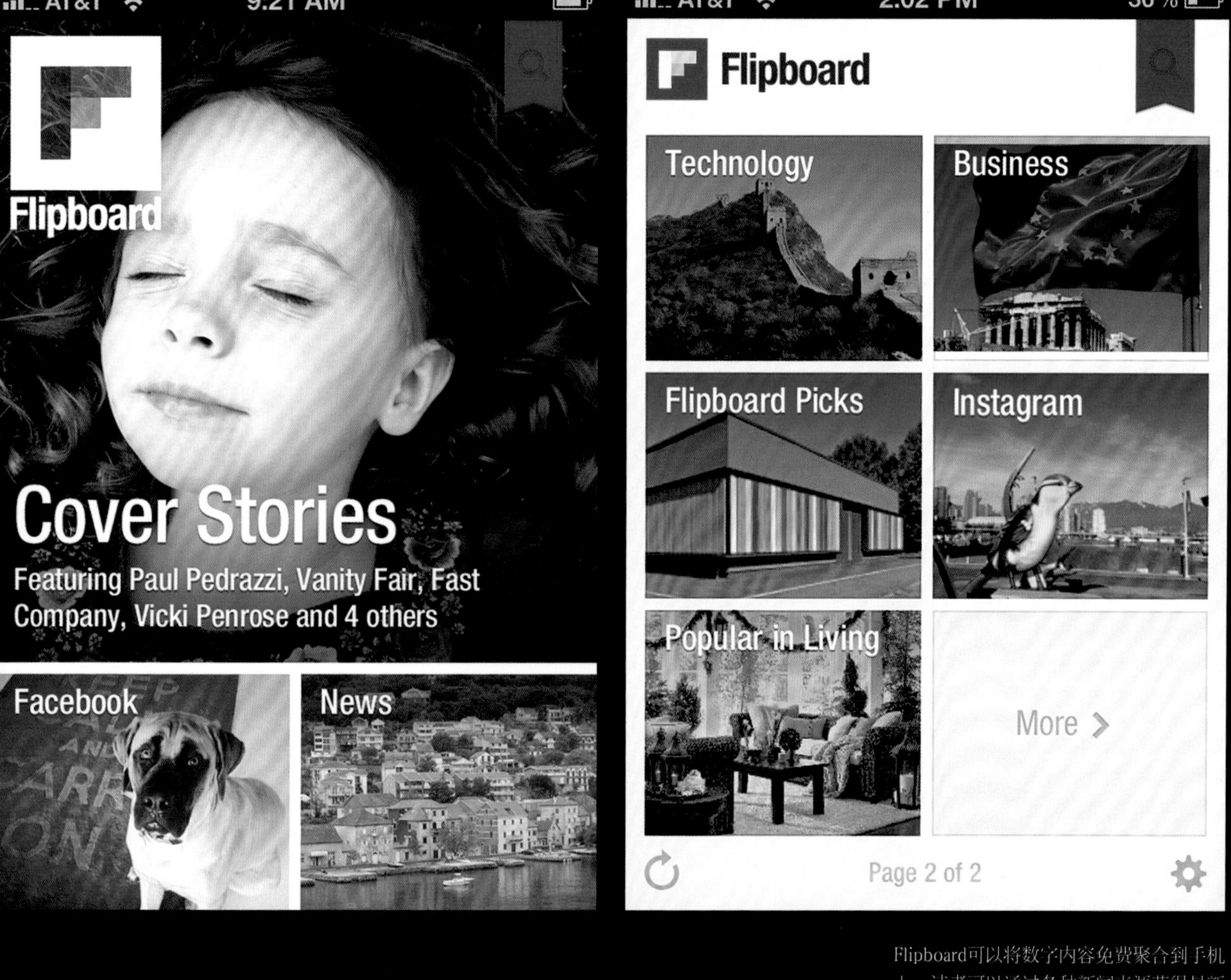

Flipboard可以将数字内容免费聚合到手机上，读者可以通过各种新闻来源获得最新讯息。用户只要滑动手指，就可以自由浏览不同页面，还能即时分享页面到社交媒体上。受到纸质媒体页面精美、翻阅轻松的启发，它将自己定位为“世界上第一本社群杂志”。

封面设计的类型

封面设计有许多不同方法，广义上封面设计可以分成三类：具象封面、抽象封面以及文案主导封面。其中文案主导封面较为少见，因为单纯使用文字或者利用双关语进行封面设计，视觉影响力不够大，编辑也就避而远之。

具象封面

一般而言，具象封面会利用一张摄影作品作为封面的主视觉，即使是传统的局部特写或者全身照，也可以通过加一些创意元素变得更具吸引力。例如，把笑脸改成愤怒、恐惧或者带一些得意的表情。当然这种创意的发挥要符合杂志调性，并且在读者的接受范围内。例如，反消费主义杂志《广告狂人》（*Adbusters*）的读者，就不太可能被负面形象吓跑，而女性周刊的读者可能就会。巧妙风趣的封面往往能吸引读者，而带有冒险感的动作镜头更能使人身临其境。即使是普通的脸部照片也可以变得有趣：时尚杂志*i-D*的Logo就是一个眨眼的形象，因此其封面上的面孔也常常是眨眼的形象。全身入境的摄影作品发挥起来空间更大，*Dazed & Confused*杂志就常常使用。*Carlos*杂志则喜欢用插图来表现封面人物，配合肆意飞溅的金属墨水，往往有着惊人的效果。

插图应用在时装类杂志的封面设计上，效果也很棒。用插图来展现服装，比照片更能传达面料带给人的感受。插图的另一个优势就是能够和文字更好地融合，而不像照片那样和文字界限分明，只有文字围绕图片或者文字压在图片上等简单形式。插图拼贴蒙太奇是惯用的手法，可以给具象的封面带来另一维度的隐喻，很适合用来表现尖锐深刻的评论。

加拿大季刊*Adbusters*（上图），这份刊物的副标题为“充满反文化干扰色彩的杂志（culture-jamming revolution）”，就是关注特殊焦点的刊物如何与目标读者培养关系的优秀范例。每期杂志都有专门的主题，就像是一本迷你书，因此每一期的外观都很不一样。

本页带来两种截然不同的对特写镜头进行诠释的方式。在*M-real*杂志（上图）上，创意总监杰里米·莱斯利（Jeremy Leslie）通过手绘面部肖像，反讽了杂志封面热衷于使用与读者有眼神接触的女性面部特写的做法。*Pop*杂志（右图）也一样，只是用了不同的方式。封面上的詹妮弗·洛佩兹，虽然也没有与观众进行眼神交流，但不同于一般封面人物冷漠、淡定的神情，这里的人物面部有着生动的表情，使得封面从众多的竞争者中脱颖而出。

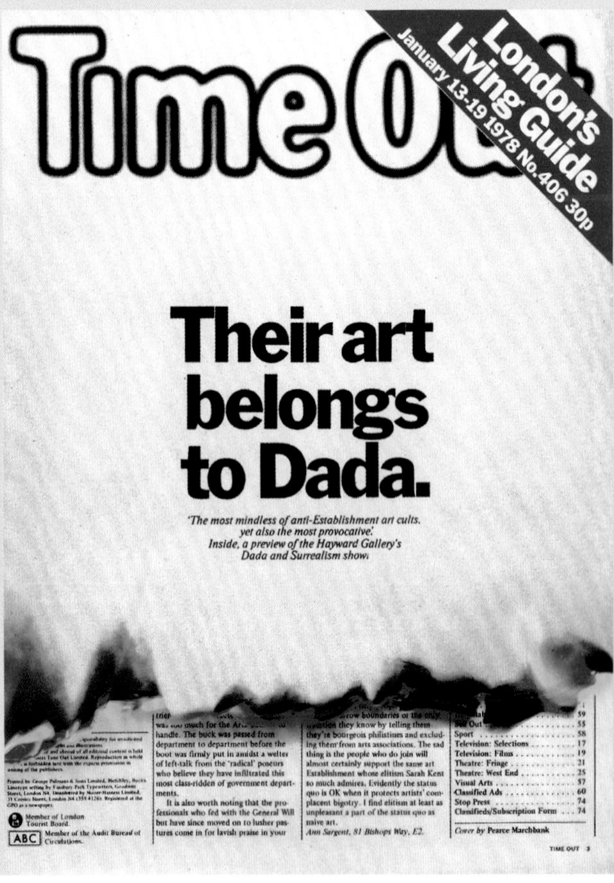

封面设计最简单的原则就是引起读者的兴趣，图像视觉是设计师最先使用的元素，但这只是其一，还有其他做法。整体而言，封面由四种元素构成：

① 开本——尺寸、形状和设计特点；

②Logo、刊名以及其他常见的封面信息，如标语、日期或条码；

③图像；

④文案和文章标题。

创刊于20世纪90年代的男性杂志*Lad Mag*，将上述四种元素使用得淋漓尽致。封面顶部的标语就说明了这是一本给“该知道更多的男性”看的杂志，它的设计和编辑手法洋溢着类似“我们疯狂了”（we' re off our heads）这种浓厚的疯狂意味，完全符合当时男性读者正在经历（或想要）的那种生活方式。编辑詹姆斯·布朗（James Brown）曾表示，这本书是为“那些相信只要不醉酒就没什么问题”的男性准备的。在设计方面，艺术总监史蒂夫·里德（Steve Read）的设计完美而巧妙地诠释了这一点，封面看似未经设计，但充满活力和动感，同时出色地运用了色彩、字体、图像和布局结构。封面和标题大而醒目，活泼又趣味盎然。

概念性的封面尤其引人注目。20世纪70年代，皮尔斯·马奇班克（Pearce Marchbank）在设计*Time Out*杂志（上图右）时，巧妙地运用了摄影、插画、拼贴和文字编排来“宣传”比较难以传达的抽象概念，如达达主义（Dadaism）和“嫉妒”（Envy）等。文斯·弗罗斯特（Vince Frost）为周六《独立报》(*The Independent*）增刊（右图）设计的封面，也很引人注目。他将剪切的摄影照片放在白色背景中，营造一种充满智慧和优雅质感的氛围，并配以宽泛故事的概念。这两位设计师都知道，要把封面当作海报来设计，因为它实际上就是海报，一定要能吸引观众。

02/09/95
Independent
Magazine
Can Britain bite back?

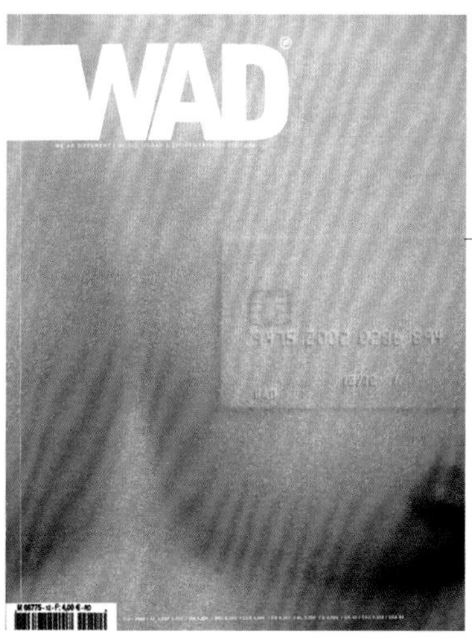

如果刊物的设计者、编辑和出版商有勇气挑战读者的接受度，勇于推翻固有的“受欢迎”“好卖”的封面设计观念，就可以设计出非常有创意的具象封面。例如*Adbusters*杂志（最左图）和法国杂志*WAD*（左图）的例子所示。*Adbusters*巧妙地破打破传统“好卖”封面的设计观念，封面上的金发女特写明显与传统审美不同。

“这是我们的第一期‘设计专刊’，我们把它看作前卫入门读物，向读者阐释哪里能找到好的设计。我们决定稍微让读者也参与一下这个过程，于是举办了一场设计比赛，邀请六位有着不同背景的优秀平面设计师，提出设计理念，并设计出六种封面方案（右上图）。一看到珍妮佛·莫拉（Jennifer Morla）的设计（上图），我们立刻就知道她是赢家。尽管这份设计非常简单，在朴素的银色背景上，用简单的Helvetica字体大胆放置“the shock of the familiar”一行文字，但《纽约时报》杂志的Logo却上下颠倒放在封面底部，读者不得不把封面倒过来看，并且意识到封面本身。解读其他五个方案也很有趣，可以了解设计师的思维以及他们解决问题的方法。”

——珍妮特·佛罗里克，《纽约时报》杂志前艺术总监

抽象封面

抽象封面很少出现在严重依赖报摊销售的零售刊物中，但在特定主题和仅靠订阅的出版物、新闻周刊或报纸增刊中，倒是比较常见。这类杂志的封面通常文案极少甚至没有文字，设计的自由度很高，连Logo也可以放在设计师认为最合适的地方，因为这些杂志不需要把自己在报刊架上的可见性问题放在首位。这种封面能够催生极具创意的设计，但一定要对设计方向和方式进行规范，以维持品牌调性。

在这方面，《连线》杂志一直是个中高手（见第68页）。在杂志创办之初，它的设计师约翰·普朗克特（John Plunkett）和芭芭拉·库尔（Barbara Kuhr）就经常使用抽象的封面插图，以简单的方式传达复杂的概念。*Adbusters*杂志也是一个很好的

简洁的文字具有的魅力和影响力，有时单靠图片是无法企及的。如同乔治·洛伊斯（George Lois）设计的《时尚先生》（*Esquire*）封面所示（下左图）。有时，它也可以作为一种概念传达工具，就像《纽约时报》（*New York Times*）杂志有关“创意”的一期封面（下右图），“创意”刊是年度最佳创意的年终总结。艺术总监珍妮特表示：“我们的想法是，按照字母顺序，把一年来最好的概念、发明和创意，以百科全书式的方式呈现出来。为此，我们制作了一个既像字典又像百科全书的版面，包括书边缘的半月形索引，营造一种页数很多的错觉，加宽栏位，文字编排上采用类似字典那样密集的排版方式。封面设计模仿布面烫金的古书封面，再用照片呈现出立体感。页面右边的假书勒口，让整本刊物看起来很厚，也让开本显得稍微窄一些。”斯科特·金（Scott King）在设计青年文化杂志*Sleazenation*时，受到了T恤设计的启发，用直接、诙谐的设计方式，好像在开读者、杂志和时尚的玩笑。

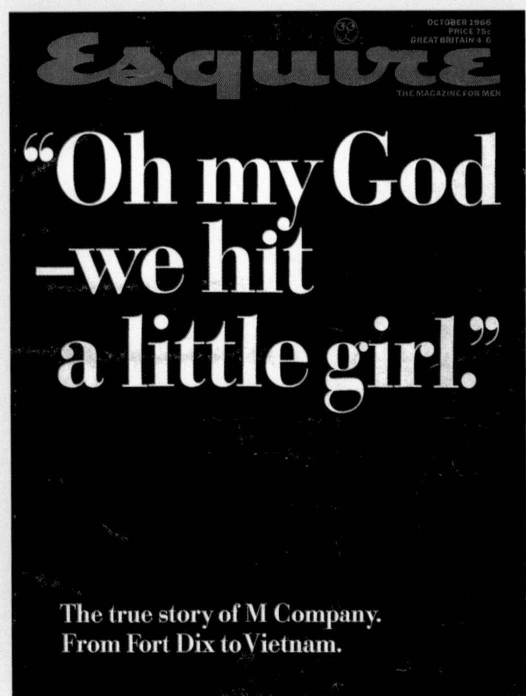

例子。费尔南多·古提耶雷斯（Fernando Gutiérrez）在设计*Tentaciones*的增刊*El País*（参见220页）时，甚至可以随性放置Logo的位置，而且有更多的封面留白空间任由他发挥创意。唯一的限制是必须使用印刷报纸的印刷机进行印制，因此无法印刷满版的出血图片，所以他让图像飘在白色的背景上，形成出血设计的错觉。

文案型封面

文案型封面在当代期刊中并不常见。不过，《时尚先生》（*Esquire*）的乔治·路易斯（George Lois）、《真相》（*Fact*）的赫伯·鲁巴林（Herb Lubalin）和《低俗》（*Sleazenation*）的斯科特·金（Scott King）等设计师，都曾采用过文案型封面，效果极佳。20世纪70年代，皮尔斯·马奇班克（Pearce Marchbank）在设计*Time Out*杂志封面时，也经常采用文案型封面。例如，在一期以日本电影为主题的刊物上，封面不用照片，而是用日本国旗的出血图，配上一行言简意赅的文案，生动形象。有一期以阿姆斯特丹旅行指南为主题的杂志上，也选择了文型封面的排版方式，艺术总监杰里米莱斯利（Jeremy Leslie）解释说：“阿姆斯特丹没有标志性的地点、建筑或活动，采用文字编排的排版方式，更能体现这座城市的喧嚣。”

文案型封面有自己的作用。但在现今视觉导向文化背景下，使用场景有限。不过，如果发生悲剧或者名人去世等事件，编辑和设计师不妨试试文案型封面，往往能够取得很大的影响力，脱颖而出。

SLEAZENATION

AN IDEAL FOR LIVING THROUGH HONEST FASHION, ART, MUSIC AND DESIGN

I'M WITH STUPID

JARVIS COCKER & BARRY 7 • ARAKI FASHION EXCLUSIVE • BIBA • THE HIVES
DJ BIRD • KILLED BY DEATH • CHARLIE LUXTON • RAPING STEVEN SPIELBERG

Time Out London

LONDON'S WEEKLY LISTINGS BIBLE
DECEMBER 14–21 2005
No.1843 £2.50

North VS

South VS

London
Time Out

LONDON'S WEEKLY LISTINGS BIBLE
DECEMBER 14–21 2005
No.1843 £2.50

每一期周刊都必须能看出不同，读者才能知道报刊架上哪本是新刊，尤其是，有时新旧两期周刊可能陈列在同一个位置（就像有一次*Time Out*的新旧两期有一天重叠了，同时陈列一样）。因此，*Time Out*就充分利用文字脱颖而出的力量，这是很多图片封面无法做到的。就像上图这个封面，这期主题讨论伦敦北岸和南岸哪里更为优越，米查·魏德曼（Micha Weidmann）的解决方案非常新颖。他没有简单地放一张泰晤士河的照片或插图，而是设计了“两种”封面：根据购买者在北岸还是南岸，可以自行选择自己的封面。在北岸的读者，可以将“北”向上放，在南岸的读者将封面倒过来即可。这样，用简单而巧妙的办法，就解决了

《连线》杂志

《连线》杂志于1993年在旧金山创刊，堪称纸质杂志中很特别的存在，其设计完全符合时代和主题。作为大众杂志,它将科技发展视为一种文化力量,与以往那种传统杂志不同，摒弃了那种严肃、稳重的设计，其版面布局、结构和美学理念，以其特有的方式挑战读者的既定观念，展现出新潮的设计、视觉表达和内容格式。《连线》杂志的色彩应用相当有特点，有时会在同色的背景上放置文本内容，让读者大开眼界，在有点儿微妙小沮丧感觉的同时，也很是兴奋。要想让读者真正感受到这种新媒体科技的神奇之处与无限潜力，杂志对读者提出了很高的要求。然而，聪明、知识渊博的读者能够解读这些设计背后的寓意，并给予热烈的回应，带动杂志销量飙升。互联网泡沫破灭时，许多杂志不得不停刊，《连线》虽然缩减规模，但存活了下来。《连线》杂志创刊后的前五年，其总体创意总监、设计和排版工作由约翰·普伦基特（John Plunkett）和他在Plunkett+Kuhr工作室的合作伙伴芭芭拉·库尔（Barbara Kuhr）负责。他们的设计师包括崔茜卡·麦吉利斯（Tricia McGillis）、托马斯·施耐德（Thomas Schneider）和埃里克·库特曼奇（Eric Courtemanche）。

2010年，时任《连线》杂志创意总监的斯科特·达迪奇（Scott Dadich），率先向纽约出版设计师协会（Society of Publication Designers）展示了一段视频，内容是为苹果平板电脑重新设计的杂志。他展示了《连线》杂志为苹果平板电脑设计的新功能，以及未来读者可以期待杂志呈现出的更多可能性。达迪奇解释说，他们希望为读者和广告商提供更多的选择，打破对杂志的固有认知。

2012年，尼曼新闻实验室（Nieman Journalism Lab）和达迪奇讨论了重新设计苹果平板电脑版杂志的问题。实验室的副主编贾斯汀·埃利斯（Justin Ellis）写道：“有些东西，老派的东西，不管你是印在纸上还是呈现在平板电脑上，是不会变的。”时任康泰纳仕集团数字杂志开发副总裁的达迪奇在封面上写道：“作为杂志制作者，我们将封面作为吸引读者花钱购买的唯一广告。是封面先触动读者拿起杂志，付出金钱和时间。

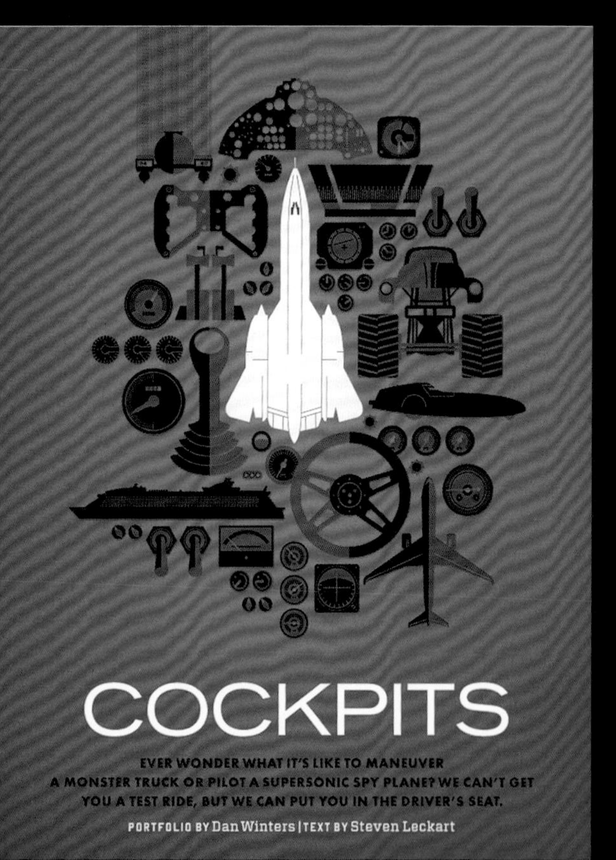

COCKPITS
EVER WONDER WHAT IT'S LIKE TO MANEUVER
A MONSTER TRUCK OR PILOT A SUPERSONIC SPY PLANE? WE CAN'T GET
YOU A TEST RIDE, BUT WE CAN PUT YOU IN THE DRIVER'S SEAT.
PORTFOLIO BY Dan Winters | TEXT BY Steven Leckart

The skipper of this massive cruise liner sits atop a cushy leather throne, naturally. Just don't expect the seat to get much use: "When we're going into port, we typically push the chairs out of the way and stand up. It makes us more agile," says Bill Wright, who was the first captain of Royal Caribbean's $1.4 billion *Oasis of the Seas*. Situated on the center line of the ship, the captain's station has two trackball-controlled 27-inch LCDs (foreground) that are used to display the electronic chart and the ship information system, which aggregates mission-critical data like radar, GPS, and sonar. Nineteen additional screens are positioned less than 10 feet away, so the captain can quickly access, say, the machinery automation system, which tracks everything from the 5.5-megawatt bow thrusters to the fore and aft ballast tanks. So how does the captain steer? "The port and starboard command chairs have built-in joysticks for controlling the ship," Wright says. But those are typically operated by other officers. "Captains should be mentoring and teaching."

Oasis of the Seas
LENGTH
1,187 FEET
CRUISING SPEED
22.6 KNOTS
CAPACITY 8,690 PEOPLE
WORLD'S LONGEST CRUISE SHIP

封面的构成要素

杂志的封面包含外封（OFC）、内封（IFC）、外封底（OBC）和内封底（IBC）。在大多数杂志中，除了外封，其余所有封面都用来刊登利润丰厚的广告，即使没有刊登广告，这些版面也比其他内部版面要珍贵，仅次于外封面。刊物识别（也称为刊头）是封面的首要元素。而指向当期内容的标题和文案，也是封面的重要组成部分。

Logo

虽然杂志刊名和外观一样重要，但对设计师而言，刊名一旦确定后就会长期固定下来。Logo则需要体现并传达刊物的调性、主题立场和态度，并且潜移默化地传达给目标读者。虽然Logo的主要功能是放在刊物封面上，但也会用于各种纸质刊物的电子版本，各种营销、宣传材料上，是品牌呈现中的元素。因此设计Logo时就要考虑到这些用途，如果杂志很成功，那么Logo就会被长期使用，必须谨慎设计和放置，还要特别注意设计出来的Logo是否足够清晰、有辨识度。

Flaunt
COMAG £3.75
FLAUNT
AUGUST 01
THE BLONDE ISSUE

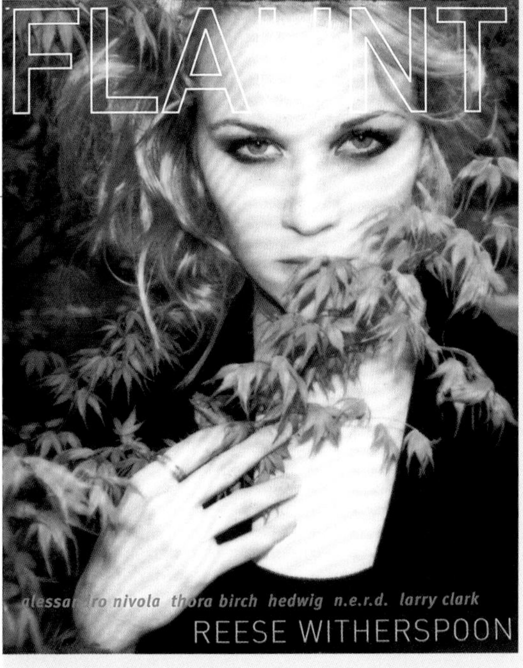

FLAUNT
thora birch hedwig n.e.r.d. larry clark
REESE WITHERSPOON

时尚杂志*Flaunt* 的封面制作和设计一直很有创意，而且分成内封和外封两种。在右边的例子中，外封是一个无法辨认是谁的填色封面，只有读者翻到内封，才会发现这是瑞茜·威瑟斯彭（Reese Witherspoon）。在其他杂志还在简单使用卡片封面的时候，*Flaunt* 已经另辟蹊径。在拍摄内封之前，我们会与摄影师多次讨论封面创意，通常我们会找那些即将或者已经有展览开幕的艺术家，或者单纯喜欢其作品风格的艺术家，进行合作拍摄。当然，每一次的设计要找到艺术总监、设计师、插画师能够完美配合的团队，实属不易，是可遇不可求的。——吉姆·特纳（Jim Turner），*Flaunt* 杂志创意总监

Logo的玩法

Logo是刊物的名片，应该清晰可见。出版社不喜欢刊物的Logo是模糊的，或者被照片和插图遮盖。但也有不少例子，就算Logo有部分或完全被遮住了，销量也很不错。诀窍就是露出的部分依然能够一眼就被读者认出来，如果遮住后所传达的概念有时比呈现完整Logo时还清楚，就可以采用。亨利·沃尔夫（Henry Wolf）为1953年3月的《时尚芭莎》设计的封面就是很好的例子（参见第51页），封面上的手套与杂志刊头交错，创造出有趣的空间感，不同元素完美搭配，非常抢眼。

美国室内设计杂志*Nest* 会不时更换Logo的设计和位置，设计师戴维·卡森（David Carson）在设计*RayGun*杂志时也经常这样做。之后，英国Substance工作室在设计姊妹刊*Blah Blah Blah*时也效仿了这一做法。英国的《金融时报》商业杂志（参见第101页）将Logo看作可以不断变化的设计元素，每周会通过对Logo进行装饰和视觉处理，让封面脱颖而出。其他杂志也佳作频出，*Nova* 采用古老的木质温莎字体（Windsor），每期封面上只放置Logo和单一主题，配上一句言简意赅但很有表现力的文案，效果很好，只使用了一条封面文案来销传推广（见第214页）。*Interview* 则使用插画家马茨·古斯塔夫森（Mats Gustafson）的手绘Logo。刊物特色得以强化，封面看上去简洁有力，也展现出大开本Logo独特的视觉风格，严谨、大胆使用名人照片。

T 杂志是《纽约时报》杂志的时尚版，该团队将Logo化作一个有趣而迷人的视觉元素。哥特式字母“T”由霓虹灯、皮草和樱桃派装饰，有时几乎无法辨认，但读者能够明白其中的趣味。艺术总监珍妮特·佛罗里克（Janet Froelich）十分注意这些图片的微小细节，传递出对刊头工艺的信心。

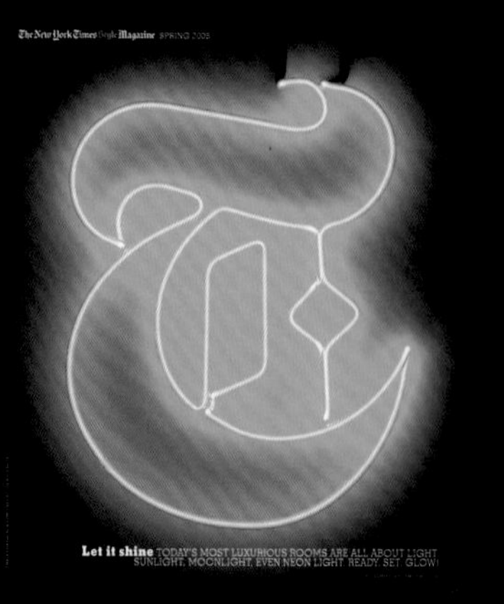

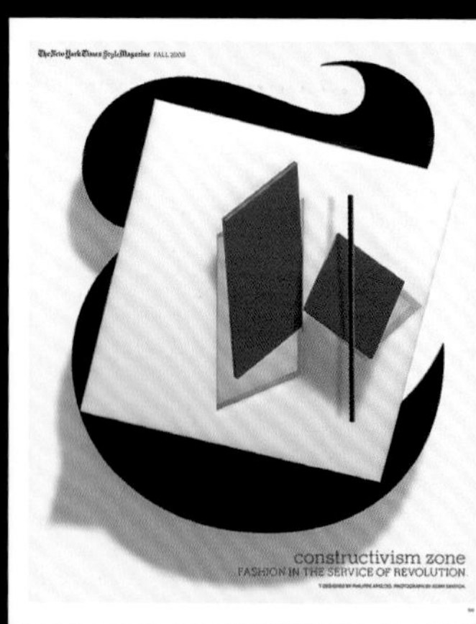

The New York Times Style Magazine FALL 2008
Fox News... AND OTHER FASHIONABLE EPHEMERA.
T DESIGNED BY FENDI. PHOTOGRAPH BY STEPHEN LEWIS.

Easy as Cherry Pie

RECIPES FOR LIVING THE GOOD LIFE.

PHOTOGRAPH BY JAMES WOJCIK

CHERRY PIE BASED ON A RECIPE FROM "PIE" (HARVARD COMMON PRESS) BY KEN HAEDRICH. FOOD STYLED BY BRETT KURZWEIL.

色彩

封面的色彩选择也很重要，绿色的Logo配蓝色的背景肯定不好卖，换成红色的接受度就会更高。金发女郎的封面往往比棕发女郎的销量高，黄色通常不太受欢迎，这些是杂志设计中既定的惯例，不过没有具体数据来支持。设计师和编辑不妨大胆依照直觉行事，发挥颜色的情感力量，别忘了任何颜色都可以用来强调或突出重点，某些用色格外具有象征意义，或是能够触发情感与记忆，不过想只靠颜色来提高刊物销量恐怕是不可行的，因为颜色引起的联想因人而异。

色彩使用

虽然刊物中色彩使用的惯例缺乏数据支持，但在文化色彩心理学领域，色彩使用确实遵循严格的规则。红色可见度高，在西方颇具吸引力。但在南非，红色与哀悼有关，它出现在封面上的概率与西方用黑色封面差不多。不管文化背景如何，蓝色使人平静下来的力量，让它在各种文化里接受度都很高，但不适合用于食物。因此用色也得视地区和环境而定。虽然用色法则并不能通用，不过记住以下禁忌倒是可以帮助设计师避开雷区。

黑色

黑色很复杂，既可以代表性感、权威、强大、威胁，也可以展现有趣、丰富、沉闷、枯燥，还可以体现光滑、质感、永恒……常常同时具有多种内涵。但由于黑色太容易让人联想到死亡和悲剧，所以要避免在封面上使用黑色，用在内页中效果则很抢眼。在色彩心理学中，许多人用黑色来暗示服从。

白色

白色的复杂程度与黑色不相上下，会让人联想到天真、纯净、富有和纯洁，但也可能代表枯燥乏味、平淡无奇。

红色

红色很有活力，有优点，也有缺点。红色很抢眼，容易让人对封面上的其他元素视而不见。但红色却能引发观看者的情感反应，使得观看者的心跳与呼吸加快。

蓝色

蓝色平和宁静，效果类似于可让身体产生镇静效果的化学物质，但要小心使用，因为蓝色也可能让人感觉冰冷和压抑。

绿色

绿色让人耳目一新，使人平静。这是属于大自然的颜色，引发的联想多半是正面的，另外深绿色有时也意味着财富与权力。

黄色

黄色是眼睛最难辨别的颜色，存在感较弱，常被其他的颜色压住，或许正因为如此，封面很少使用黄色。

紫色

若使用得当，紫色会让人联想到奢华、财富、浪漫和成熟，但也可能显得过于女性化或笨拙。

橘色

橘色会引发正面的联想：令人兴奋、充满活力和欢乐。但它可能是一种很难使用的颜色——太红，可能过于鲜艳；颜色太黄，可能会过于平淡。

棕色

棕色是另一种带来正面联想的“自然”的颜色：浅棕色意味着真诚，深色可以联想到木材或皮革。是很适合男性相关主题的色系。

封面文案

封面文案往往用于定期发行的刊物，报摊上的零售刊物通常更会用大量的文案向读者宣告自己的内容比竞争者更多、更好。若用字体大小来区分重要性，最粗最大的文案通常是封面故事。*GQ*、《名利场》（*Vanity Fair*）以及《美丽佳人》（*Marie Claire*）杂志.封面文案的内容、使用和放置位置，通常由编辑和艺术总监共同决定，决策过程中会考虑发行和竞争因素，通常封面文案会放在封面左边1/3的位置，这样是在地面的书架上能见度最高的。然而封面文案的外观与调性，包括色彩如何从竞争者中脱颖而出，数字长度和文字都在共同阐释着刊物的个性和特质，这些是设计师的重要责任。设计报纸版面时也是一样，编辑设计师喜欢利用横幅上方的空间将刊登文章的重点逐条列出来，突出报道的特色。

刊脊

书籍设计师都清楚书脊很重要，需要用心设计。杂志出版商常常忽略刊脊这个狭长的空间，只来放置刊名和出版日期。这是不对的，原因有二：其一，刊脊能够帮助销售，如果杂志放在一起竖着排放，刊脊的可见度甚至比封面还高；其二，这个区域也很适合强化刊物的品牌和风格。例如，*Arena*、*Loaded*、*Vanidad*和*Wallpaper**等的设计师都不会小看这个空间，他们不是简单地在刊脊列出刊物信息。前两家刊物会利用刊脊进行文字叙述，营造出一种读者只是购买了系列中的一期的感觉，从而吸引读者，引起他们整套购买的欲望，建立忠诚度。*Wallpaper**会在刊脊列出主要内容，成为很好的索引。要区分资讯的重要程度，可通过字体的粗细大小来体现。期刊的Logo和日期，必须远远就能看得出来，以吸引目标读者靠近翻阅。

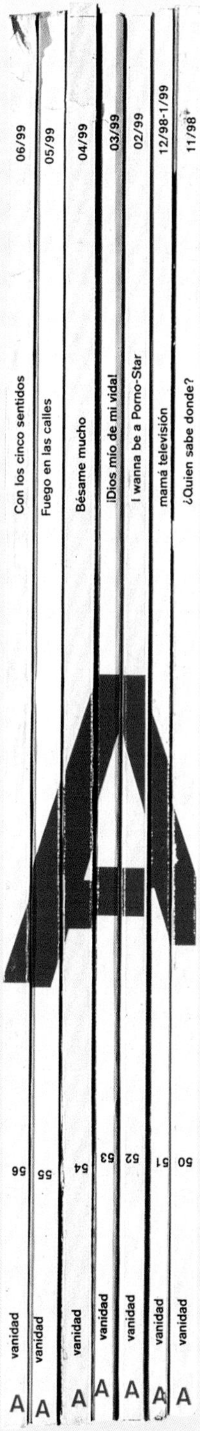

费尔南多·古提耶雷兹（Fernando Gutierrez）为*Wallpaper** 设计的刊脊，整个期刊竖着排列起来之后，可以成一个V字，倒过来放又会变成A字。

工作坊2
刊头和封面

目标

为你的杂志设计一个能够体现刊物理念的刊头和三份封面。

操作步骤

首先，运用工作坊1中的杂志情绪版（Mood board）（见第38页）构思三个封面。运用刊头（Logo）的文字编排，反映杂志的视觉哲学。寻找合适的字体或自行绘制字体。先确定刊头是否要占据整个页面的焦点，还是处理得含蓄一些。要考虑杂志的目标读者是谁，想想看你使用的视觉语言是否足够清晰，是否能够传达你想要传达的内容。

其次，利用排版软件里提供的模板，帮杂志的封面设计版面。先选一个开本，但不要用A4，因为A4太高，不适合翻页。杂志一般比A4纸略宽、略短。找一张喜欢的图片扫描过来，也可以用手边现有的照片或自己做一张图。接下来决定封面上的文字字体，文字编排会影响整本杂志的视觉识别，因此要仔细考虑字体的大小和样式。

假设你的第一份封面是创刊号，接下来两张封面就是后续的月刊或季刊的封面。如果你的刊物是A5大小，印制时记得不要缩放，如果是小报尺寸，则需要再在印刷时分开印刷，再组合起来。总之一定要制作出刊物实际大小的页面。

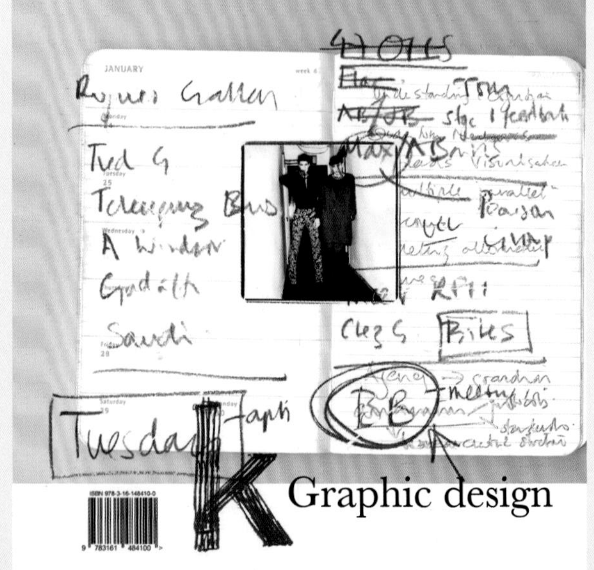

以下是中央圣马丁艺术与设计学院（Central Saint Martins College）平面设计系学生杰迈尔·德沃拉尼（Jetmire Dvorani），为自己的杂志*Sketch*设计的封面。这是一本关于素描草图的杂志，刊头的设计风格强烈，自成一个图标。我们可以看出他的设计思想发展的过程。刊头的设计看似很简单，用非常粗的字体加上横线，有点类似涂鸦。谨记，若在封面用摄影师的作品，一定要先取得摄影师正式的许可。

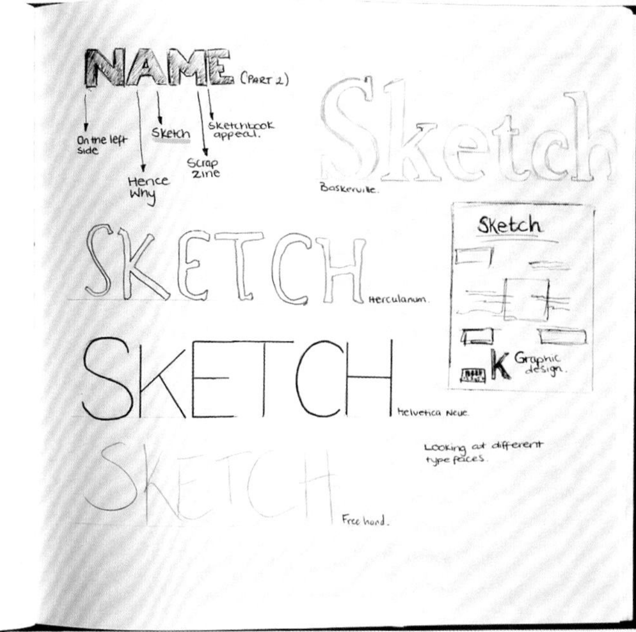

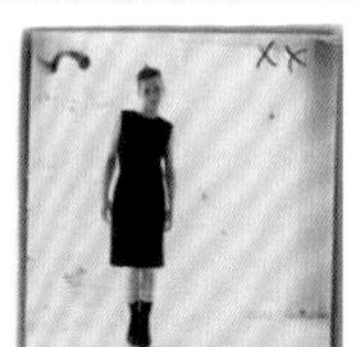

Photography:Gary Wallis

"magical, treasuring a special moment"

Onseri conseque volum repel est, solum saectae enda voloreh enianis essi volo beatur molupicipis animo quo is repratus experitat dolestia velluptae repratincte illabo. Et odi to corisquam, quatia nosanime molupta tionseribus nis con provid quia poratia nones exerepe llendisquia serferchil et, conseque offictiate quasitiae quia nis ressinia sitaqui veliqui cusdae sum eat omniend usamet elest ma nonsequi del id eum rest imo tem aute perum quia voloribusa dolut es doloria quianih iliqui corit laboreperum eum dit et lanihilicius quibus.

Turest opta quatem nonsequia cum hit aliam verist, que necabor epudant omnihil ipient ommodit atempor roviti nimi, si simincil maion consenimenis nonseriae volenis doluptatem. Nam, te parum eatur?

Alicte num iditi as es accusam, nus is commolo rehendae rempor alique etur, nonsecepuda precea dolut aut officiureped utas everum entotatus dolupta tempor reium ipienti te ped quam quae. Itaecaes repe vollit idelitam re consequi ipid unt autaqui doluptatur?

Photography: Gray Wallis

VOGUE
JUNE
£2.90
SPECIAL ISSUE
BEST OF BRITISH
HOT, HAPPENING AND HERE

第四章　刊物内容

对编辑设计有了初步理解之后，就要进入更复杂的阶段——对出版物内容价值的理解，并将其有策略地应用到刊物或报纸内部的细节设计中。谨记，不仅仅是停留在表面去制定策略，更重要的是理解策略背后的深层知识和动机。

刊物解析

期刊一般被分成多个区域，每份期刊都有大致固定的版式。一般可以分成三个部分：前三分之一以新闻为主，称为“前段”（front of the book），中间三分之一为专题报道，称为“专题报道区”（feature well），后三分之一称为“后段”（the back of the book），主要是资讯内容，包括评论、娱乐活动清单、目录等。同样，报纸也可以依据内容分成几个区域：重要新闻（难以预测，包括国际新闻和商业新闻）；新闻分析与时事评论；其他固定内容（电视、股市信息、评论、天气、体育报道等）；不定期进行的专题报道。

翻看任何报纸或杂志都会发现，不同的单元往往由不同的版式设计和网格来区分：栏宽、标题、字体及文字粗细、图像等都略有差别，以方便读者能够区分清楚和阅读。设计师当然可以根据内容表改动格式，但是一定要有明确的指示和导航，以保证不熟悉变动的读者能够顺畅地阅读。特别是对于那些固定的内容版块，如电视节目列表、天气预报、读者来信、字谜、占星术等，更是如此。

目录

杂志的目录页有什么作用?当代读者使用目录页的方式有几种：查找封面故事、浏览刊物内容、找到自己最喜欢的栏目，或者找到他们依稀记得多年前读过的故事。有些人根本不用目录页，有些人从后往前阅读或翻页，让前面的目录页变得多余。但目录页仍然很重要，位置在封面之后，它是唯一能真正引导读者深入阅读出版物、提示如何浏览内容的页面。目录页通常位于版面右侧，因为这个位置可见度高。但也正因如此，目录页右侧，尤其是靠近杂志封面的页面，对广告商也很有吸引力，因此可能会被卖给广告商，迫使目录页出现在页面左边。

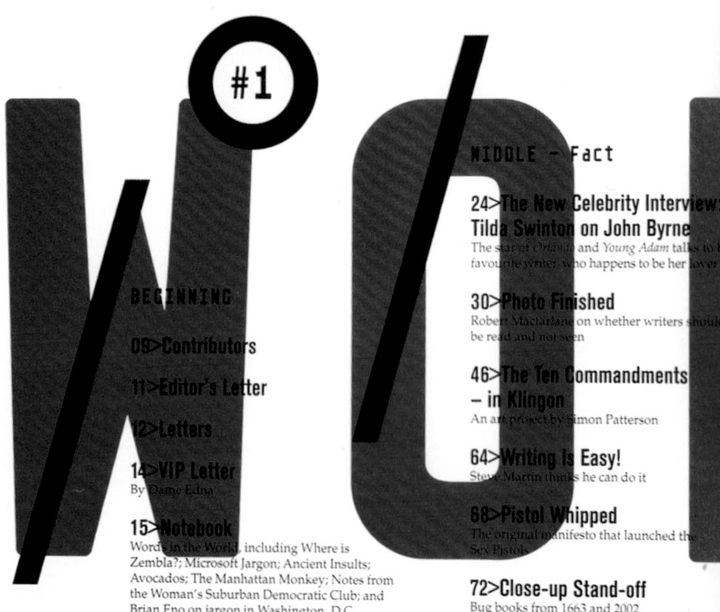
Zembla magazine >contents

#1

BEGINNING

09>Contributors

11>Editor's Letter

12>Letters

14>VIP Letter
By Dame Edna

15>Notebook
Words in the World, including Where is Zembla?; Microsoft Jargon; Ancient Insults; Avocados; The Manhattan Monkey; Notes from the Woman's Suburban Democratic Club; and Brian Eno on jargon in Washington, D.C.

MIDDLE - Fact

24>The New Celebrity Interview: Tilda Swinton on John Byrne
The star of *Orlando* and *Young Adam* talks to he
favourite writer, who happens to be her lover

30>Photo Finished
Robert Macfarlane on whether writers should be read and not seen

46>The Ten Commandments – in Klingon
An art project by Simon Patterson

64>Writing Is Easy!
Steve Martin thinks he can do it

68>Pistol Whipped
The original manifesto that launched the Sex Pistols

72>Close-up Stand-off
Bug books from 1663 and 2002

88>The Dead Celebrity Interview
Michel Faber interviews Marcel Duchamp
Illustration by Manolo Blahnik

[04] zembla magazine *september_two thousand and three*

目录页必须列出刊物中的全部内容，但在设计时可以发挥创意。如*Zembla*和《纽约时报》杂志的例子所示。后面一本是2003年11月《纽约时报》杂志的年度特刊《灵感特刊》(*Inspiration*)，前艺术总监珍妮特·佛罗里克在Cooper-Hewitt设计博物馆的双年展上找到了灵感。她说："在展会上，我看到设计师兼作家保罗·埃利曼（Paul Elliman）用工业废弃物设计的作品*Alphabits*，受到启发。于是利用瓶盖、电脑元件、发动机零部件等来做字母，巧妙呈现出灵感从何而来的问题。封面的字母图像性很强，让封面看起来新奇又充满启发性，制作过程也充满趣味。"

RD/S

MIDDLE - Fiction

38>Sanchakou
By Matthew Kneale

54>Under the Weather
By James Hopkin

66>The Marcus Van Heller Legend
An exclusive excerpt by John Stevenson

81>Sometimes the Daughter Says the Things Her Mother Thinks
By Donna Daley-Clarke

87>Dr Mortimer's Observations
The problem with 'nice' people

END

92>Writer's Shock
Toby Litt remembers the writer and educator Malcolm Bradbury

95>Global Literary Calendar
Upcoming events around the world

96>Reviews
Authors on their own books; real kids on kids' books; and Zemblans on what to read now, damn it

102>Big Philosophy, Big Talk
By Alexander Bard and Jan Södergvist

112>Cartoon
By R. Crumb and Gavarni

115>Crossword: Crass Words
By Francis Heaney

116>Competition: Sex and Death
Who is the best-qualified person to write on sex and death? Tell Zembla publisher Simon Finch and win a first-edition classic novel

120>Lost Page
Never-before published pictures

september_two thousand and three zembla magazine [05]

The New York Times Magazine 11.30.03

69 Inspiration:
Where Does It Come From?
By Arthur Lubow

71 The Seeds of Inspiration:
A Design Portfolio
The design team of Hjalti Karlsson and Jan Wilker, principals in the firm KarlssonWilker, were asked to investigate the evolution of seven designs they found curious and intriguing. Some of the subjects will be familiar, others less so, but all illuminate the mysteries of the creative process.
Text by Camille Sweeney

78 The Guts of a New Machine
The iPod became an instant classic by combining high design and powerful technology. But as Apple has learned before, that formula alone doesn't keep you on top. By Rob Walker

86 Necessity Is the Mother of Invention
The greatest design achievements, according to an innovative teacher at M.I.T., ought to save lives and cost next to nothing. By Pagan Kennedy

INSPIRATION

Design 2003

NOT YOUR AVERAGE JOE

71 78 86 90 96

90 Vote Getters
In posters and buttons, leading designers make a visual statement for each of the major Democratic presidential candidates. By Gail Collins

96 The Shape of Jackets to Come
To create a new world of clothing, Olivier Theyskens started with the gentlest of notions — a new silhouette.
By Lynn Hirschberg

Continued on Page 20

How Do You Spell Inspiration?

目录页包含重要信息，它的第一设计法则就是要清晰易读、容易理解、便于查找。以前目录页通常尽量放在靠近封面的位置，但只要每期目录页的位置保持一致，放在哪里就没那么重要了。位置固定，读者阅读起来就会感觉亲切熟悉。目录页的组织和布局要清晰易懂，能够帮助读者快速定位和浏览，例如找到封面故事或者某个喜欢的固定版块。同时，要用字体、图片、线条和图标等图像元素，概括并突出重要的专题和报道，吸引读者阅读。毕竟，看到目录页时，读者可能还没买下这本杂志。最后，目录页应该如实呈现内容版块的前后顺序。如果在新闻后面才是专题报道（参见第83页）和名录，那就应该这样安排目录页。

4 Geht die Kultur der Liebe kaputt?
Ein Mädchen beklagt sich: Die Männer sind wie Kaninchen

8 Die Lust, ein Polizist zu sein
Polizeibeamte schreiben in twen über ihre Erfahrungen mit Demonstranten

15 Feliciano. Die neue twen-Platte
Die aufregendste Stimme, die in den letzten Jahren entdeckt wurde

22 Werden Gehirne manipuliert?
Gehirn mit elektronischer Ladung. Wissenschaftler bauen den mechanisierten Menschen

26 Witze

28 Machen Sie Ihr schönstes Farbfoto
Startschuß zum zweiten großen Farbfotowettbewerb

30 Die Jungen machen das Geschäft
Vier Rezepte, wie man einen eigenen Laden auf die Beine stellt

41 Wenn ich ein junger Deutscher wäre
William Saroyan schreibt für twen

45 Auch an der Garderobe abzugeben: Citybikes
Die kleinsten Motorflitzer auf zwei Rädern

51 Leserbriefe

53 Die Verfolgung
Kurzgeschichte von Henry Miller

60 Sex und Schule
Die Klasse der schwangeren Mädchen

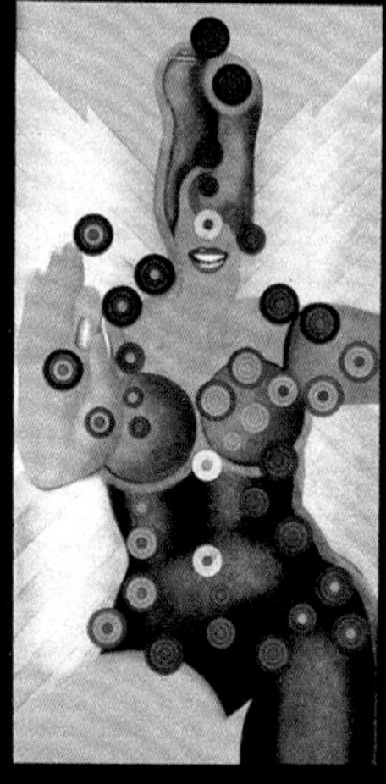

80 Kopf, Herz oder Bauch?
Das twen-Testspiel verrät alles über Ihr Temperament

90 Touching ist schöner als Sex
Die Kunst, zu berühren und zu fühlen

78 Steve McQueen ist nicht zu fassen
Hollywoods erfolgreichster Außenseiter

85 Warum malt Hockney am liebsten Knaben?

88 Dieser Sommer wird noch heißer!
Werden die Studenten den politischen Kampf auf die Spitze treiben?

100 twen fordert einen deutschen Pop-Sender

102 Morgen bin ich ein anderer Mensch
Wie ein Mädchen mehr aus sich machte

108 Fiesta der Zigeuner
Fahrendes Volk trifft sich in Saintes-Maries-de-la-Mer

121 Supermann in Unterhosen
Gordon Mitchell in den neuen Geblümten

131 Das heiße Lied
Georg Stefan Troller schreibt über französische Chansons

138 twen-Magazin
Neue Bücher, neue Filme, neue Platten

166 Sterne und Zitronen

70 Bikinis in Israel
twen flog mit EL AL ans Rote Meer

目录页也可以采用列表式设计，例如本页的*Twen*，右页下方的*Sleazenation*和右页左上方的*About Town*都是很好的示范。它们严格遵循网格系统和平面设计的原则，如实呈现文字内容，信息丰富又保持了整体的优雅质感。不过，还有很多其他方式来设计杂志的目录浏览方式。例如右页右上角的《大都会》（*Metropolis*）杂志，前《大都会》艺术总监克里斯韦尔·拉滨（Criswell　LAppin）说："目录布局图是葆拉·舍尔（Paula Scher）在1999年刊物改版时提出的设计方法，这种组织结构能够让读者明白不同文章之间的相互关系，虽然在过去五年杂志目录出现了很多设计上的变化，但是这种地图式形式依然保留，并且成为设计中的重要元素。

EDITOR Clive Labovitch
EXECUTIVE DIRECTOR Michael Heseltine
ART DIRECTOR Thomas Wolsey
FEATURES EDITOR David Hughes
PRODUCTION MANAGER Denis Curtis
ADVERTISEMENT MANAGER Lindsay Masters
ASSISTANTS:
EDITORIAL Morfudd Davies
PRODUCTION Jeanette Collins, Gillian Humphrys
ADVERTISING Peter Evans, Stephen Hollis, Anthony Reader
MERCHANDISING Gordon Deighton

Cover photographed specially for About Town by Terence Donovan

volume 2 number 3 MARCH 1961

ABOUT TOWN'S EXAMINATION OF THE TORIES BEGINS ON PAGE 20 WITH AN EXCLUSIVE INTERVIEW WITH THE PRIME MINISTER, THE MAN WHO TODAY REPRESENTS THE PUBLIC IMAGE OF THE PARTY

about town

PERSONALITIES	20 Tory leaders: a series of interviews by *Godfrey Smith* and *Michael Heseltine* 46 Political Fraserology: *Liz Fraser* demonstrates catchphrases of the platform
CLOTHES	32 Man about the Black Country, photographed by *Terence Donovan* 78 Address book: tailors and services throughout the country
MODERN LIVING	58 London's wine houses: a drinking tour by *Anthony Haden-Guest* supported by *Bill Brandt* and *John Donaldson* 40 Are we more bored than we used to be? asks *Elizabeth Smart*
BRITAIN TODAY	20 The Tory Party: *David Hughes* explains its phenomenal success 43 The Midlands for me: *John Wain* defines the peculiar charm of his home country 50 England's hard centre: *George Lamming* and *John Bulmer* investigate the black spot on the affluent English conscience
CARS	62 The open roadsters: a comparison of sports cars under £1,100, by *Richard Bensted-Smith*
FOOD AND DRINK	48 Splitting the apple: a guide to the which, how and when of cider-drinking 70 On the town: where to drink, dine or dance in London
COMMENT	31 The junkies come to town: *Nat Hentoff* on *The Connection*, a new play 15 Tip-off on share tipsters: this month's guide by *Henry Spencer*

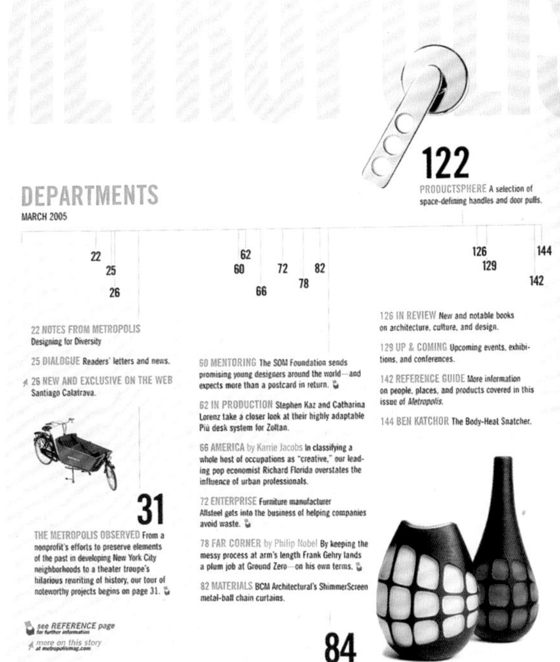

DEPARTMENTS
MARCH 2005

22 25 26 60 62 66 72 78 82 126 129 142 144

122 PRODUCTSPHERE A selection of space-defining handles and door pulls.

22 NOTES FROM METROPOLIS Designing for Diversity

25 DIALOGUE Readers' letters and news.

26 NEW AND EXCLUSIVE ON THE WEB Santiago Calatrava.

31 THE METROPOLIS OBSERVED From a nonprofit's efforts to preserve elements of the past in developing New York City neighborhoods to a theater troupe's hilarious rewriting of history, our tour of noteworthy projects begins on page 31.

60 MENTORING The SOM Foundation sends promising young designers around the world—and expects more than a postcard in return.

62 IN PRODUCTION Stephen Kaz and Catharina Lorenz take a closer look at their highly adaptable Più desk system for Zoltan.

66 AMERICA by Karrie Jacobs In classifying a whole host of occupations as "creative," our leading pop economist Richard Florida overstates the influence of urban professionals.

72 ENTERPRISE Furniture manufacturer Allsteel gets into the business of helping companies avoid waste.

78 FAR CORNER by Philip Nobel By keeping the messy process at arm's length Frank Gehry lands a plum job at Ground Zero—on his own terms.

82 MATERIALS BCM Architectural's ShimmerScreen metal-ball chain curtains.

84 PORTFOLIO Finnish designer Anu Penttinen's glass work defies classification.

126 IN REVIEW New and notable books on architecture, culture, and design.

129 UP & COMING Upcoming events, exhibitions, and conferences.

142 REFERENCE GUIDE More information on people, places, and products covered in this issue of *Metropolis*.

144 BEN KATCHOR The Body-Heat Snatcher.

see REFERENCE page for further information
more on this story at metropolismag.com

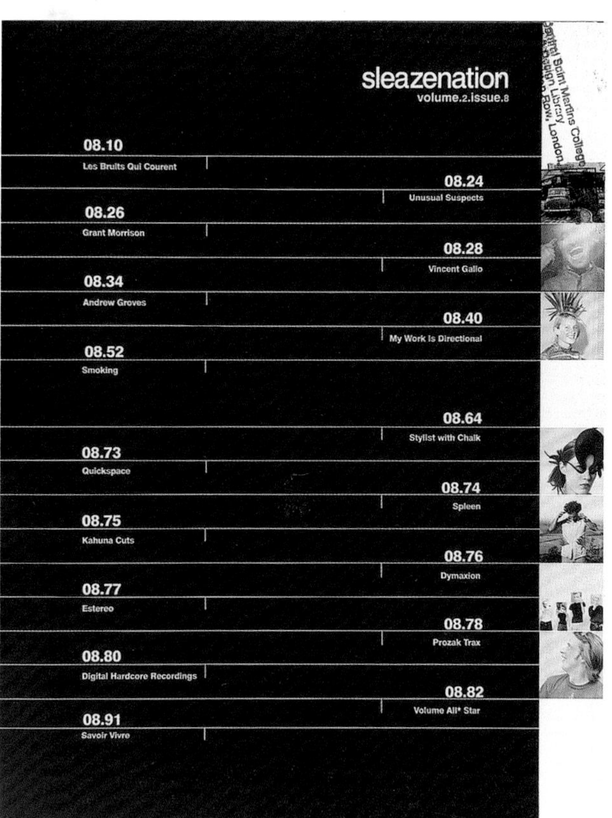

sleazenation
volume.2.issue.8

08.10 Les Bruits Qui Courent
08.24 Unusual Suspects
08.26 Grant Morrison
08.28 Vincent Gallo
08.34 Andrew Groves
08.40 My Work Is Directional
08.52 Smoking
08.64 Stylist with Chalk
08.73 Quickspace
08.74 Spleen
08.75 Kahuna Cuts
08.76 Dymaxion
08.77 Estereo
08.78 Prozak Trax
08.80 Digital Hardcore Recordings
08.82 Volume All* Star
08.91 Savoir Vivre

有些刊物没有目录页。*Adbusters* 的前艺术编辑克里斯・迪克森（Chris Dixon）就取消了刊物的目录页，他和编辑卡勒・拉森（Kalle Lasn）一致认为，目录页过度分割杂志，限制设计。他们另辟蹊径，通过线条和色块进行导览，通过线条和色块的长度、宽度和不同颜色区分，来表示单元内容的长度和类型。这当然是一种极端的解决方案，因此只适用于某些固定类型和开本的刊物。如果新闻周刊报道娱乐消息的杂志没有目录页，那很快会惹恼读者。另外，要是刊物超过三百页，是肯定需要实用的导航目录的。

CON/TE/N/TS

'WAS a game at first. A publisher had been tonewalling my proposal for a new magazine o I started to toy with the idea of publishing it ıyself. Independence – now there's a thought o set any journalist's pulse racing. But I never onestly thought I'd go through with it. On the ther hand, what if I *could* actually escape the ustrations and constrictions imposed by having to conform to somebody else's notions of ow a magazine should be run and what it hould be? So I started to check it out, juggled few figures using some £4,500 of savings, and alculated that – yes – I could just about cover ıe costs in the event of total public rejection. hen there arrived the day when I telephoned ın order for £7,500 of Finnish paper, literally eeling queasy in my stomach as I replaced the eceiver and contemplated the commitment. I ad the same feeling at 5am one April 1980 norning in South Wales when the first issue ame off the presses. The initial idea was, in a ense, self-indulgent. To hell with market research! This would be the magazine that I vould enjoy reading – visually attractive, urprising, diverse, colourful, stimulating. Vind it up and let it go! Profits (sic) would be pent improving the product; behind-the-cenes costs cut to the bone. To survive during he first year I was the only full-time staffer, vorking first from a corner of a room in omebody else's office, then in a musty one-oomed basement. In that period there were at east two occasions when disaster was imminent. The fact is that it's relatively easy to start magazine, a major act of endurance and faith o keep it going when things don't immediately click. But there has always been enormous goodwill towards THE FACE and it was this, with equal measures of stubbornness, pride and my wife's support, that kept us afloat. Today we have a rising circulation and an nternational reputation. On top of UK sales, every month we export 24,000 magazines to some 36 countries. Quality of content improves with every issue; there's no room for complacency. Last year we were elected Magazine Of The Year in two polls. This year (so far) we've put out two 96-page issues, expanded the scope of coverage, tried to stick our necks even further out. Profits are still being reinvested. Not just me, but all of us at THE FACE would like to thank the readers, contributors, advertisers, associates and friends of the magazine for their generous and continuing support. *And* we're still independent!

NICK LOGAN

00 ABC

AMERICA • AUSTRALIA • CANADA • DENMARK FRANCE • GERMANY • GREECE • HOLLAND HONG KONG • ITALY • JAPAN • NEW ZEALAND

The World's Best Dressed Magazine

INTRO

- 8 "Chaka Khan meets Eurythmics" **Helen Terry**
- 11 **Sara Sugarman**/Olympic shirts/Sunday Best
- 15 The Black Theatre Coop/Richard 'Popcorn' Wylie
- 16 **Yellowman**/Grace Jones/**Was Not Was**
- 18 Alternative Country – High noon in NW1!
- 20 **Jermaine Jackson**/Intro's record selections
- 22 **Absolute Beginners**/Loose Ends/**Bronski Beat**

FEATURES

- 24 Hippie as Capitalist – profile of **Richard Branson**
- 28 **Mods!** The Class of '84 hit the Trail of '64
- 38 The General Revisited – **Jerry Dammers**
- 44 **Gang Wars!** Sudden death in Olympic City
- 52 Photospread **The funeral of Marvin Gaye**
- 58 Saint **Joseph** the Tastemaker!
- 62 Exclusive! **Stevie Wonder has a dream**
- 67 EXPO **Night People** by Derek Ridgers
- 79 All for Art – **Gilbert & George**

SECTIONS

- 42 **Julie Burchill** How to get *It!* How to lose *it!*
- 50 **Music** for June reviewed by Geoff Dean
- 55 **Letters** The readers bite back!
- 66 **Back Issues** (going fast!)
- 74 **Films** New on the circuits
- 76 **Nightlife** THE FACE in Turin/Club selections
- 82 **Bodylicious!** 12 pages of **Style** start here...
- 94 **DisINFORMATION** for THE FACE's 50th
- 95 Subscriptions advice **THE FACE by post!**

The Face • 4th Floor, 5/11 Mortimer Street, London W1, England

Publisher/Editor • **Nick Logan**
Assistant Editor • **Paul Rambali**
Designer • **Neville Brody**

Features • **Paul Rambali**
Intro/Features assistant • **Lesley White**
Accounts/subs • **Julie Logan**
Design assistant • **Ben Murphy**

New York Editor • **James Truman** (212 989 4579)
Ad Manager • **Rod Sopp** (01-580 6756)

+ Peter Ashworth/Janette Beckman/Max Bell/Ian Birch/Chris Burkham/Julie Burchill/David Corio/Kevin Cummins/Giovanni Dadomo/Chalkie Davies/Anthony Denselow/Robert Elms/Anthony Fawcett/Jill Furmanovsky/Laura Hardy/David Johnson/Marek Kohn/Neil Matthews/John May/Joe McKenna/Jamie Morgan/Neil Norman/Steve Pyke/Derek Ridgers/Dave Rimmer/Helen Roberts/Sheila Rock/Fiona Russell Powell/Chris Salewicz/Jon Savage/Kate Simon/Carol Starr/Jay Strongman/Kevin Sutcliffe/Steve Taylor/David Thomas/Paul Tickell/Steve Tynan/Elissa Van Poznak/Jane Withers/Patrick Zerbib

THE FACE 7

在另类时尚杂志*The Face*中，编辑尼克·洛根（Nick Logan）和设计师兼字体设计内维尔·布罗迪（Neville Brody），共同探索后朋克（post-punk）风格的设计，利用俄罗斯建构主义（Russian Constructivism）的政治和视觉美学，引领字体、版面和整体设计进入新方向。即使这样，目录页的编排设计依然能够清晰明了，概念上也算传统。

前段

杂志的前段主要以文字专题为主，充分刊登文化、时尚、运动、音乐、旅行或室内设计等方面的内容。前段通常会有明确且相对固定的设计规范，包括字体大小和粗细、颜色和页面元素（包括图示、线条等），都会在网格系统上结构清晰地排列好。通常，设计师也要参与设计刊头和社论版块，因为这里往往是展现刊物风格的重要部分，所以要格外重视，才能清晰传达刊物的调性。

至于时事新闻版块，现在经常学习设计网页的方式，利用框线、底色和各种大小、粗细不一的字体，让页面变得生动活泼。《连线》和《商业2.0》杂志率将网页设计方式用在了纸质刊物上，英国《金融时报》商业版的加里·库克（Gary Cook）也随之效仿。为了让《连线》杂志的新闻页面看起来充实有活力，设计师使用了大量重叠的方块、色彩、字体、图片和图形。留白空间带来的宁静感不适合新闻页面使用。

报纸的头版与杂志的前段有些相似，头版一般是最新消息，也是利用网格系统制作一些可以进行灵活处理的模板，用来刊载突发新闻故事等无法事先预测的内容。

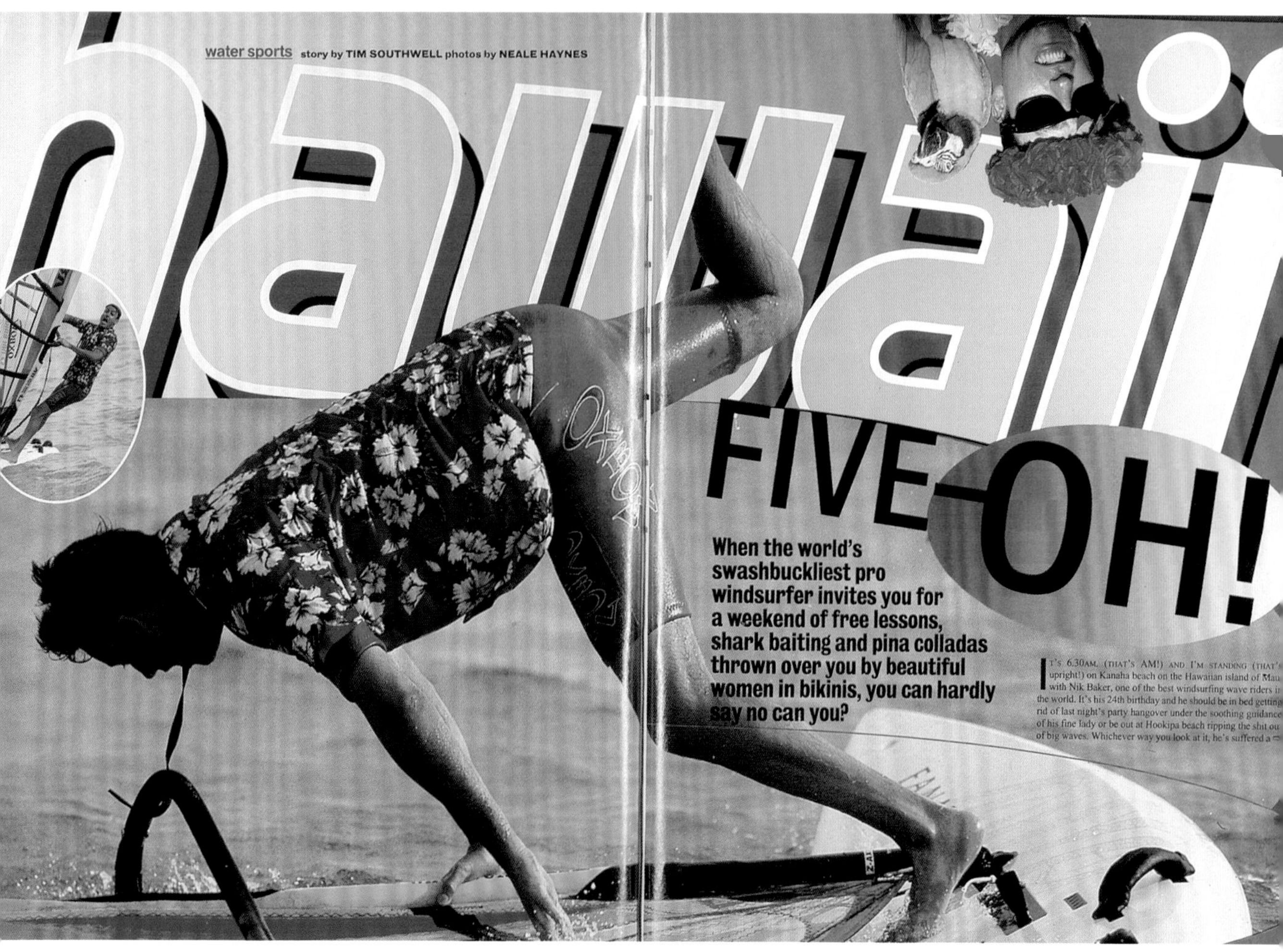

water sports story by TIM SOUTHWELL photos by NEALE HAYNES
hawaii
FIVE-OH!
When the world's swashbuckliest pro windsurfer invites you for a weekend of free lessons, shark baiting and pina colladas thrown over you by beautiful women in bikinis, you can hardly say no can you?
IT'S 6.30AM (THAT'S AM!) AND I'M STANDING (THAT'
upright!) on Kanaha beach on the Hawaiian island of Mau
with Nik Baker, one of the best windsurfing wave riders i
the world. It's his 24th birthday and he should be in bed gettin
rid of last night's party hangover under the soothing guidance
of his fine lady or be out at Hookipa beach ripping the shit ou
of big waves. Whichever way you look at it, he's suffered a

专题报道区

专题报道是对杂志品牌影响最大的文字版块。可能刊登名人访谈报道，也可能是对某个事件、主题和热点的深入分析和报道，抑或是独家新闻报道。正是这些内容的写作风格、调性和排版布局，让杂志从竞争刊物中脱颖而出。

很多专题报道区会采用刊物的标准视觉风格，并通过细节差异将该板块与其他内容版块进行区分，例如加大栏宽、增加留白空间、变化字体、放大标题、加长导语等。如果专题报道以跨页开始，那么通常会利用一张满版的出血图片（大头照、图表或插图），搭配单元刊头。专题报道单元的刊头通常包含文字（标题、导语、正文和重要引文），或许还有其他小图，使整个出血主图更契合专题特点（如果不用图片来做

20世纪90年代的*Loaded*杂志，号称是世界上第一本男性杂志，虽然目标锁定的是新市场，但其专题报道的内容和文风非常契合目标读者的喜好，一上市就取得成功。杂志的作者充分表达了读者想做、想想、想看的内容，拥有或者正在经历的状态。设计也出色地传达了文章中那种精力旺盛、混乱和反体制精神。刊物的设计总监史蒂夫·里德（Steve Read）的设计手法看似“没有设计”，仿佛只是将各种报道和元素凑在一起，却有出色的表现力。运用鲜活的色彩、满版大图、充满动感的照片和巨大标题，传递内容的活力和深度。直接印在图片上的问题，也营造出具有纯粹生活乐趣的氛围。

"J'ai le caractère *obstiné*, *tenace*, la *motivation* de mon *père*. De ma mère, j'ai hérité la *discrétion*. Elle *préfère* être *observatrice plutôt* que le centre *d'attention*."

Elle a choisi l'Hôtel, rue des Beaux Arts, QG d'Oscar Wilde, à quelques pas du Café de Flore où elle se régale de croque-monsieur avec les doigts, tout contre l'immense appartement dépouillé qu'elle occupe le temps d'un film. Le clin d'œil du réceptionniste, son «bonjour mademoiselle» connivent, laissent à penser que Sofia Coppola a ici quelques habitudes. Peut-être vient-elle, dans ce cadre «rococo-Art déco-bordelo» rêver à Marie-Antoinette, tête d'affiche de l'Histoire, couronnée reine de son troisième long métrage? Silhouette frêle parfumée à la fleur d'oranger, teint transparent semblable à celui de ses jeunes héroïnes indécises, lèvres pleines, comme un rideau de chair tendre sur un émail éblouissant, nez de caractère, phrasé sucré, hérissé de *«you know what I mean»*, Sofia Coppola, interrogée sur sa devise, répond du tac au tac «tais-toi et agis». Une ligne de conduite talonnée par la réalisatrice qui, en l'espace de deux films, a su tracer sa route très personnelle, se soustraire à l'ombre d'un père dont le nom charrie à lui seul le mythe Hollywood. Avec *Virgin Suicides* (adaptation du livre de Jeffrey Eugenides) puis *Lost in Translation* (Oscar du meilleur scénario, trente millions de dollars de recette), Miss Coppola s'est imposée fascinante «fée-ciné» du cool et de l'innocence dont les films, comme des albums intimes, des collages photos, recèlent une sensibilité visuelle chic et moderne. Un halo soyeux qui prend dans la toile de sa langueur diaphane, de ses couleurs crème glacée, de ses ralentissements abyssaux, de ses lumières tamisées, de ses mélodies planantes, les supplices et les délices d'un âge impossible, d'une passion platonique, l'essence de l'essentiel comme la substance des petits riens. «C'est drôle, dit Sofia Coppola, quand j'étais jeune, ma mère s'inquiétait parce qu'avec mes copines, on avait un avis sur tout. Aujourd'hui, j'ai un boulot où tout ce qu'on me demande c'est justement d'avoir une opinion. Un cinéaste doit être curieux de toutes sortes d'images et de sons. Je suis une généraliste, sensible à la mode, au design, à la photo, à la musique et je jette dans mes films tout ce que j'aime, sans la moindre restriction. Je rends les choses le plus personnel possible avec tout ce que ça comporte d'effrayant.»

D'où l'irrépressible tentation de cerner l'autobiographie, de démêler les fils du soi de Sofia sur le tapis de la fiction, ceux-là mêmes qui suturent le huis clos moiré avec un quinquagénaire au sommet d'un cinq étoiles tokyoïte ou l'odyssée laconique d'une fratrie de blondes évanescentes qui se soustraient au monde des vivants. «J'ai adoré le livre d'Eugenides, au point de m'angoisser quand j'ai appris qu'un réalisateur avait l'intention de l'adapter, dit-elle. Je me sentais comme investie de la mission de protéger ce bouquin. La plupart des films américains qui abordent l'adolescence sont affligeants de caricature et de niaiserie. Ce qui m'intéressait, moi, dans cette étape cruciale de la vie, c'est la collision entre l'obsession des grandes questions, la mort, l'amour, et l'infatuation du futile, l'océan du dérisoire. J'ai donc commencé à écrire un script de mon côté. Bien plus tard, j'ai compris que ce livre a sans doute fait écho à la disparition accidentelle de mon frère aîné, Gio, alors que j'avais une quinzaine d'années. Mon adolescence, qui aurait dû être la période la plus insouciante de mon existence, a été la plus triste. Un poison dont on ne se débarrasse jamais.»

Sofia Coppola est née à New York, le 14 mai 1971, en plein tournage du *Parrain*, «moment où, pour la première fois, mes parents ont gagné beaucoup d'argent». L'enfance est vécue entre la grande demeure victorienne de San Francisco – décorée de néons et de canapés en forme de lèvres par Eleanor, la mère éprise d'art conceptuel, où défile le tout- cinéma, de George Lucas à Wim Wenders en passant par Marlon Brando, et les différents plateaux de tournage aux quatre coins du globe où Francis Ford, alias «le Sultan», débarque invariablement avec son clan. Pour *Apocalypse Now*, Sofia est exilée trois ans à Manille, elle s'éveille sur les bancs d'une école chinoise, s'amuse de tours en hélicoptère sur les genoux du patriarche, de pâtés de boue au bord de la rivière, et croise avec la même fascination mandarins de Hollywood et serpents venimeux de deux mètres de long. «Mon père nous voulait toujours autour de lui, dit-elle. J'étais constamment entourée d'adultes ce qui, du coup, m'a rapprochée de mon second frère, Roman. Et de mes parents, bien entendu. Je ne crois pas qu'il y ait beaucoup de pères qui donnent des conseils de réalisation ou de scénario à une petite fille. Le mien n'arrêtait pas. Je crois que j'ai hérité de son caractère obstiné, tenace, de sa motivation. Et de la discrétion de ma mère. Elle a toujours préféré être observatrice plutôt que centre d'attention.» L'adolescence de Sofia a eu pour cadre Napa Valley, ses somptueux jardins, ses vignes vallonnées, le dernier refuge de l'ogre Coppola qui prend le large, se protège d'un Los Angeles sous perfusion d'un «cinéma business» dans lequel il se sent à l'étroit. Études d'art, premiers flirts sur fond d'Elvis Costello, de Siouxsie and the Banshees, des Clash, escapades à répétition en grosse américaine, direction San Francisco ou Seattle, pour se frotter en live aux prémisses du grunge… composent le «moodboard teenager» d'une jeune boudeuse aux yeux noirs qu'aurait pu imaginer Salinger, vie capitonnée, destin en suspens. «J'ai été torturée par l'idée de ne pas savoir ce que j'allais faire, dit-elle. De n'être géniale en rien.» Sans avoir l'air d'y toucher, elle s'essaiera à la comédie (sur injonction du père, qui l'impose sur le *Parrain III*), à la photo, à la peinture *(Suite page 295)*

"C*ela* m'a sans doute *arrangée* qu'au *départ*, on ne m'ait pas prise au *sérieux*. Ça m'a laissé la *latitude* de *surprendre*."

Vogue Paris 的前艺术总监法比安·巴伦（Fabien Baron），擅长用黑色和留白空间来设计跨页，配合严谨的格线系统，让文字变得格外醒目。右页上方，是维利·弗莱克豪斯（Willy Fleckhaus）在1968年设计的*Twen*杂志，大胆使用留白空间，以出其不意的方式来处理主题，在视觉和概念上呈现出令人惊喜的优雅感。.

主视觉，标题文字也可以承担这一功能）。如果专题报道的起始页为单页，而且上一页是全彩广告，专题页则可用黑白主图或者适当留白，与旁边的广告页形成对比，引导读者的视线从广告转向刊物内容。留白设计在新世纪成为设计主流，但20世纪90年代的刊物偏向花哨的设计，后来才转向利用留白营造空间的简洁风格。要让留白设计具有一致性，就要事先在设计网格或模板中进行明确标注，例如天头地脚的留白，标题和导语之间的留白，或文字框周围的留白。

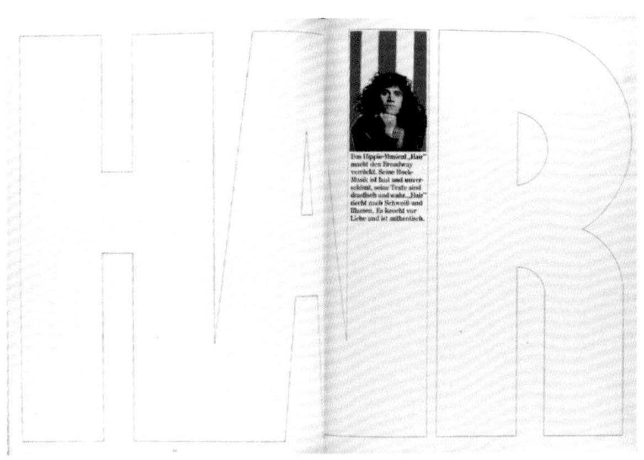

后段：评论、信息列表与投稿

和刊物的前段一样，专题报道区之后的部分（如各种评论、列表、投稿与星座）一般由初级设计师设计。这部分有相对固定的结构和网格，固定的配色、字体、字号和风格。图片是这些页面的重中之重，能够利用好插画和照片，决定了这些页面是否活泼、吸引人。图文编排也是如此，在空白背景上放置图片，不仅醒目，还能留给页面呼吸空间。前段的页面就不一样了，因为前段往往被图文排得满满当当。很少有人会阅读所有报道，但页面设计出彩，往往可以吸引读者的目光。这对于所有类型的版面都适用，即使是无趣的资讯页面，也要充分利用字体与线条，带来更加生动的效果。杂志的后段，最珍贵的就是最后一页，读者如果从后向前翻，最先看到的就是这一页。正因如此，有些杂志会将最流行、最受欢迎的内容放在这一页上，例如星座、读者来信，或者把刊头再重复一次。

纽约时报旅游杂志*T：Tavel*这一章节的开篇，就通过手绘的歌德字体（Fraktur）强化品牌形象，而在沙子上写下的字母则传达出杂志的冒险精神。这画面不仅有创意，也很吸引人。

单元开始页

单元开始页在刊物中可说是相当奢侈，但读者却很喜欢。通常单元开始页会占用一整个跨页，并以抢眼图片搭配极少文字，让视觉有喘息的空间。跨页是难得可使用横图的机会——横图的设计比一个人物肖像的直图更有效果（但要确保图片质量佳、强而有力，能持续找这种图并不容易）。单元开始页若设计得当，能给人留下深刻印象，因此可以当成“引导标志”使用，引导读者顺利找到想看的版块。这一页面可以单独建立专用版式，无论是老读者还是新读者，都能够分辨与查找自己想看的单元。

文字编排的作用

版式是编辑设计工作的骨架，也是设计师务必掌握的基本知识。数字刊物和纸质刊物一样，应依循最基本的文字编排原则，再根据具体情况调整。只要能充分了解字体，悉心处理复杂的文字素材，无论是什么样的出版形式，都能设计出好看的版面。

文稿（copy）

在编辑发排的文稿中，常会使用许多关于文稿名称的术语，对于不习惯使用这些术语的设计师来说，会感到很困惑。同样的内容有时候会有各种各样的叫法（参见88页图），这就更让人一头雾水了。然而提到文稿，有四件事情设计师不可不知。

- 文稿中的各个术语名称。
- 不同形式的文稿结构。
- 报刊写作（各类文稿）和其他写作类型的区别。
- 这些区别对设计有何影响。

标语（tagline）

刊头下方代表刊物精神的标语或口号能为出版物增加巨大价值，标语写得好，不仅能告诉读者文章内容，还能表明刊物的调性与读者对象。标语能够强化读者的感受与认知，例如下面这些标语：“该知道更多的男性”（*loaded*）、“关心周围事物的人（*Wallpaper**），或是“我们是与众不同的行家”*WAD*），彰显出特立独行的时尚主张。对于新读者来说，通过标语，可以快速了解杂志内容。

大标题（headline）

文稿编辑会说，在说服读者阅读一篇文章时，大标题和版面一样重要。标题在出版物与读者群间建立牢固联系，好像在说：“我们了解你、喜欢你、我们有共同的幽默感、兴趣以及文化背景。我们知道你够聪明，能够理解这个标题与报道。”处理大标题要采用适当的尺寸、位置与方式，以文字为主的报纸尤其重视大标题，毕竟报纸无法通过漂亮的图片吸引读者购买，文字标题便成为重中之重。

导语（stand-first）

导语通常放在标题后，从文本内容上来说，甚至比标题还重要。它能够决定文章调性，告诉读者文章的主要内容，成为标题和主文间文字与视觉的桥梁。因此导语一定要与标题一脉相承，必须以简洁、吸引人的方式总结并向读者推销故事。

引文（pull quote）

引文对设计师来说是很好用的元素，不仅能引导读者的视线，还能将文章分段，制造出阅读节奏，让专题报道更引人入胜。引文的内容可直接从文章里摘取，也可将一段文字精简、重组。

小标题（subhead或cross-head）

小标题在冗长的新闻报道中最常用，可将冗长的文章分段显示。文章内容太长，读者没耐心读完时，可以通过小标题找到想看的内容。小标题也很适合用来提示新起段落与章节，或主题转换。如果读者无法一次读完文章，小标题能帮助读者快速找到上次读到的位置。

署名与图片出处说明（byline and credit）

署名与图片出处说明的处理方式与放置位置会因具体刊物的重视程度而有不同。杂志通常会列出文章作者与编辑人员（尤其请知名作家撰稿时）、摄影师或插画师。报纸更重视新闻内容，不太强调记者是谁，因此常没有署名；而在杂志中，最新资讯类内容的署名，也比专题报道的要小。

“设计师要愿意读文字材料，经过深思熟虑和热烈讨论，才能在刊物的预设架构之下构思出正确的视觉元素，丰富阅读体验。”

——马丁·威尼茨基，*Speak*艺术总监

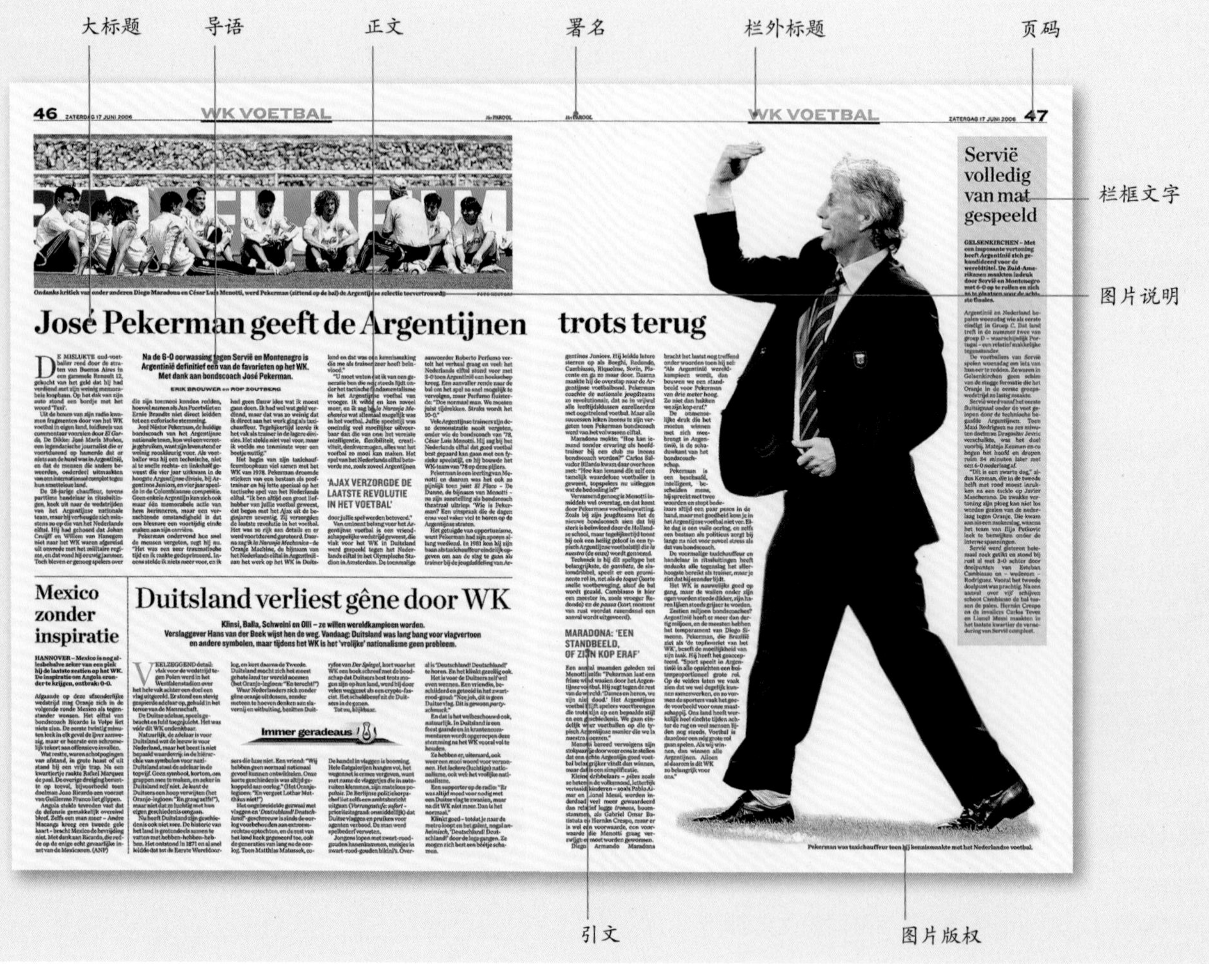

正文（body copy）

虽然刊物可以靠设计来吸引读者，但读者如果发现文字内容或正文不符合期待，销量还是会下降，进而流失广告，最后沦落到停刊。刊物内容会随着潮流变化而变化，但同时要忠于品牌与品牌传达的信息。要做到这一点，就必须有扎实的内容、优质的作者和编辑团队。在正文部分，设计师主要有两项工作要做：一是根据正文的需求和特色来选择字体和栏位设计，传达品牌精神；二是设计师也要把自己对文化趋势的想法表达出来，让内容在视觉上更有活力。

当然，出版物的内容会随着潮流而变化，并与读者保持联系，但这种变化的关键是保持对品牌和品牌信息的忠实，这其中最重要的是内容、作者和全体员工的力量。

资讯框、文本框、边栏和信息图表（Panels，box copy，sidebars and info-graphics）

资讯框主要用来放短篇新闻或者长篇报道的补充材料，例如相关事实、统计数据、案例研究或者其他与

Black shirt by Tom Ford for Yves Saint Laurent Rive Gauche Homme.

NOT HALF BRAD

by Jim Turner photographed by Tony Duran

12 NACIONAL EL PAÍS, DOMINGO 11 DE JUNIO DE 2006

ANÁLISIS. La resaca del debate sobre el estado de la nación. Por Javier Pradera

Ruptura intempestiva

Rajoy aduce como razón para retirar el apoyo de su partido a la política antiterrorista del Gobierno el anuncio de que los socialistas vascos se disponen a reunirse con la ilegalizada Batasuna

CAJERO AUTOMÁTICO EL ROTO

El indiferente

LA COLUMNA. El PP se ha convertido en el partido del no. No al Estatut, no al proceso de fin de la violencia. Por Josep Ramoneda

Cataluña, con Euskadi al fondo

EL PAÍS, DOMINGO 11 DE JUNIO DE 2006 ECONOMÍA 13

La discreción del PNV

Los nacionalistas vascos, inquietos por la ruptura anunciada por Rajoy

LA CRÓNICA por Soledad Gallego-Díaz

Dirigentes nacionalistas estiman que las mesas de diálogo político servirán difícilmente para algo si no está presente el PP, tercer partido político de Euskadi

Ángel Acebes, en un mitin contra el Estatut.

AGENDA GLOBAL Los tipos de interés, al alza. Por Joaquín Estefanía

Precio del dinero: no pasarse de frenada

Los bancos centrales, que deben garantizar la estabilidad de los precios, se encuentran en una incómoda encrucijada: estar atentos a la subida del petróleo y a la enorme liquidez, mirando al crecimiento con el rabillo del ojo

上图*Flaunt* 的标题比较传统，但仍然非常抢眼、巧妙、精彩。这个标题不但不会抢了超级巨星布拉德·皮特的风头，反而暗示读者文章内容会有更精彩的见解，不是那种半知半解的浅显报道。

越来越多报纸仿效消费类杂志，用引语来吸引读者目光，从而将密集的正文分割开来。左侧西班牙的《国家报》就是很好的例子。

in the suburbs,' Burton says, 'and somehow if you are deprived of certain feelings, there is a desire to get them out. Otherwise, you feel like you're going to explode.'

Burton has the manner of a precocious teenager who has spent a great deal of time happily alone. He is enthusiastic, but not effortlessly so — his sentences start and stop and cut back on themselves in the manner of someone who is used to an internal monologue and resists outside analysis. But Burton is engaging; he makes himself understood. He isn't shy about his art. His drawings are pinned to the walls of the studio, and there are toys of his own design on the coffee table, atop the fireplace and stuck in nooks in the kitchen. Burton is surrounded by friends.

"I think best when I'm drawing," Burton said as he folded himself into a large red velvet sofa. "I always loved drawing, but that's true of all kids. When children draw, they all draw the same. There's a certain kind of passion that all kids share. And then it gets beaten out of them. At that point, most people stop drawing. I almost succumbed to the idea that I couldn't draw. I wasn't that good at life drawing, and I almost gave up. But I thought, I don't care if I can't draw in this traditional way. To punch through, that was what saved me. It was a catharsis. I realized that a kind of dark, dramatic, depressed, sad, moody thing was kind of healthy."

For most of his career, Burton, 45, has communicated through his artwork. His explanations or descriptions came in the form of sketches. "Tim showed me several drawings of his Edward," said Johnny Depp, who starred in "Edward Scissorhands." "I'd read the script, of course, but Tim's drawings said everything. I instantly fell for the character — he made his way into my body. I was traumatized by Tim's drawings. They haunted me."

Since none of his past movies — from "Pee-Wee's Big Adventure" to "Batman" to "Planet of the Apes" — have been realistic or particularly plot-driven, Burton's visual brilliance has dominated. For years, he told his collaborators that he wasn't really comfortable with dialogue. Burton preferred the immediacy — the emotion, in a sense — of a drawing come to life. Edward Scissorhands, with his pinprick eyes, kabuki face and shears for fingers, barely speaks, yet is extraordinarily expressive. "He approaches filmmaking from a really artistic place," said Denise Di Novi, who produced five of Burton's movies. "He could do one drawing, and you knew what the whole movie was. Tim comes from a deeply emotional place, and the drawings combine with that emotion to create something poetic, whether it's the Bat Cave or the topiary in 'Edward Scissorhands.' It's a coherent vision."

Burton explained it this way: "For most of my career, I have not found most dialogue to be meaningful. It doesn't ring true to me or feel honest in the way that a fully realized character can. Most movie talk is just not my experience of how people actually talk."

Despite his originality and visual acuity, Burton's movies have often been classified as too quirky or macabre to be taken seriously. No matter how fully realized, movies like Burton's that are based on comic books or children's stories, or feature aliens or monkeys, are often discounted when awards are handed out. "I don't like when people say of my movies, 'Well, they look nice,'" Burton said. "I'm not an interior decorator. Do you want me to come over and do your apartment? 'Big Fish' is about what's real and what's fantastic, what's true and what's not true, what's partially true and how, in the end, it's all true." Burton smiled. "To show those contrasting elements was the challenge."

Lynn Hirschberg is a contributing writer for the magazine. She last wrote about Sofia Coppola.

52

"Big Fish" tells the story of Edward Bloom, who is played by Albert Finney. (Edwards figure prominently in the Burton oeuvre — there's Edward Scissorhands and the director Ed Wood and now Edward Bloom.) Bloom is a wild fabricator who has turned every incident in his life — from the birth of his child to his stint in the military to the wooing of his wife — into an elaborate tale. His estranged son, played by Billy Crudup, long tired of his father's yarns, has nonetheless returned home to Alabama to reconcile with Bloom before he dies. "Big Fish" jumps back and forth in time as Bloom, played as a young man by Ewan McGregor, spins his tales of wild adventures, some of which involve a giant, a pair of Siamese twins and a witch. Bloom is a mix of Odysseus, Candide and your long-winded uncle, and just when the movie seems in danger of veering off into the sentimental, Burton pulls it back. The extraordinary ending of "Big Fish," in which the son, finally, gets to tell his own story, is a perfect melding of the visual and the verbal: the words are bolstered, rather than overwhelmed, by Burton's artistry.

Burton's attraction to a strong narrative, to a tying together of the emotional and physical world through plot, seems to mirror a larger trend on display in this year's holiday movies. There is an abundance of films based on densely plotted novels — from "Cold Mountain" to "Master and Commander" to "Mystic River" — stories in which, despite violence and tragedy, a kind of order prevails. "Big Fish" is not violent, but it is about the need for a coherent story that represents the truth. "With this movie," Burton said, still twisted on the sofa, "it was so important not to be maudlin. That was the challenge. I wanted to hit that right note — sadness and humor and a little bit of horror. I've realized that really seems for me to be what comprises the essence of life."

Burton was interrupted by a knock at the window. His girlfriend, a very pregnant Helena Bonham-Carter, was standing in the courtyard. Burton, who met Bonham-Carter when she starred in "Planet of the Apes," sprung up from his seat to let her in. Bonham-Carter was dressed in a flower dress and had attached large cloth roses to her pink sweater and her piled-up curly hair. "I dreamed I gave birth to a frozen chicken," she told Burton. He laughed and hugged her. The baby was due in a week, and although they did not know the sex, both assumed their child would be a boy. (It was — born on Oct. 4.) "In my dream," Bonham-Carter said, "I was very pleased with a frozen chicken." Burton smiled, aware, perhaps, that this dream could be a scene from one of his films.

It would be easy to assume that Burton's attraction to "Big Fish," with its central paternal conflict, would be due to his imminent fatherhood. But his interest in the film is more closely linked to his own childhood in Burbank, Calif. "The clearest thing for me is where I came from," Burton said. "I had a kind of sensory deprivation growing up in Burbank. I said once that my childhood was like a kind of surreal, bright depression, and it was. I grew up in the suburbs, and somehow if you are deprived of certain feelings, there is a desire to get them out. Otherwise, you feel like you're going to explode."

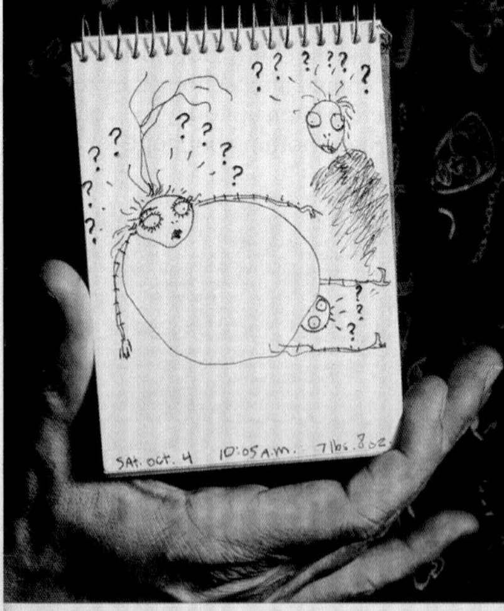

For Burton, *who once worked as an animator for Disney, life is very much a sketch. Here, he bears witness to the recent arrival of his son.*

Burton was never close to his father, who had played baseball and worked for the city's parks and recreation department, or his mother, who at one point had a gift store devoted to cats. When he was 12, he went to live with his grandmother. "My grandmother used to tell me that before I could walk, I would crawl out of the house," Burton said. "Whenever anybody came over, I wanted to leave with them. Maybe that's why my parents had a complex. But my grandmother left me alone. At some level, I was understood, and her house felt like a sanctuary." Burton paused. "My parents were very stiff. Until around 15 years ago, I would flinch if anyone touched me. Those memories of my childhood are the most vivid for me. Whenever anyone asks me what I find scary, it's never something like a monster movie. I would have to say my mother. I was scared of her mood swings. Certain relatives, perhaps, also scared me. If I had a nightmare, it was something normal, like I didn't take out the trash. That would make me nervous and edgy. Monsters always felt understandable, while people were terrifying."

In ninth grade, Burton was already heralded for his art: he drew an antilittering poster that was displayed on the garbage trucks of Burbank. After high school, he was accepted at the California Institute of the Arts on a Disney fellowship, and in 1979 he went to work for Disney as an apprentice animator. "It was like army recruitment," Burton recalled. "I was accepted when I was 21, but as soon as I got to Disney, I realized I couldn't do animation. I was drawing foxes. That's where I broke down. They were supposed to be cute foxes, but my foxes looked like they had rabies, like they had been hit by a car. Then they put me on 'The Black Cauldron.' I felt like they'd look at my stuff and say, 'That's great, that's great,' and then they'd leave the room and say, 'What was that all about?' I think they felt they had to watch me or I would do something strange." Burton shook his head. "As soon as I got to Disney, I got . . . weird. I'd sleep all the time — I learned how to sleep sitting up, holding a pencil. It was terrifying: I was turning into a kind of zombie factory worker."

Eventually, Burton's talents were recognized, and in 1982 he was allowed to make "Vincent," a five-minute animated short that detailed the fantasies of a deeply unhappy suburban boy. "My mother was disturbed by 'Vincent,'" Burton recalled. "She said, 'It's a shame you didn't have a *worse* childhood.'" Two years later, he made "Frankenweenie," a live-action short about a boy who brings his dead dog, Sparky, back to life. "I remember a test screening where kids started crying when Sparky turns into a monster," Burton recalled. "Disney felt that was too intense, but adults forget that being scared isn't a bad thing for kids. It's actually helpful in your life.

"After 'Frankenweenie,' I thought I would never make it out of the basement at Disney. Out of my room I could see the hospital where I was born and the cemetery where my family was buried. It was like this Bermuda Triangle. I remember thinking, I have to get out of here."

In 1984, Stephen King recommended "Frankenweenie" to a Warner Brothers executive, who showed it to Paul Reubens, aka Pee-wee Herman, who soon after signed Burton up as his director. The box-office success of "Pee-Wee's Big Adventure" led to Burton's directing "Beetlejuice," which firmly established his sensibility. The plot of "Beetlejuice" was nonsensical, but that didn't really matter — the film looked like nothing that had been made before. "'Beetlejuice' involved a lot of throwing away of conventions," says Bo Welch, who was the production designer of "Beetlejuice" and other Burton films. "Design-wise, the movie was an epiphany for me. Tim would communicate through his drawings of the characters. His drawings were the movie. Through them, Tim led me toward a more expressionistic, rather than literal, way of looking at things."

"Beetlejuice" was also a box-office smash, and Burton next signed on to direct "Batman." In 1988, Burton's idea of exploring the dark side of a comic-book hero was an original approach. Critics were dazzled by "Batman"'s production design — its rendering of Gotham City as a dark, menacing metropolis — but less impressed by the storytelling. This became the standard Tim Burton criticism: he had a brilliant eye and no ear.

Audiences didn't care. "Batman" was an enormous financial success, earning $425 million worldwide. After completing "Edward Scissor-

PHOTOGRAPH BY DAN WINTERS FOR THE NEW YORK TIMES

THE NEW YORK TIMES MAGAZINE / NOVEMBER 9, 2003 53

上图的两份刊物，处理正文的方式各有不同。《纽约时报》的正文虽然排满整个页面，但通过谨慎选择字体，营造出轻盈易读的空间感：这里使用的字体有强纳森·霍夫乐（Jonathan Hoefler）重新设计的Cheltenham、赛勒斯·海史密斯（Cyrus Highsmith）与马修·卡特（Matthew Carter）重新设计的Stymie，还有Garamond与Helvetica。右页的例子是*Speak*杂志，威尼茨基通过运用图形，打造出正文的空间感，将各种不同元素整合起来，为跨页版面创造出视觉上的连贯性。

正文不同却相关的元素。因此，资讯框通常不像正文那样深入复杂，而是更为活泼。句子更短，采用陈述事实性的语气，有大量的信息片段和元素，把连续的文章分解成列表和要点。在进行设计时，要考虑到版式能够易于读者阅读和理解。

图片说明

图片说明是重要的设计环节，正如导语连接标题与正文一样，图片说明也连接着图像与文字，需要精心设计。图片说明的设计方式与摆放位置有很多种（详见第五章），设计师必须了解图片说明在刊物中的角色与调性，才能做出合适的设计。

页码（folio）

页首和页尾会放置页码、书名，有时也会加上单元名或章节名。页首和页尾是页面中不可或缺的元素，能

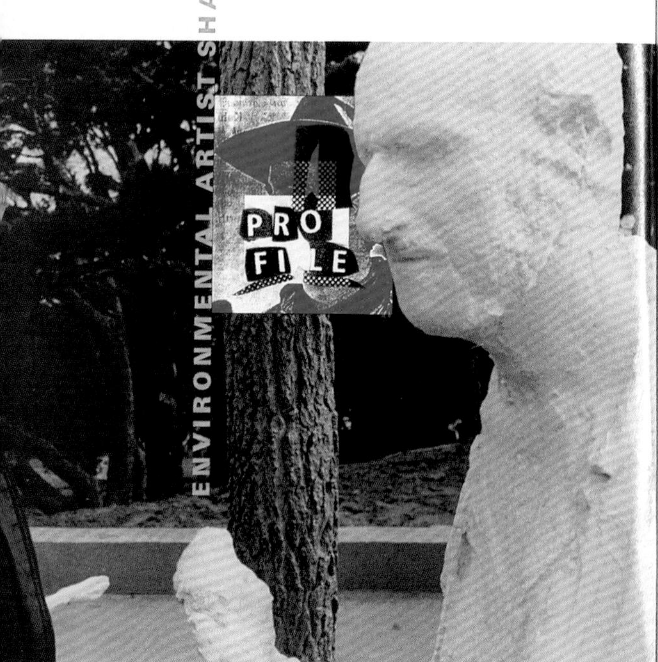
ENVIRONMENTAL ARTIST SHAPES HISTORY
PRO
FI LE

george segal

by Amanda Nowinski George Segal's sculptural environments allow the viewer a voyeuristic glimpse into the muted, introspective moments of everyday life. Augmented by a realistic backdrop—sitting in a restaurant, talking on the phone, operating a dry cleaning store, riding the bus, driving the bus—the narrative is seemingly clear and straight-forward. But upon further inspection, it becomes evident that the tableaus represent more than just a fleeting snippet of tedium; many of Segal's characters, who appear unwittingly tied to urban life and capitalist drudgery, are caught in the only form of escape they know—daydreaming. Conversationless, the characters impel viewers to fill the silence with their own psychological interpretations, which are infinite and remain unresolved. Unlike an isolated video frame, Segal's narratives are heavily organic, weighed down in their environments with layers of hospital bandaging and white plaster. Here the temporally mundane is transformed into a complex, unsettling permanence.

I wanted to deal with *reality*, with *psychology*, *with all the interior stuff* plus the vivid reality of the world.

Born in 1924 in New York City, Segal is the son of Jewish immigrant intellectuals who made their living on a chicken farm in New Jersey. Although he pursued a degree in arts education at Cooper Union after graduating high school, his escape from the fourteen-hour workday was cut short when his older brother was drafted. It was through his life on the farm that Segal developed his views on the hardships of the labor class.

Beginning as a painter in the late 1950s, Segal's early canvases evoke the grand simplicity of Matisse, but are rendered less playfully through his use of muddied primary colors and non-figurative lines. Segal was an artist emerging from the tail end of Abstract Expressionism and he and his New York peers were faced with the daunting task of creating something entirely new. To rebel against the anti-figural and material gestures of the Abstract Expressionists, however, meant that the new art must not only provide the antithesis to the previous movement, but that it must also avoid the Renaissance perspective that the Abstract Expressionist school worked to destroy. Robert Rauschenberg and Jasper Johns, leaders of the New York school, provided the link between pure abstraction and the advent of Pop with their Dadaist-inspired Assemblages—the use of real objects was the logical divergence from the splatters of Jackson Pollock.

Inspired by the three-dimensionality of Rauschenberg and Johns, Segal expanded his vision to plaster casting—a move which would change the direction of his career forever. Beginning with sparse tableaus, normally of people sitting, Segal soon embraced more complex backdrops, including the quintessential Gas Station (1963-1964) and Times Square at Night (1970), a neon-lit setting for two disengaged pedestrians. His later work transgresses the quiet narratives, depicting more politically engaged scenarios, including the controversial In Memory of May 4, 1970: Kent State—Abraham and Isaac (1978),

2 5

引导读者的阅读方向，强化版式架构，从而巩固品牌形象。如果刊物内容简单，通常不会太注重页码设计。但如果刊物读者很在意视觉效果，那么刊物就需要运用字体、粗细与位置等，让页码更突出，成为独立的设计元素。*RayGun*的大卫·卡森（David Carson）、*Speak*的马丁·威尼茨基（Martin Venezky）与*Émigré*的鲁迪·凡德蓝斯（Rudy VanderLans）都很重视页码的设计。许多杂志为了满版出血的图而舍弃页码，虽然无关紧要，但如果缺少页码的连续页面太多，可能导致设计团队与印刷厂工作时很困惑，读者在浏览时也觉得不太舒服。所以如果遇到左页和右页只能选一页放置页码时，优先选择右页，因为可见度更高。

屏幕字体和印刷字体（font versus print font）

仅靠眼睛看就能发现字体印在纸上与在数字屏幕显示时差异很大。屏幕是有背光的，字体会随着屏幕的亮度与校准程度不同而产生很大差异。尤其是当屏幕对比很强时，字体可能会看起来很暗淡。而印刷在白纸上的黑字，阅读起来是否舒适，和纸张的反光程度有关。使用报纸或再生纸印刷时，油墨网点通常会晕开，如果用放大镜观看，就会发现字体的边缘比较模糊。

因此，设计师需要了解屏幕与纸质印刷字体在设计上的差异，并且谨慎选择字体。在数字排版时，设计师通常都会购买知名字体公司的字体产品。然而网络上也有丰富多样的免费字体可供下载，如果设计新人能够了解字体，会对设计选择和判断很有帮助。也有许多字体热爱者组成的社群以及字体协会的活动与演讲。经验老到的字体设计师都知道，字体与文字编排值得终身学习，所以专业的设计师要与时俱进，不断学习最新知识。

约翰·希尔

像《泰晤士报》这样的大型新闻机构，能利用其庞大的历史档案库，将这些内容通过纸质版和数字版的形式在当前新闻报道中引用。通过这种方式，出版社过去的内容资产也能在新时代发挥价值。《泰晤士报》的目标受众是那些想读到可靠、值得信赖的新闻产品的读者。如果现在读者不得不付费阅读网站上的内容，他们就更期待一个内容精致、完善的产品。收费的数字产品意味着更高的制作标准。

约翰·希尔（Jon Hill）是伦敦《泰晤士报》的设计编辑，曾负责报纸、网站与苹果平板电脑版本改版，也曾主导《泰晤士报》旗下的*Eureka*月刊的推出。他在以下访谈中谈到对文字编排的看法，以及数字环境对设计的影响。

你如何运用《泰晤士报》的品牌，让它保持新鲜感？

我认为与其要设计师遵照设定好的模板工作，不如设定大的工作原则。这些原则可以适用刊物设计的所有部分，连字体设计也不例外。我们使用Times Modern字体，这是2006年、2007年布洛狄在改版时开发的。

我们努力确保这个字体在纸质版、数字版等各种平台上都能适用。Times Modern是《泰晤士报》的DNA，也可以说是我们的品牌。因为《纽约时报》的大部分版面的标题用的就是这一字体。它几乎无处不在。

众所周知，你是一位严格的字体设计师，非常了解刊物页面的深层结构及视觉语言。但在制作《泰晤士报》的数字版本时，这些字体编排的技巧还能派上用场吗？

我认为最大的挑战，在于和程序开发者及设计内容管理系统的工程师合作。我发现他们认为成功的数字刊物就是要能自动出版，最好只要把新闻内容汇入，就能直接转换成网页和数字刊物。然而艺术总监对外观更有要求，例如想让标题更好看，或调整一下苹果平板电脑版的图片裁剪方式等，不可能固定设计版面，只让对方灌进内容就行了。这样就会和技术团队发生冲突，然而正确的做法是介于两者之间的。我们尽量让系统自动排好数字版本，设计师多少仍有调整空间，能修改文字和版面。

谁来决定数字刊物的最终版式？

因为《泰晤士报》设有付费墙，所以我们的目标是保证为付费读者提供精致、完美的产品。这么一来，制作时就要有详细的规范，例如如何设定首字母大写或段落间距，好好打磨细节，才能体现我们的用心，也能让产品从免费新闻网站中脱颖而出。我会和开发者讨论这些细节，确保他们重视并优先考虑这些细节。

是否觉得数字字体的质量已经大幅提高，能媲美印刷字体的效果？

现在比较满意了。我们仍在与Monotype公司合作，想办法改善网站上的字体，因为现在浏览器种类很多，每种呈现出的字体效果都不一样。在平板电脑上，可以控制字体呈现效果，因为平板电脑屏幕的分辨率是固定的。2012年之后，有些屏幕的显示效果已和纸质版一样好。我有信心，我们比五年前更有能力在各种数字版本上，掌控色彩与品质。

可否请你说说新的数字环境对你的工作有何影响？

没有任何出版商敢断言自己完全掌握了在苹果平板电脑出版数字刊物，以及在这个平台丰富读者阅读体验的真谛。现在谈影响还言之过早，但迟早会有影响。设计师务必学习交互图片以及动态效果的设计，还要能应付大量的数据，也必须知道在数字环境下如何讲好故事。如今编辑设计不只是处理文字与图像，数字刊物的设计师需要学习全新的知识和全新的设计方式。

THE JACKSON FAMILY TREE

Kendall Brown (great-great-grandfather)
b circa 1828
Once a slave for an Alabama family named Scruse, a name which he later adopted, calling himself Prince Scruse. Renowned for his voice, he sang in a Russell Country Church.

Mary Etta (great-great-grandmother)
b circa 1850 - 1920

Prince Scruse (great-grandfather)
b circa 1880
Tenant cotton farmer in Alabama.

Julia Belle Jordan (great-grandmother)
b circa 1883

Crystal Lee King (grandmother)
b circa 1900

Samuel Jackson (grandfather)
b circa 1900
Raised on a farm, Fountain Hill Arkansas, Ashley County. Became a highschool teacher and had a reputation for being a strict and distant father.

Prince Albert Scruse (grandfather)
Oct 16, 1907 - Jan 21, 1997
Born in Alabama, he worked for the Seminole Railroad and also as a tenant cotton farmer before moving to East Chicago, Indiana to work as a Pullman porter with the Illinois Central Railroad.

Martha "Matti" Upshaw (grandmother)
Dec 14, 1907 - Apr 29, 1990
Later changed her name to Martha Bridgett.

Joseph Walter Jackson (father)
b July 26, 1929
Eldest of four children. He had a short career as a boxer before becoming a blue collar Crane Operator for a steel company in Gary, Indiana. He had a brief marriage to another woman, which was annulled, and at 20, married Katherine Esther Scruse with whom he had ten children. He became manager of the Jackson 5 and was honoured by the Rock & Roll Hall of Fame as "the greatest musical manager of all time".

Katherine Esther Scruse (mother)
b May 4, 1930
Born in Alabama, she suffered from polio as a child. She was a country and western fan and played clarinet and piano. Her parents separated when she and her sister, Hattie, were still young. She married Joseph aged 19 in Indiana and became a Jehovah's Witness around 1966.

Maureen Reillette "Rebbie" Jackson (sister)
b May 29, 1950
The oldest Jackson child was a backing singer for acts including Chaka Khan before her 1984 solo album Centipede became a hit, kickstarting a moderately successful solo career. Her daughters sing together in the group Geneva.
Husband Nathaniel Brown
Daughter Stacee Brown, b 1971 (grand-daughter London Blue, b 2005)
Daughter Yashi Brown, b 1977
Son Austin "Aussie" Brown, b 1985

Sigmund Esco "Jackie" Jackson (brother)
b May 4, 1951
Released a self-titled solo album of Motown ballads in 1973. Now manages the career of his rapper son Sigmund, aka DealZ.
Wife Enid Spann
Son Sigmund Esco Jackson Jr, b 1977
Daughter Brandi Jackson, b 1982 (adopted)

Toriano Adaryll "Tito" Jackson (brother)
b Oct 15, 1953
His guitar skills inspired his father to form, with Jackie and Jermaine, The Jackson Brothers - the group that later became the Jackson 5 - although the record label, Motown, did not allow Tito to play on their recordings. Tito extended the Jackson family brand in the mid 1990s by managing his three sons in the R&B group 3T.
Wife Delores "Dee Dee" Martez
Son Toriano Adaryll Jackson Jr, b 1973
Son Taryll Adren Jackson, b 1975
Son Tito Joe "TJ" Jackson, b 1978 (grand-son Royalty, b 2000)

Jermaine LaJaune Jackson (brother)
b Dec 11, 1954
Shared lead vocals with Michael in the Jackson 5. Converted to Islam in 1989 after a trip to Bahrain. Appeared in Celebrity Big Brother in 2007, where he performed the Jackson 5 song *I Want You Back* and finished in second place. In 2008 he was sued by a family friend for an alleged debt. In the 1990s Jermaine married the former wife of his brother Randy, Alejandra Oaziaza.
Wife 1973 Hazel Gordy (Motown CEO Berry Gordy's daughter)
Son Jermaine La Jaune Jackson Jr, b 1977
Daughter Autumn Joy Jackson, b 1978
Son Jaimy Jackson, b 1987
Partner Margaret Maldonado
Son Jeremy Jackson, b 1986
Son Jourdynn Jackson, b 1989
Wife No 2 Alejandra Oaziaza
Son Jaffar Jackson, b 1996
Son Jermajesty Jackson, b 2000
Devo (adopted, sex unknown)

MICHAEL JOSEPH JACKSON

August 29, 1958 - June 25, 2009

Wife Lisa Marie Presley, married 1994, divorced 1996
Wife No 2 Deborah Jeanne Rowe, married 1996, divorced 1999
Son Prince Michael Joseph Jackson Jr, b 1997
Daughter Paris Katherine Jackson, b 1998
Unidentified mother
Son Prince Michael Joseph Jackson II, "Blanket", b 2002

Joh'Vonnie Jackson (half-sister)
b Aug 30, 1974
From Joseph Jackson's affair with Cheryl Terrell.

Janet Damita Jo Jackson (sister)
b May 16, 1966
The second most successful Jackson solo artist, Janet has had several chart-topping singles and won five Grammy awards. During a performance with Justin Timberlake at the 2004 Superbowl her right breast was accidentally exposed - an incident that became known as "nipplegate". At Michael Jackson's 2005 trial for child molestation she wore a T-shirt emblazoned with the slogan "I'm a pervert too".
Husband James DeBarge
Husband No 2 Rene Elizondo Jr (now divorced)

La Toya Yvonne Jackson (sister)
b May 29, 1956
Released debut, self-titled, solo album in 1980, which peaked at No 116 on the Billboard 200. She has released ten more albums since then, several produced by her husband Jack Gordon. She became estranged from the rest of the Jacksons after posing nude for *Playboy* and publishing an autobiography in which she claims that her father, Joseph, abused her.
Husband Jack Gordon (now divorced)

Steve Randall "Randy" Jackson (brother)
b Oct 29, 1961
Joined the family band in 1976, when Michael left and they changed their name to the Jacksons. Formed his own band, Randy and the Gypsies, in 1989. Last year the *New York Post* reported that he now does odd jobs, including fixing cars in a Los Angeles garage owned by a family friend.
Wife Eliza Shaffel
Daughter Steavanna Jackson, b 1990
Wife No 2 Alejandra Oaziaza
Daughter Genevieve Jackson, b 1989
Son Donte Jackson, b 1991
Son Steven Randall Jackson, b 1991

Marlon David Jackson (brother)
b March 12, 1957
After a brief solo career Marlon became a businessman involved in real estate and TV. His latest ventures is a controversial plan to develop a $3.4bn (£2.4bn) slavery memorial and luxury resort in Badogry, Nigeria, including "animatronic vignettes" of the Jackson 5.
Wife Carol Parker
Daughter Valencia Jackson, b 1977 (grandson Noah, b 2006)
Daughter Brittany Jackson, b 1978
Son Marlon David Jackson Jr, b 1985

Brandon Jackson (brother)
b March 12, 1957
Stillborn after an emergency caesarian section. His twin Marlon survived.

[Illu]stration by Edward McGowan

A LONELY LIFE IN REVERSE

Michael Jackson lived his childhood as an adult, then, when he grew up, retreated into infantilism and a dangerous fascination with children

WILL PAVIA

The most ordinary years in the life of Michael Jackson were his first half dozen in the rough industrial town of Gary, Indiana. Then, like Tibetan monks fighting on a young Dalai Lama, the future King was recognised by the agents of pop music and groomed for a life apart.

Watching the ten-year-old Jackson reorganise the publicity shoot for the Jackson 5's first single, placing himself in front of his brothers and on one knee, a record executive was sure that they had found one who belonged at the top of the industry.

"How did he know that, how to take a publicity photo," said Ben Brown, then a high-ranking executive at Steeltown Records, albeit speaking with a great deal of hindsight. "He was an old soul, as if he had been a superstar in another life."

Born on August 29, 1958, the seventh of nine children, his mother, Katherine, was a shy woman from rural Alabama who walked with a limp because of her polio.

His father, Joseph, was a crane operator for a steel firm who earned $65 a week and aspired to work in the music industry. His own band never amounted to much but when he heard his second son, Tito, playing his own discarded guitar another avenue into that world opened before him.

"I envisioned these kids making audiences happy by sharing their talent," he said later. "Talent they'd maybe inherited from me."

The Jackson children practised before and after school. The neighbours complained: rocks were hurled through their windows and passers-by gave abrupt assessments of their talents, shouting: "You ain't nothing, you Jacksons."

At the age of 3 Michael began singing and at the age of 4 it became apparent that he ought to be the lead. He was quick on his feet too: Tito recalls him dodging blows whenever his father "took a swing at him": the first signs of the fleet-footedness that was a hallmark of his career came as he avoided violence.

His father was a strict disciplinarian with some odd ideas of how to enforce order. When his children left their bedroom window open at night, he would apparently climb up and burst in screaming and wearing a mask to teach them to keep it shut. It gave them nightmares.

Later Jackson would recall the thrashings his father considered necessary to keep the young band in line, a cycle of violence that intensified as he, alone among the brothers, began to stand up to him. The only threat that worked on his father was the threat to stop singing.

His mother's parents had divorced and she was determined to hold her own family together even though she and her husband were now moving in radically different directions.

She had become a devout Jehovah's Witness, he was busy booking the brothers to play strip clubs. Aged 9, Jackson was quite used to watching striptease. As part of one of their early numbers, he was encouraged to leave the stage and crawl beneath the tables, looking up ladies' skirts, flashing lewd smiles.

"I may be young," he said while introducing a song by Smokey Robinson, "but I do know what the blues are all about." Robinson himself would say that the child had an "old soul".

"I remember hearing that all the time when I was little," Jackson told Oprah Winfrey in an interview in 1993. "They used to call me a 45-year-old midget wherever I went."

Commentators have since noted the paradox at the heart of Jackson's strange life: a childhood spent being an adult, followed by an adulthood spent as a child.

The Jackson 5 graduated from strip clubs and talent contests to the top of the charts via a record deal with the Motown label in 1969. They were brought to Los Angeles, and by November 1969 the single *I Want You Back* had topped the charts. In the next seven years the brothers were worked for all they were worth, releasing 13 albums and touring venues and television studios with "Baby Michael" at the centre of the whirlwind.

"I would do my schooling, which was three hours with a tutor, and right after that I would go to the recording studio and record, and I'd record for hours and hours until it's time to go to sleep," he said later. "I was most comfortable on stage but once I got off I was . . . very sad."

Jackson appeared to be following the parabolic trajectory of the child star who betrays his fans by growing up: adolescence brought pimples; fans who saw him up close were shocked; his father told him he was ugly. He was certain this was true: he felt his skin was too dark and his nose too wide.

He did not follow the example of his brothers and sleep with the band's groupies. His first love, he said, was Diana Ross, who at other times had served as his surrogate mother in show business. He claimed the teen actress Tatum O'Neal as his first real girlfriend. She held his hand at a party in 1977. "It was serious stuff," he said. "She touched me."

For her part, Ms O'Neal, like Diana Ross and, later, Brook Shields, insisted that any relationship was strictly platonic.

The brothers had left Motown in 1976; in 1977, Jackson made a break for freedom, following Ross to New York to appear alongside her in a film of *The Wizard of Oz*, playing the scarecrow.

His first massively successful solo album, *Off the Wall*, was born out of his friendship with the film's musical director Quincy Jones. *Off the Wall* was released in 1979. Jackson the songwriter was emerging from the chrysalis of Jackson the child star: he wrote a hit single, *Don't Stop ('Til You Get Enough)* and then, after more work with his brothers, started on what would become *Thriller*, the most successful album of all time.

Children all over the world gasped at the album cover and the $10 million video of Jackson shaking and popping before a support act of zombies. A global brand was established: the moonwalk, the single sequined glove, the unlaced sneakers.

An incident in 1984 buffered the legend: performing before a crowd in Los Angeles for a Pepsi commercial, a fireball exploded across the stage and set his hair alight. He remained calm as it burnt — some concluded it was part of the act until they saw him carried away on a stretcher.

Sales of *Thriller* jumped by 150,000 week on week. It was at the height of his success that Jackson began to retreat from the world that was so fascinated by him. He bought a ranch in California and called it Neverland. He built his own amusement park, stocked it with a menagerie of animals, including a chimpanzee called Bubbles, and spent afternoons alone riding his own Ferris wheel.

The stories multiplied: it was said that he slept in an oxygen tent and was determined to purchase the remains of the Elephant Man, Joseph Merrick. The blaze of rumours was stoked further by the singer's own changing appearance. Some trace an addiction to plastic surgery back to the late 1970s. Susan Blond, a publicist who worked for Jackson during those years, found herself promoting a changing face. "By *Thriller* he had already gone overboard," she said.

Jackson denied that he had undergone any operations but for work on his nose to help with his singing, and said the cause of his whitening skin was a dermatological condition called vitiligo.

His global commercial success continued with the album *Bad*; the video was directed by Martin Scorsese; the concerts incorporated lasers and the sight of Jackson flying like his hero Peter Pan on a wire. Concurrent with this sell-out tour, his autobiography was published. Jackson declared himself "one of the loneliest people in the world".

Critics felt he was losing his touch, however. The music that always hauled him above the mire of stories about his sexuality, his appearance and his close friendships with child actors, like a giant crane at a stage show, was losing some of its power.

Rumours of child molestation hardened into accusations that threatened to land the singer in court. The police raided his home in 1993 after an allegation by the family of 13-year-old Jordy Chandler. He later arrived at an out-of-court settlement with the family for around $20 million. Marriage, to Lisa Marie Presley, was widely regarded as an attempt to restore his reputation. The marriage failed, as did a second, to Debbie Rowe, his former nurse.

The albums of the 1990s were poorly received and his ability to raise his own children was questioned after he dangled the 11-month-old Prince Michael II from the window of a German hotel in 2001. Astonishingly for a man who had sold so many records, his finances were also said to be in a state of perpetual crisis, brought on by unrestrained spending.

His next appearance on the world stage followed a controversial television interview with the British journalist Martin Bashir, in which he admitted sharing a bed with a boy.

Police subsequently issued a warrant for his arrest, and he surrendered in a blaze of publicity, to stand trial on charges of molesting a 13-year-old boy. Though he was acquitted after his defence team highlighted the extraordinary past life of his accuser, the boy's mother, it was a rather partial victory. Two jurors said they thought him almost certainly guilty of child molestation and remarked that they hoped that he would stop sleeping with young boys.

His final years somewhat resembled his own peripatetic youth, and were equally bizarre. Forced out of Neverland, which he mortgaged to the hilt, he was for a time a guest of the son of the King of Bahrain, Sheikh Abdullah. He went out in public dressed as a Muslim woman, wearing the veil.

Running from his own financial difficulties, he stumbled into new ones: the sheikh believed Jackson had agreed to record a new album and to allow him to contribute to the material, and to write an autobiography. For his part, Jackson, interpreted the money and hospitality that the sheikh had afforded him as gifts. The case, that was due to come before the High Court in London and in which Jackson was due to appear as a witness, was settled, at the last moment, by an amicable deal out of court.

For all his troubles, his poor health and the chaos of his life, his music still seemed to offer a means of salvation. A new tour, announced in March, sold out in days. He insisted he could still moonwalk.

All that the tour might have been, is now left, like most of the rest of his barely believable life, as a subject for speculation.

"Why not just tell people I'm an alien from Mars," he told one of his biographers, J. Randy Taraborelli, in 1995. "Tell them I eat live chickens and do a voodoo dance at midnight. They'll believe anything you say."

From the archive June 18, 1984

Michael Jackson is a swindle, according to the Moscow paper *Sovetskaya Kultura*. Both the album and the video of *Thriller* amount to "great show-business swindles," the paper said. The Kremlin is increasingly worried that what a Ukrainian paper recently called "empty and senseless Western music" is crowding out "glorious, inspiring, Soviet songs".
Richard Owen

From the archive November 14, 1972

The Jackson Five at the Empire Pool. Shrill teenage squeals echoed round the cavernous hall as skinny yellow suited Michael Jackson launched into the show's first ballad "Let me fill your heart with joy and laughter," he sang in that inimitably sexy and provocative voice. The 14 year old has an astonishing command of gesture, and his twinkling feet scarcely seemed to touch the stage.
Myles Palmer

End of the road for Jacko caravan

CAITLIN MORAN

I was at Jackson's last-ever UK appearance — the press conference to launch his residency at the O2 centre. Indeed, by the time he came on stage, I was ten feet away from him — standing in a scrum of screaming fans and tense, scribbling journalists. Ninety minutes late in appearing, Jackson had already worked the room into an insecure frenzy — 7,000 people who were starting to believe that of course Michael Jackson wasn't going to appear in this freezing, incongruously unglamourous hallway at the O2 centre with a Starbucks opposite. He doesn't do SE22. He's Michael Jackson.

So when he did finally appear on stage — waving awkwardly, oddly disengaged — it was almost alarming. The crowd surged and screamed — mothers pushing children forward, shrieking "Michael!", as if offering them to Jackson — while I stood, staring at arguably the most famous person in the world without a nuclear capability. The three most famous residences in the world are, after all, the Vatican, the White House and Neverland.

"His appearance is shocking," I wrote, at the time. "He looks absolutely surreal. In profile, his nose has a sharp, skull-like point, and his cheekbones protrude like child's elbows. His hair is a black lacquered curtain, and aviator shades cover fully half his face. In 2009, what you get from a personal appearance for Michael Jackson is his nose, mouth and little else."

His face was paper-pale — as if he'd been caught in the blitzkrieg of flashbulbs so often that it had finally bleached him white. And of course, in a way, they had. For, mask-faced and frail, he looked as if the onerous task of being "Michael Jackson" had exhausted him. That absurd, escalating, pharaoh-like existence of his later life — with the chimps and the gold statues and the Presley wife and the veiled children — had finally left him too weary to sing, or dance, or write another *Thriller*.

When he walked on stage, I involuntarily said "Oh!"

It was the same sound I made when I heard that he'd died. Put simply, three months ago the news that he was alive was as shocking as the news that he had died.

THE EARLY YEARS

August 29, 1958
Born Michael Joseph in Gary, Indiana to Joseph and Katherine. Lives with his five brothers (Jackie, Tito, Jermaine, Marlon, Randy) and three sisters (Rebbie, LaToya, Janet) in a two bedroom clapboard house at 2300 Jackson Street.

1963
His mother Katherine becomes a Jehovah's Witness. Enrols at Garnett Elementary School. Also becomes "the lead guy" in the Jackson singing group.

1964
The Jackson Five perform their first non-competition show, at a Big Top supermarket.

1968
The Jackson Five release their debut single, *Big Boy*, on Steeltown Records. Later that year they sign a contract with Motown Records.

October 1969
They release their first Motown single, *I Want You Back*

October 18, 1969
They make their first television appearance on *The Hollywood Palace*

December 1969
The release of their debut album, *Diana Ross Presents the Jackson 5*.

January 31, 1970
I Want You Back reaches No 1 in the Billboard chart, and goes on to sell more than six million copies worldwide.

GOING SOLO

October 1971
Jackson releases his debut solo single, *Got to Be There*. It fails to reach No 1 but still sells 1,583,650 copies.

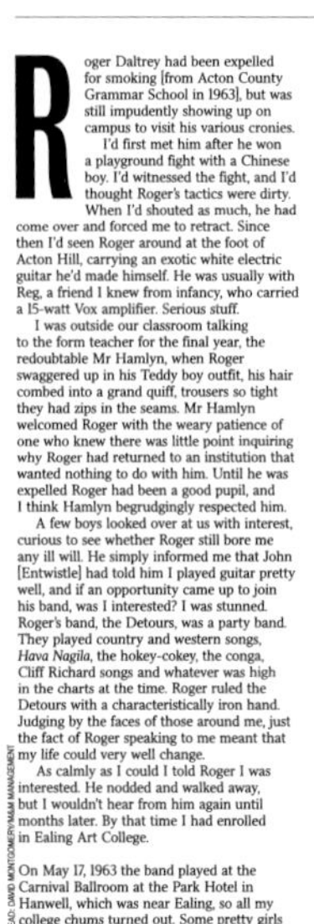

Roger Daltrey had been expelled for smoking [from Acton County Grammar School in 1963], but was still impudently showing up on campus to visit his various cronies.

I'd first met him after he won a playground fight with a Chinese boy. I'd witnessed the fight, and I'd thought Roger's tactics were dirty. When I'd shouted as much, he had come over and forced me to retract. Since then I'd seen Roger around at the foot of Acton Hill, carrying an exotic white electric guitar he'd made himself. He was usually with Reg, a friend I knew from infancy, who carried a 15-watt Vox amplifier. Serious stuff.

I was outside our classroom talking to the form teacher for the final year, the redoubtable Mr Hamlyn, when Roger swaggered up in his Teddy boy outfit, his hair combed into a grand quiff, trousers so tight they had zips in the seams. Mr Hamlyn welcomed Roger with the weary patience of one who knew there was little point inquiring why Roger had returned to an institution that wanted nothing to do with him. Until he was expelled Roger had been a good pupil, and I think Hamlyn begrudgingly respected him.

A few boys looked over at us with interest, curious to see whether Roger still bore me any ill will. He simply informed me that John [Entwistle] had told him I played guitar pretty well, and if an opportunity came up to join his band, was I interested? I was stunned. Roger's band, the Detours, was a party band. They played country and western songs, *Hava Nagila*, the hokey-cokey, the conga, Cliff Richard songs and whatever was high in the charts at the time. Roger ruled the Detours with a characteristically iron hand. Judging by the faces of those around me, just the fact of Roger speaking to me meant that my life could very well change.

As calmly as I could I told Roger I was interested. He nodded and walked away, but I wouldn't hear from him again until months later. By that time I had enrolled in Ealing Art College.

On May 17, 1963 the band played at the Carnival Ballroom at the Park Hotel in Hanwell, which was near Ealing, so all my college chums turned out. Some pretty girls from the fashion school stood at the front of the stage, pretending to scream at me like Beatles fans; they were teasing, but everyone was impressed, especially when we played the slightly funkier R&B tunes I'd managed to sneak into our otherwise catholic repertoire. This was a formative moment for me. My friends from college could see the band I had been so reluctant to talk about; John, Roger and Doug [Sandom, the drummer at the time] could see my art-school friends, and how ➡

PREVIOUS PAGE: ART KANE ARCHIVE. THIS SPREAD: DAVID MONTGOMERY/M&M MANAGEMENT

30 The Times Magazine

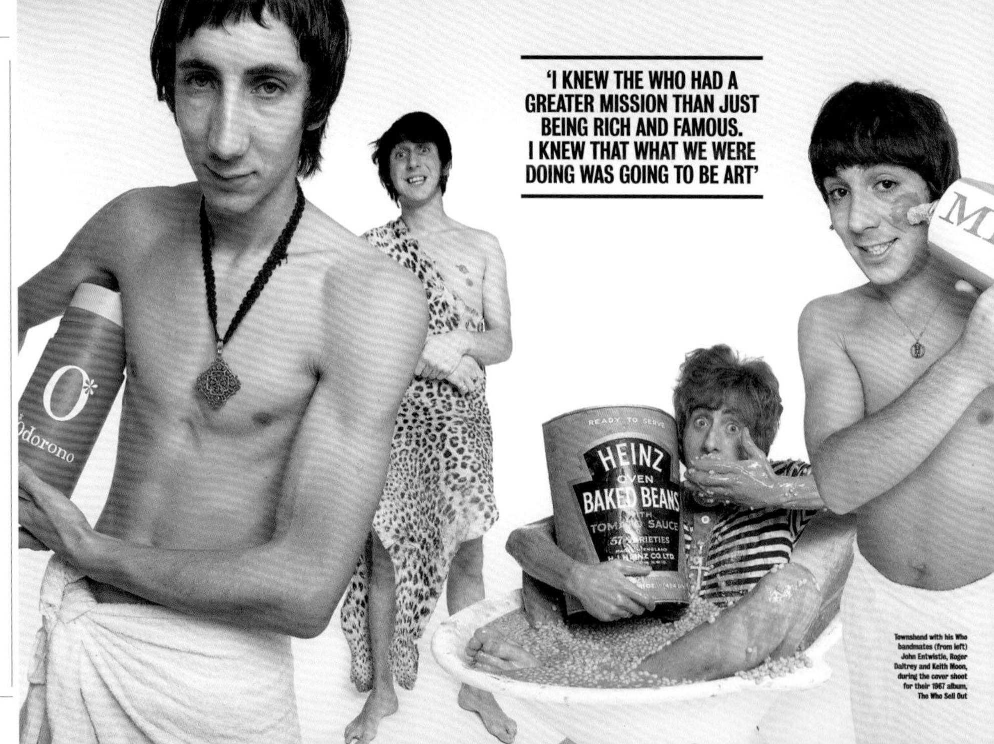

'I KNEW THE WHO HAD A GREATER MISSION THAN JUST BEING RICH AND FAMOUS. I KNEW THAT WHAT WE WERE DOING WAS GOING TO BE ART'
HEINZ
OVEN
BAKED BEANS
Odorono

Townshend with his Who bandmates (from left) John Entwistle, Roger Daltrey and Keith Moon, during the cover shoot for their 1967 album, The Who Sell Out

THE TIMES
MAGAZINE
29.09.12

I can explain.
Pete Townshend in his own words.
Exclusive extracts from his memoir

PLUS
KOFI ANNAN'S VERDICT ON TONY BLAIR
By James Harding

聊聊苹果平板电脑吧。你们在平板电脑上使用的版本，是否和网站版本差异很大？

先说一点儿背景吧：鲁珀特·默多克（Rupert Murdoch）曾向史蒂夫·乔布斯（Steve Jobs）承诺，我们将是首家在苹果平板电脑上发布新闻的报纸，而且是在苹果平板电脑登录英国后三个月内就要出刊。于是我们在平板电脑上市后三个月内推出《泰晤士报》的App，这是每天一刊的版本，不是不断更新的网站。我们的编辑口号是它“就和报纸一样，但是更好”。

《泰晤士报》App背后的理念是，它不像网站一样是“实时”产品，而是和报纸一样一天出刊一次，除了拥有和纸质版报纸一模一样的内容，还增加交互式图像、视频与更多的新闻档案资料，为更长的报道提供背景信息。读者可以有一个基于文本的线性体验。

相信读者会更青睐这种产品，现在读者能在不同版本上享受线性阅读，而不是像浏览新闻网站时那样，无休无止地阅读。

!

希尔给有志成为编辑设计师的年轻人的一些建议

你认为年轻设计师需要了解文字编排的规则与基本知识吗？

是的，年轻人必须对设计与文字编排有热情，也要证明自己能处理复杂的信息与大量文字。

毕业生必须对文字编排与信息设计有基本知识，还要能够自学新软件。在职场上的设计师都是利用互联网随时补充新知识。要多少懂一点编程知识，才能理解程序开发者所做的事，并用他们的语言来沟通。

至于专业出版软件，可以边学边做。许多大型出版商都是使用内部的专用系统，在市面上根本找不到，因此只能在公司里才学得到。这意味着，快速学习的能力很重要，应聘者如果有很强的学习能力，是加分项。

学生能不能一边工作，一边回头学习文字编排的基本规则？

我们常要招聘有专业背景的人，因为文字编排得花更多时间才能解释清楚。我们会找有技术能力的学生，再花时间系统训练他们如何处理字体。他们的作品集如果能够体现出自己对编辑设计的热情和兴趣最好，这样未来在工作上才能讨论得更深入。

我们寻找的人才都有不少文字编排经验，或至少能对字体展现出热情（即使是木活字），这样才能和他们讨论细节。毕竟在《泰晤士报》，无论是纸质版或数字版、信息设计或动画都离不开字体设计。

你希望应聘者具备什么专业能力？

最好要了解或至少能够清楚网站的基本程序代码如何实现，目前需要掌握的程序代码包括HTML5与CSS。我们不会要求设计师亲自处理程序代码，但必须知道网页是怎么做出来的。知道如何处理有大量数据、视觉效果与图像的新闻报道也很重要。现在信息图表是很常见的讲故事工具，各机关单位与政府机构都喜欢用这种方式来发布数据，这对设计师来说是好事。设计师要能独立分析新闻数据，并利用布局图或动画等，用简单的方式讲述复杂的故事，在设计数字版时尤其需要这项能力。编辑设计师无法仅靠会编排图文就打遍天下，还得能够运用很多方式讲好故事。如今我们必须十八般武艺样样精通，动画、处理数据与程序代码都要懂，同时，版面设计、文字编排与图片制作的传统能力也要扎实。

关于字体设计的迷思

cultuur

'Dromen zitten je geluk in de weg'

Misja M.: muzikale anarchist 30

Belangrijke prijs v

Amsterdamse gal

HOLLAND FESTIVAL

De wortels en de vruchten van de meester

Geef mij maa

ORANJE

对还是错？刊头一定要放在封面的最上方。

这一原则源自杂志只在店面或报摊贩售的时代，当时杂志在展示时会层层重叠，因此刊头必须出现在封面的最上方，才不会被挡到。如今许多杂志是依靠订阅而非零售，情况已有不同。刊头放在最上方，符合阅读逻辑，并有利于展现权威，所以封面最上方仍是刊头最理想的位置。

对还是错？栏宽和x-字高（x-height）要能搭配，一行的字数不能过多。

这是个关于“易读性”的问题，不妨阅读相关书籍，例如贝蒂・宾斯（Betty Binns）的著作《改善字体》（*Better type*，Roundtable出版社出版，Waston-Guptill，1989年）。x-字高是指某小写字母从基线（baseline）到中线（meanline）之间的高度。不同字体即使字号相同，x-字高也不一定一样。如果一行的字数太多，就不容易拉回原来的位置。因此一般来说，每一行要保持适当的字数才好。

对还是错？标题应该尽量避免压在图片或者任何图像上。

如果可以避免的话，标题尽量不要压到新闻摄影照片或者其他任何图像，尤其是当新闻图片和艺术品一样珍贵时。想想看，艺术家的作品如果被标题压住，艺术家肯定很不高兴。图片上有文字原本是无可厚非的，文字如果能让图片更抢眼，会为画面加分，这样的文图结合才是精彩的设计。可参见乔治・路易斯《六十年代封面设计》（*Covering the sixties*，The MonacelliPress出版，1996年）。

对还是错？别用黑底和反白字，读者不爱读。

这是因为精美刊物往往采用CMYK四色印刷（分别指青、洋红、黄与黑），图片的“黑”至少加了点蓝色油墨，有时还会加上洋红与黄。在印制过程中，由于印刷机套色（套色是彩色印刷机的印刷方法，以确保颜色对齐，印出清晰锐利的效果）可能不准，小小的白字体就会显得模糊，导致读者无法阅读。

对还是错？标题应占据页面最上方，但不能跨到装订线的留白区。

这个规定起源于文字编排非常讲究层次结构的时代，方便帮助读者一步步消化内容。然而现在阅读渠道很多，非传统刊物没必要墨守成规。平板电脑版本的页面可在横向和纵向之间转换，标题也可以移动，可以当成浏览内容的链接，就算标题位于页面底部，也可以自由缩放。纸质杂志因为需要装订，因此厚的铜版纸刊物会有1.2厘米的订口空白，这地方不该放内容。多数杂志使用胶装（用热熔胶把装订区黏合），如果刊物很厚，标题就不能太靠近装订线，以免文字因订口封装而被遮盖。但数字版本或采用骑马钉装订的杂志，标题延伸到订口空白处也没关系。这种问题可以站在读者的立场思考。

字体设计的参考资源

网站：

关于如何使用印刷字体与数字屏幕使用字体

艾莉萨·费登与埃伦·勒普顿在普林斯顿建筑出版社（Princeton Architectural Press）出版了*Opinionated font facts, research and writing*一书。在www.thinkingwithtype.com这个网站可以看到他们的著作。

想了解阅读心理学

《辨识文字的科学》（*The science of word recognition*），凯文·拉森（Kevin Larson）著，Advanced Reading Technology，Mircrosoft Corporation出版,2004年7月）。网站：www.microsoft.comtypography

参考书：

《阅读的旅程》（*While you are reading*）,格拉尔德·翁格尔（Gerard Unger）著,Mark Batty Publisher出版,2007年。

《排版细节》（*Detail in Typography*），约斯特·贺朱利（Jost Hochuli）著,Hyphen Press出版,2008年。

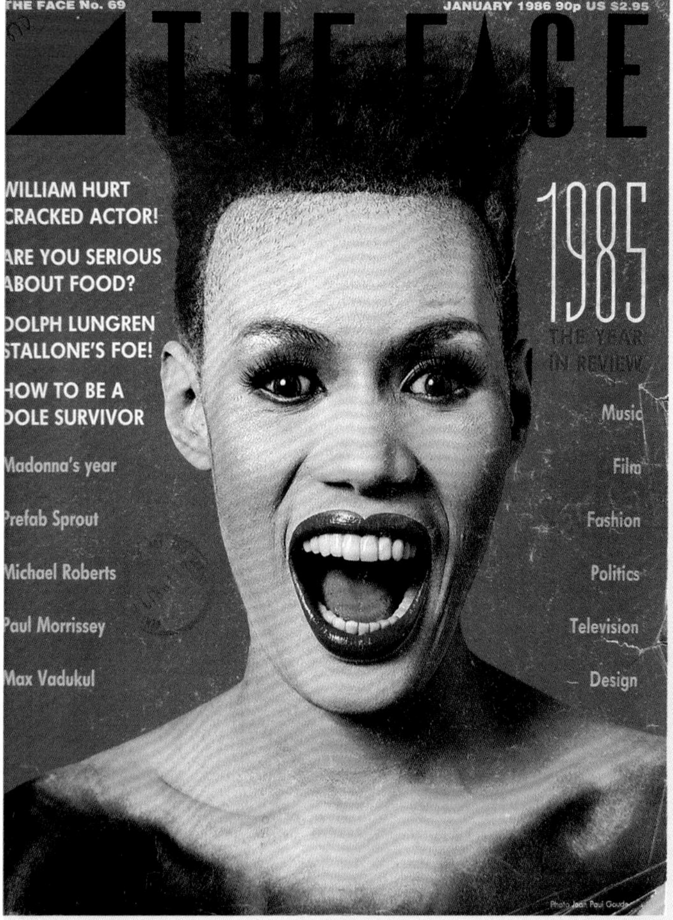

在1981年至1986年为*The Face*设计的作品中，奈维尔·布罗迪（Neville Brody）曾在排版和布局上进行了激进的尝试。受构成主义和达达主义的影响，他打破了文字编排的诸多传统，凸显个别单词和短语，在同一个单词中使用不同大小的字母。上面是1986年1月的封面，是歌手格蕾丝·琼斯（Grace Jones）的肖像，由让-保罗·古德（Jean-Paul Goude）拍摄。

图片处理

设计师如何处理图片对刊物质感影响很大。以前报纸以文字内容为主，现在越来越依赖利用图片来讲述故事，即便是新闻报道对图像（包括图表、插图与图像工具）的需求量也大幅提升。不过报纸与杂志在使用图片时，依然有根本上的差异。马里奥·加西亚（Mario Garcia）

说：“刊物性质不同，用图重点也不同。报纸用图要传达现场感与实时性，杂志用图则比较注重趣味与放松效果。”制作过程与预算也是一个重要因素。和杂志相比，报纸花在图片上的时间与经费要少得多。

关于艺术总监的迷思

对或错？封面是杂志中最重要的页面。

封面确实越来越重要。如果封面无法吸引读者，那么之前所有的工作都是浪费。电子杂志的封面更是刊物的入口，也是强化品牌的主页。在网站上，杂志封面会被缩得很小，所以封面必须格外用心制作，即使是小尺寸的封面缩略图，也要留下令人难忘的印象。

对或错？编辑设计师只能听命于文字编辑。

在创作过程中，设计师是编辑的伙伴，和编辑处于平等地位，有助于碰撞出更多火花，让平凡的刊物内容变得好看。最佳的编辑设计团队组合是要彼此尊重合作，而非谁听命于谁。

对或错？每张图都需要图片说明。

如果希望读者了解图片的含义，就需要附上图片说明。但如果图片只是装饰，就不需要。图片要标示出摄影师姓名或图库名称。设计再好，也不能侵犯他人版权，这一点务必要注意。

对或错？花钱制作原创摄影或插画是有必要的。

设计师都乐于制作有原创性的设计作品。如果预算充足，允许找寻新素材，不妨为杂志制作专属的素材。此外，付给原创作品费用，应与权利人协商，允许拥有日后再使用该图片的权利。因此，出版商可着手建立能重复使用的图像数据库，时尚与生活类杂志甚至可以利用自家图库来赚钱。

对或错？运用网络上的免费图片，是未来的趋势。

首先应该是要做出独一无二的产品。原创图片虽然成本不低，但好处很多。网络上的免费图片可能到处都

这是*Wallpaper**的*Reborn in India*（在印度重生）特刊中所刊登的时装照片，画面充满色彩与动感。照片于印度色彩节（Holi）期间拍摄，在*Wallpaper**网站上还可以看到影片。满版照片极大强化了杂志的特性——在这个例子中这系列摄影彰显了杂志对不同文化与国度的好奇。

在苹果平板电脑的单页（非跨页）浏览模式下，单张静物摄影表现力很强。在这张*Port*杂志的时装照片中，呈对角线摆放的长椅与苹果平板电脑产生有趣的视觉互动。这张精彩的服装摄影照片，没有采用常见的去掉图片背景的方法，而是通过场景安排，如实呈现出大衣的硬挺质感与故事性。

有，缺乏新意。此外，使用网络上的图片时需要标明出处，也要仔细确认是否真的没有版权问题。

对或错？留白空间在杂志上不受欢迎。

如果留白空间能增加氛围，营造视觉张力，提高内容的故事性，那么留白空间就有价值。如果留白空间只是拖长文章或者为了充页数，那就没有必要。本书179页的*Émigré*杂志与*RayGun*杂志，是运用留白空间创造活泼视觉的优秀典范。

照片运用

新闻摄影具有视觉报道和讲述故事的功能。如今摄影技巧丰富，风格五花八门，编辑设计师们的选择很多。即使出于预算原因，没办法委托摄影师专门拍照，公关公司或是报道对象所提供的照片在编辑、后期制作后也很好用。委托摄影师帮新闻报道拍出特定风格的照片只是开始（本书第190页会谈到如何成功委托）。虽然照片的长宽比多半是固定的，但设计师可裁剪图像、改变形状、调整色调或用其他方式处理图片。这都是图像编辑工作，就像文字编辑会删改文章，同样地，艺术总监也可调整刊物的图像内容。

不仅如此，如果编辑设计师希望通过照片呈现不同的报道角度，可以故意不告诉摄影师任何信息。这样，可能带来完全不同的图文解读方式，发挥一加一大于二的效果。正如编辑可以通过文字编辑来传递观点，设计师也可以通过设计来实现。图文的选择、结合方式与相对位置，以及向读者解说图片内容的图片说明，都能强烈暗示出或许并不存在的“事实”。此外，针对照片的处理技术越来越复杂（无论是拍照时还是后期制作过程），甚至能将真相变成谎言。因

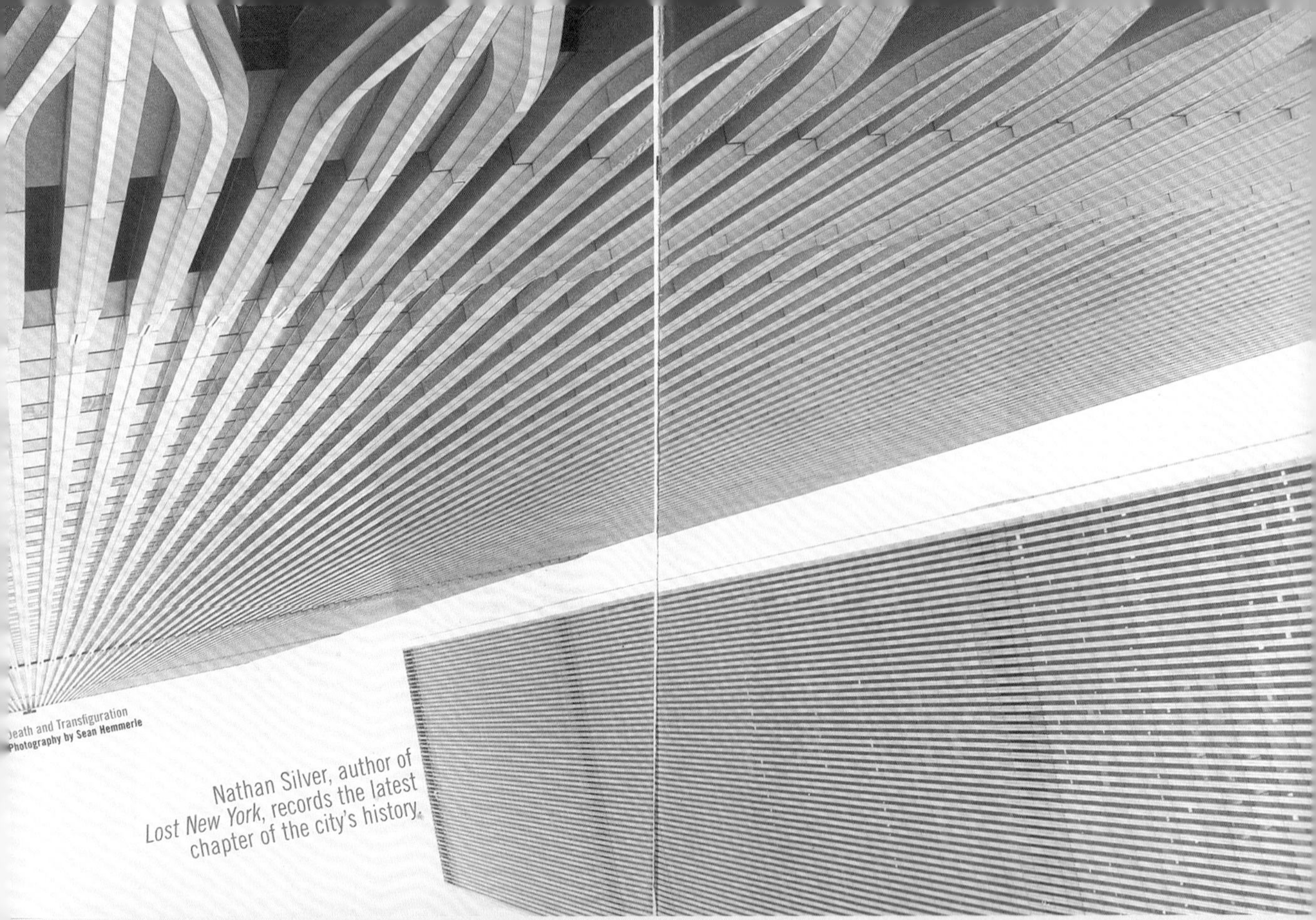

Death and Transfiguration
Photography by Sean Hemmerle

Nathan Silver, author of *Lost New York*, records the latest chapter of the city's history.

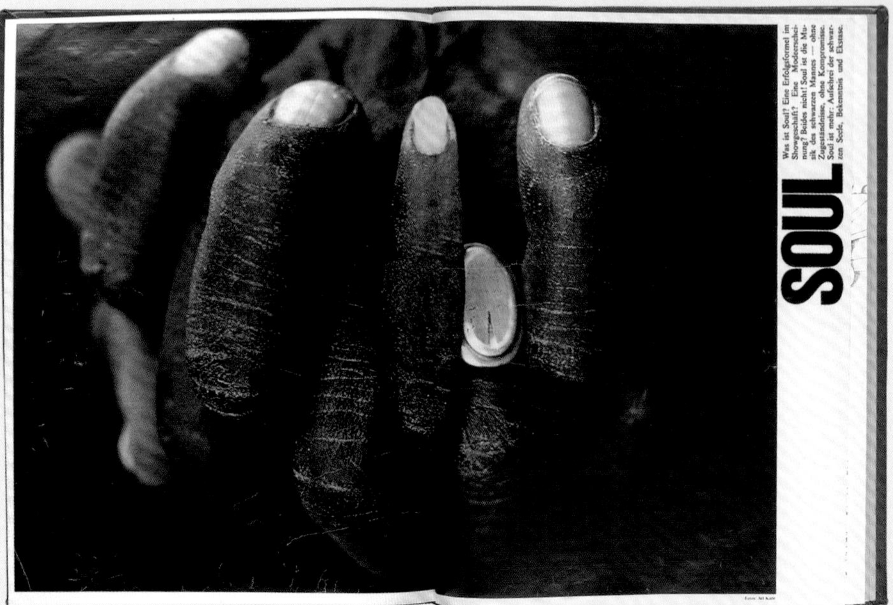

《大都会》的前艺术总监克里斯韦尔·拉滨（Criswell LAppin）说：“2001年12月刊，是我们在‘9·11事件’之后首度出刊。由于刊物有两到三个月的准备时间，送印前，双子大楼遭毁的照片已经满世界都是。我们决定不再刊登任何双子大楼遭毁的照片，而是把这些版面空间用来纪念过往，展望未来。”这张由肖恩·汉莫瑞（Sean Hemmerie）拍摄的照片完成了这项任务。

好的摄影作品是杂志的生命线，但适当的裁剪、缩放、定位、色彩与纸张使用，也能进一步引发共鸣，这些必不可少。左图选自*Twen*杂志，就是一张裁剪出彩的好图。

The Guardian

A declaration of war

此，美术编辑、文字编辑、图片编辑与摄影师都能使一则报道呈现出千变万化的形态。迈克尔·沃茨（Michael Watts）曾长期负责《金融时报》周末版的编辑工作，他回忆起当时的经历时表示：

> “有时我和负责这份周末报刊的莉娅·卡斯伯特森（Julia Cuthbertson）会意见不合。她希望杂志保留经典传统的内容，例如美食与餐厅评论，但我们想要一眼看上去就充满动态感的新颖的东西，以呼应互联网的繁荣和新的、炫酷的商业状态。所以我们会耍点小手段从视觉上来颠覆传统，例如在美食版，别人都是拍煮好的料理，但我们拍生的食材。我们的图片编辑卡洛琳·梅特卡夫（Caroline Metcalfe）为此找了几位优秀的食物摄影师，尤其是罗伯·怀特（Rob White）。有一次，我们把玛丽恩·麦吉华瑞（Marion MCGilvary）风趣的评论，搭配上餐厅地图及插画风格的菜单，这些元素结合之后，杂志就呈现出与众不同的样貌，虽然这也惹恼了一些人，但是喜欢的人更多。《金融时报》周末版非常有特色，我在职期间的销售量冲到了25万份，增长了20%。平常没看《金融时报》的人，也会在周末买来看。”

杂志里的照片可发挥和文字一样强大的力量，而报纸的图片却往往是文字的辅助。然而，报纸设计已潜移默化地向杂志设计靠拢，越来越依赖图像的表现力。现在的报纸头版常用一张抢眼的图片来吸引读者目光，这有着仅靠语言无法传递的效果。2001年9月12日《卫报》的头版封面就很有力地体现了这一点。

1967年7月的《时尚芭莎》（*Harper's Bazaar*）（左图）刊登了一篇关于诗歌的文章，搭配德国作曲家卡尔海因茨·施托克豪森（Karlheinz Stockhausen）的乐谱《第十一号演奏者的副歌》。艺术总监鲁思·安塞尔（Ruth Ansel）与比阿·费特勒（Bea Feitler）让五线谱与乐曲的优雅细致与抒情画面，和诗的概念结合起来，为文章提供了一个绝妙的视觉隐喻。现在报纸比以往更经常使用插图，让页面更活泼，也为报道提供文字以外的解读。下图为马里恩·杜查尔斯（Marion Deuchars）为《卫报》创作的插画，让页面充满生机和活力，构图和色彩与下方的文字形成很好的对比，插画形式给故事报道带来了图片无法带来的质感。左下图选取的平板电脑页面中，插画家丹·威廉姆斯（Dan Williams）的笔触和利落的文字相互映衬，达成完美的平衡。

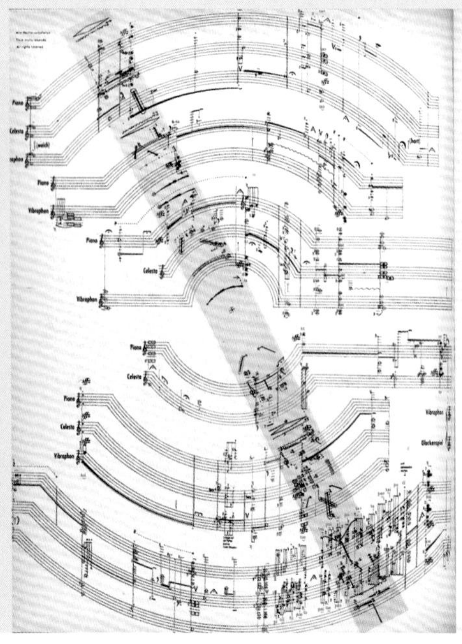

ROBERT CREELEY

TALKS ABOUT POETRY

I'm *given* to write poems. I cannot anticipate their occasion. I have used all the intelligence that I can muster to follow the possibilities that the poem "underhand," as Olson would say, is declaring, but I cannot anticipate the necessary conclusions of the activity, nor can I judge in any sense, *in* moments of writing, the significance of that writing more than to recognize that it is being permitted to continue. I'm trying to say that, in writing, at least as I have experienced it, one is *in* the activity, and that fact itself is what I feel so deeply, the significance of anything that we call poetry.

For some sense, then, of how it was I came to be involved with poetry, at the outset I was much more interested in *writing* apart from its designated modes, and perhaps I am characteristically American in that respect. To begin with, I was shy of the word "poet" and all its associations in the world I was then intimate with. It was not, in short, a fit attention for a young man raised in the New England manner, compact of Puritanically deprived senses of speech and sensuality. Life was real and life was earnest, and one had best get on with it. The insistent preoccupation with words did begin for me early, just that I did want so much to know what people were saying, and what, more precisely, they meant by it.

I think the most significant encounter for me as a young man trying to write was that found in the work of William Carlos Williams. He engaged language at a level both familiar and active to my own senses, and made of his poems an intensively emotional perception, however evident his intelligence. Despite his insistence on his Mediterranean connections, so to speak, he was as Puritan as I—or Lawrence, or Thoreau or the Melville of *Pierre*.

Otherwise, the Forties—the time in which I came of age—were complicated in many bitter ways indeed. Not the least of the problems then evident for someone trying to realize him or herself in the world was the confusion about the very nature of "literature" itself. Coming from New England, I felt awkwardness about books to begin with, because they were for *me* often instances of social mark or measure, even at times a privilege of intellectual order—just as Hardy speaks of them in *Jude the Obscure*. I was very shy about communicating my own commitments in reading, and yet I used books as a very real *place* to be. Not merely an escape from the world—the difficulty was how to get *into* it, not away—books proved a place very deeply open to me, at moments of reading, in a sense few others were ever to be.

Thinking of that, let me note kinship with another writer—Robert Duncan—who has played a very important role in my life, both as mentor, very often, and as one whom I feel to share with me this particular sense of world, and writing, and poetry, which I most deeply respect. In a collection of his called *The Opening of the Field*, significantly enough, the first poem begins:

Often I am permitted to return to a meadow

Then continues:

as if it were a scene made-up by the mind
that is not mine, but is a made place,

that is mine, it is so near to the heart,
an eternal pasture folded in all thought
so that there is a hall therein

that is a made place, created by light
wherefrom the shadows that are forms fall.

This sense of a poem—that *place*, that *meadow*—has echoes of so many things that are intimate to my own sense of the reality experienced in writing. One would find that field or "meadow" in Whitman also, and it would be equally the sense of place I feel Allen Ginsberg many times to be entering, to be speaking of or longing for. Charles Olson too possesses its occasion in his sense of "open" verse or that open field, as he insists upon it, in composition. I have found it deeply in (Continued on page 120)

Karlheinz Stockhausen has, as have many contemporary composers, been preoccupied with the aleatoric, or chance, elements in music. "Refrain for Three Players, no. 11" (left) employs a circular score and a plastic strip which varies the succession and intervals of entry of the notes written on it, according to its position. Stockhausen observes on the record jacket (s8001) "to understand . . . (one) must read the score."

Secret City
Lorin Stein

The Tailor
New York

"Wearing clothes is a pleasure," wrote the photographer Edouard Levé, "buying them a trial." My feelings exactly, until a friend sent me to Kirk Miller of Miller's Oath. Wearing Kirk's clothes is indeed a pleasure. His suits, shirts and coats are beautifully cut, and the place is fun too. Kirk takes a sort of editorial joy in solving unusual problems in the most elegantly simple way. Last month I asked him for a suit that I could wear over a two-week train journey. The result, a light two-button corduroy, had strangers stopping me on the street from New York to San Francisco, and yet there is nothing dandified about it. It is the most comfortable thing I own – and was, from the moment I put it on.

INITIATIONS　　[Map]　　[Miller's Oath]　　**ILLUSTRATIONS BY DAN WILLIAMS**

The Guardian | Saturday September 17 2005　　23

Saturday

Marina Hyde: Lost in Showbiz 25 | Marian Salzman, doyenne of trends 27 | Mark Lawson on scriptwriters 28 | Norman Johnson joins the Guardian 29

An American on Exmoor grapples with an eccentric British tradition

Goodwill hunting

Essay by PJ O'Rourke Illustrations by Marion Deuchars

A stag flanked by two female red deer, or "hinds," trotted down a steep moorland pasture toward a wood. Two mounted hunters were behind them, and staghounds were in between. The deer's trot was faster than the hunters' canter and as fast as the all-out run of the hounds. A horn was blown. We were off – over a soaked, slippery sheep meadow, between the stone posts of a narrow gate, down a muddy track perilous with ruts, into a country lane barely an arm's length wide, through the tiny streets of an old village with tourists hopping out of our way, and then making a hairpin turn on to a paved road, speeding uphill around blind curves, and narrowly avoiding several head-on collisions with trucks. It was a thrilling ride, even if it was in a Suzuki SUV driven by a retired grocer – an enthusiast of stag hunting who had volunteered to show me the Devon and Somerset Staghounds hunt on Exmoor.

We arrived at a hilltop above the woods. From there I could look across a valley at ... not much. On the crest of the far slope several dozen members of the hunt were sitting on their horses. They watched two dogs sniffing the underbrush below. Two men in scarlet coats were with the dogs, more closely watching them sniff.

The idea of a stag hunt evokes chivalry – knights in jerkins and hose, ladies on sidesaddles with wimples and billowing dresses, a white stag symbolising something-or-other, and Robin Hood getting in the way. An actual stag hunt is more like a horseback meeting of a county planning commission. The equestrian committee is responding to deer-population-growth issues. Red deer are noble animals – big, anyway. But red deer are also pests. A farmer may find a hundred deer in his pasture, each eating as much grass as three sheep.

Each stag hunt has a "harbourer", a specialist whose job is to watch the herds and select a specific quarry, chosen for its lack of Darwinian promise, the herd. It was this singling-out that I was watching. Although it wasn't working perfectly. As unpromising as that young stag may have been from the harbourer's point of view, the two hinds thought he was worth running away with.

Once the stag is solitary, the huntsman's assistant, the "whipper-in," is supposed to bring up the full pack, and the hunt's members and guests fall in behind the hounds. Miles and miles of furious cross-country galloping ensues.

Unless it doesn't. As it seemed not to be doing from my vantage point across the valley.

When the chase does happen, the usual outcome is that the stag, at last, turns and "stands at bay", facing the hounds. Then (rather disappointingly for those whose imaginations run to tenderhearted indignation or to bloodlust) the hounds do not tear the stag to shreds. They bark. There's probably not much else they could do with an irked and antler-waving stag. Staghounds are not giant Scottish deerhounds or hulking, red-eyed mastiffs. They're just foxhounds, happy and hound-doggy and friendly if you aren't prey. "You can set your baby down in the middle of a pack," a hunter told me, "and they'll lick him silly."

What happens to the stag is that the huntsman walks over to it and prosaically shoots it in the head with a special short-barrelled, folding-stock shotgun. This is an illegal weapon in Britain. But on stag hunts it's legally required. Speaking of Britain's laws, killing wild mammals with the aid of dogs, as the Exmoor hunt was trying to do, is forbidden. Except when – as I understand the Hunting Act of 2004 – it is mandatory. The act contains certain conditions for "exempt hunting" that allow the killing of wild mammals with the aid of dogs – if "as soon as possible after being found or flushed out the wild mammal is shot dead by a competent person". No letting it go, even if it's Bambi's mother.

The Hunting Act came into effect on February 18 2005, a few weeks before this Exmoor meet. I got in touch with Adrian Dangar, the hunting correspondent for the Field magazine. He said that I shouldn't write about foxhunting, which is the most notable activity banned under the act. He said that the stag hunters were a doughty and resolute lot, and stag hunting was more of a way of life.

I went to Exmoor with Adrian. We stayed with the chairman of the stag hunt, Tom Yandle. He was doughty and resolute, the owner of a family sheep farm of centuries' standing. I went to the meet expecting a scene of American seething, full of the half-suppressed violence that Americans thwarted in their beliefs or their hobbies half-suppress so well. What I found was a cheerful, natty crowd on horseback listening to a talk from the hunt secretary about strict adherence to the Hunting Act, especially in the matter of using just two dogs.

Having just two dogs in the field was exactly the problem. So I was told by the retired grocer and other hunt followers gathered on the hilltop vantage point. Two hounds were not enough to break the stag away from the hinds. Or two hounds were not enough to make the stag stand at bay. Two hounds were certainly not enough to make what I was told was the music of a pack in full cry.

The hunt was moving. Horses were trotting over the far hill. The two hounds did their best in the music department. There was a spate of elderly, excited driving as hunt followers hurried to find a better view. We parked by a tributary of the Exe. The stag either did or didn't go into a strip of woods from a van. There was a tense murmuring, as from a golf gallery.

The staghounds and stag hunters trotted through a farmyard, and I followed on foot. Some local farmers are not hospitable to the traffic through their property – not hospitable, specifically, to the traffic of me. I was trying to take notes and make haste and avoid deep puddles and horse droppings, and I wasn't wearing a necktie. "Is he all right?" I heard a farmer ask. "He thought you were an 'anti'," the retired grocer explained later. "They come around bothering the hunts."

Wind, rain, and temperature grew worse. The hunt descended into a precipitous dell where I'd have thought the riders would have to walk their mounts. They didn't. The stag and every trace of it had vanished, and the hunters decided to "pack it in, to spare the horses".

Tom, Adrian, and I headed back to Tom's farm in his horse van, a bit disappointed. And then came that music I'd been told about: the full cry of a pack. It is a bouillabaisse of a noise, with something in it of happy kids on a playground, honking geese headed for your decoys, and the "wheeee" of a deep-sea fishing reel when you've hooked something huge.

A beagle pack, 30-some strong, was bounding across a pasture. Beagling is like foxhunting or stag hunting except that the quarry is hare, and it's done without horses. Beaglers follow the pack at a very brisk pace – on foot. Hunting hares with beagles is banned by the Hunting Act. But rabbits can still be hunted. "Because they're considered pests," Tom said. "Because a lot of Labour voters hunt rabbits," Adrian said. Also, for some reason, "the hunting of a hare which has been shot" is permitted.

The pack arced away from us across a broad field. "The hare went that way!" Tom shouted to the master of the beagle hounds.

RED DEER ARE NOBLE ANIMALS – BIG, ANYWAY BUT RED DEER

维珍航空发行的头等舱上的杂志*Carlos*除了中间两三个跨页是铜版纸印制的全彩广告之外，其他页面都用手绘插画取代了摄影图片（右图）。这一做法“旨在摆脱由名人照片主导封面，让视觉呈现方式变得单一”。用当时艺术总监杰里米·莱斯利的话来说：*Carlos*是‘后摄影时代’风格的杂志!”

“如果没有把织品和插画结合起来（右图），就只是展示地毯样布，会非常呆板乏味。但画家克里斯托弗·尼尔（Christopher Neal）把布料和线图结合之后，则起到了一箭双雕的效果。读者除了能看到赏心悦目的样品，还能了解更多的产品信息，例如不同材质的布料可以适用的不同场所（学校、办公室、机场等）。我只提供给克里斯托弗基本版面要求、每张插图的大略尺寸及每种布料该搭配什么环境，他就用有限的信息创作出这样优秀的内容。他将插画成品寄给我之后，我就决定标题‘EDGES’（意为边缘）一定要用手写字体，来和他的整个设计做呼应。”

——克里斯韦尔·拉滨（Criswell LAppin），《大都会》（*Metropolis*）前艺术总监

SOFTENING THE EDGES

The textile industry is reaching out to an intriguing group of designers that you wouldn't necessarily associate with its often staid and traditional products. Can the likes of Bruce Mau and Shashi Caan give contract carpet and fabrics a long overdue makeover?

Illustrations by Christopher Silas Neal for *Metropolis*

Translations
Shashi Caan Collective
Karastan Contract, The Mohawk Group

No one ever accused Shashi Caan of lacking ambition. When Tom Lape, president of the Mohawk Group, approached her about creating a new line of contract carpets, she saw an opportunity. "I was intrigued because the company reaches a huge segment of the market. And I believe designed environments that impact the most people have the most responsibility. With a company like Mohawk, a small shift in emphasis—toward design, for instance—creates a large ripple effect on the whole industry."

The four-pattern collection Translations is a collaboration between Caan, Mohawk's in-house design team, and the company's technical staff. "They were essential," Caan says of the technicians. "They understand the art of making carpet on a massive scale." Using transparent and metallic yarns, the loom technicians experimented with different pile heights to create almost three-dimensional textures. "Walking on a woven fabric is a tactile experience," Caan says. "My goal here was to push the design envelope incrementally and provide an incentive for them to work with design in the future. The craft of weaving is an ancient art. Personally I think hard flooring is trendy. I'm convinced that our ancestors threw furs on their cave floors. That essential connection will always be with us."

Theory
Cannon Design
MILLIKEN

If you've ever sat in a class lecture and stared at the rug until the seemingly random patterns started to play tricks with your eyes, then you'll appreciate Milliken's latest carpet line, Theory, a family of eight modular carpets that may turn zoning out into an acceptable form of scholarship. Half of the patterns are comprised of collaged letters that range in scale from small and densely packed (in Pop Quiz and Cram) to grand but still indecipherable (Mad Professor).

According Cannon's interiors director, Kate Wendt, who led the effort to develop the line with Milliken, Theory is intended for educational environments like college dorms and classrooms. "We like to pair our in-house team with an architecture-and-design firm doing specialized work in the industry to develop products that are useful to them," says Ronnie Magee, Milliken's design director. (With HDR, Magee also codeveloped Sense, which is for health-care interiors and launches later this year.) During a charette the Cannon and Milliken design teams brainstormed words and images associated with school. The result was carpets with coordinating patterns and scales representing varying degrees of seeing and understanding, notions that both teams felt were central to the learning process.
—Laurie Manfra

插画运用

马克·波特在《卫报》工作时常采用插画，他认为“插画是《卫报》重要的视觉元素。通过引入更当代的插画家，确保报纸能散发新鲜感和现代感”。如果某篇报道需要概念性或较隐晦的诠释，缺乏好的摄影照片或只是需要更有趣味性的视觉呈现时，编辑们就会改用插画，以展现不同风貌。插画比照片更能传达抽象感受，尤其是当插画颇具隐喻性时，读者也更喜爱主观地“解读”它们。但当读者只看到一张照片，就不会有如此乐趣，往往只能解读表面含义，如照片中的是什么人、穿什么衣服、拍摄地点是哪里等。但是插画通常能引发较有表达性与抽象性的联想，甚至比照片更能传达出时代精神，还更能有弹性的支撑和塑造品牌形象。英国的*Illustrated Ape*就以只用插画、不用照片而闻名，*Carlos*除了广告之外也只用插画。

许多刊物都会使用插图，但只有*Illustrated Ape*是完全以插画为主角的。这份刊物的编辑是克里斯蒂安・帕特森（Christian Patterson）与迈克尔・西姆斯（Michael Sims），设计师则是看见工作室（See Studio）的达伦・埃利斯（Darren Ellis），刊登的插画作品很多出自顶尖插画家之手，例如保罗・戴维斯（Paul Davis）。这份刊载了许多诗歌和小说的文艺季刊的插画通过作家、艺术家与插画家投稿而来，图像均是插画作品，就连文字也常常图像化，跨页设计相当丰富、有质感，一气呵成，充满新意。

图片裁剪

图片裁剪、放大、重复，或是以特殊方式拍摄，都会对版面布局产生巨大影响，营造出具有独创性和意想不到的透视效果。图片经过裁剪和局部放大后，会将读者的视线聚焦在照片最关键的部分，或在图文之间创造有意义的对话，使文字与版面达到一种颇具动态感的融合效果。如果图片数量比较多，营造这种和谐的图文关系会更加复杂，需要在图像之间、图像和文本之间实现叙事交互的效果。美国工业设计杂志*I.D.*是裁剪图片的高手：通过把产品在页面上以巨大比例放大，让即使是牙刷这样的日常用品也能呈现出超现实的美感，获得雕塑一般的特质，视觉上也很吸引人。微距摄影也有同样的效果，能彰显某物体的奇特曲线、形状及花纹等平时不易察觉的特色。

希代の珍菜か、強力な媚薬か
——アーティチョークを解剖する

scary monsters

It was a brave diner indeed who, upon first contemplating this grotesquerie, then chose to eat it. Yet the rich, nutty artichoke earned the love of the Greeks, the lust of the French and the greed of the Mo…

Trailblazers

Balancing comfortable economics with functional innovation, Copenhagen bicycle-maker Biomega taps high-profile designers for new concepts.

By David Pescovitz

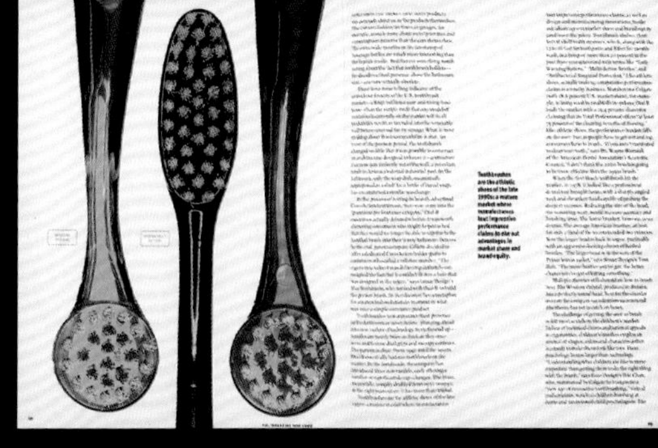

最上图选自日文杂志*Eat*，近距离拍摄倒挂朝鲜蓟的鳞片，把平凡植物拍出了雕塑般的艺术质感，令人大开眼界。在*I.D.*杂志中，单车图（左图）大胆以两个跨页呈现，传达出流动性与延展性：在另一期中，牙刷成为抽象的透明物体（上图）。

工作坊3

文字排版样式表

目标

为你的杂志制作文字排版样式表单。

操作步骤

为你的杂志设计有视觉识别性的排版样式，想想看，该如何利用字体来反映杂志内容的理念。首先，用两三种相互关联的字体，搭配出标题字体、副标题字体与内文字体。其他设计元素要成为这三种主要字体的辅助。虽然之后你可以把字体减少为一到两种，但一开始保持开放的想法更好。切记，如果你的杂志风格是大胆有力的，就要选用能让页面充满力道的字体。如果你的杂志风格比较文静内敛，就要选视觉上较为柔和安稳的字体。

写几条标题与导语。探索一下首字下沉的比例，看看哪种效果最好，导语用哪种字体读起来最顺畅。然后打印出来，贴在墙上，从不同的角度观察。将设计成果贴到素描本上，供日后参考。

微调字距设定。几乎所有的默认字体都需要调整才更适合，计算机不懂人眼的审美，你得学着自己去感受字体，体验不同字体的实际使用情况。观察文字的构成和实际效果。

要对自己有信心。你的目标是做一份文字排版样式表单，以反映并强化杂志的理念，设计过程中要对自己有信心。

下面是学生制作的杂志范例。在这个例子中，学生利用图像化的语言，探索以文字编排为主的版式，反映刊物观点，这是对艺术学校经历的一种略带愤世嫉俗的看法，也是对生活方式的一种批判。设计师乔登・哈里森-特威斯特（Jordan Harrison-Twist）采用锐利的Futura斜体字当作标题字体，以挑衅性的封面来吸引读者。

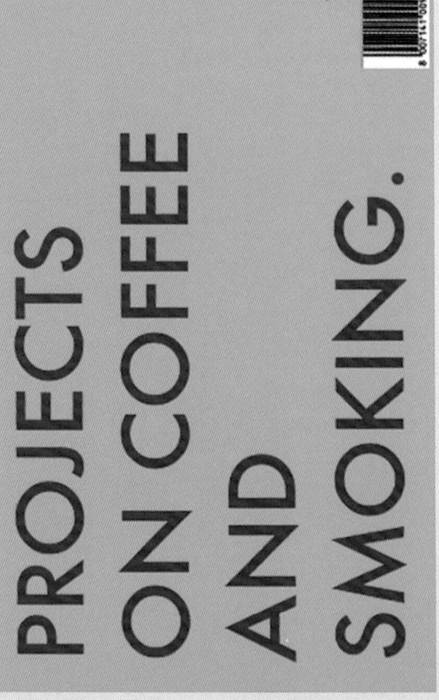

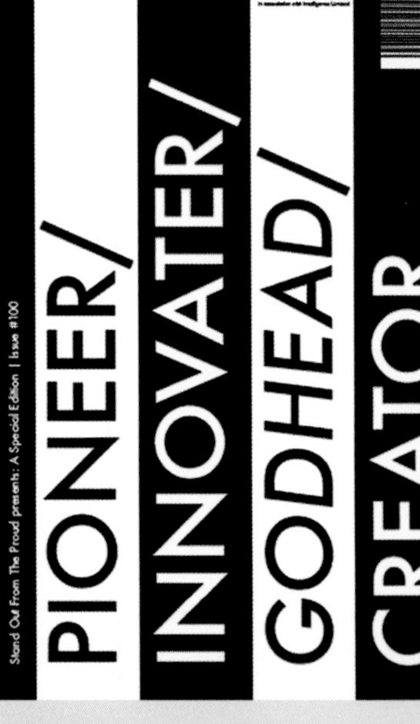

08 | Stand Out From The Proud | Issue #01

"IT'S AMAZING, APPARENTLY ALL HOMELESS PEOPLE LIKE TO BE PHOTO-GRAPHED."

07 | Stand Out From The Proud | Issue #01

Weirdly sparse page gives eyes less daunting task of actually reading something as opposed to nodding sycophantically at pictures.

It contrasts to every other grid system previously used which makes the publicist seem either rebellious, or consciously aware of his/her choice of breaking convention, which makes the page seem even more important.

Even though it is only a few paragraphs and it is probably the least visually impressive page on display, you can't help keep reading as it shows you are interested, and everybody can see you're interested and hey it ain't too long, so it's not tiresome, and you can move on fairly swiftly after finishing.

This section may even be tilted slightly to one side which is utterly mental. Though, there is a fairly simple uniform grid-system used, it is used in a dynamic way to sustain interest, regardless of the content.

written by Jorgen Harrison-Tier

"You see, it's all about thought. Art can't be anything that's been done before. And, you know, not everything is, like, beautiful. It's about capturing a moment that inspires or... or excites. Like, I did a project on homeless people last year and I think it kinda, really opened everyone's eyes to, to the plight of these people."

"Like, sometimes you just don't notice homeless people, but if you take a photograph of them, it kinda forces people to think about it. It's like Live Aid or Children in Need."

"You know, it's all about perspective. What we thought was ugly years ago, now isn't necessarily ugly. It's just different. You put your personal philosophies into what you make, and that's beautiful."

-24/02/12 - anonymous

02 | Stand Out From The Proud | Issue #01

"I FIND YOU CAN OFTEN GET TOO *TIED DOWN* WITH WHAT THE WORK ACTUALLY *MEANS*."

03 | Stand Out From The Proud | Issue #01

Lorem ipsum dolor sit amet, consectetur adipiscing elit. Fusce suscipit, nibh at fermentum tincidunt, orci velit convallis nibh, at tincidunt nulla nisl quis odio. Cum sociis natoque penatibus et magnis dis parturient montes, nascetur ridiculus mus. Sed lobortis mollis sapien non convallis. Phasellus dictum ultricies ullamcorper. Praesent erat libero, rhoncus nec tempus at, accumsan a libero. Nunc lacinia volutpat tellus vitae mollis. Duis ullamcorper laoreet dictum. Praesent ultricies ligula a diam lacinia tempus. Mauris eget molestie velit. Cras congue, elit a porttitor congue, arcu arcu sollicitudin magna, a suscipit nisl diam nec leo. Aenean eu lorem at ipsum tempor luctus. Cras id augue a ante interdum fringilla. Suspendisse ligula lectus, posuere quis aliquam laoreet, vehicula eget enim. In hac habitasse platea dictumst. Morbi nunc massa, tempor eu pretium in, ornare cursus erat. Sed leo risus, volutpat et malesuada ut, elementum sed nibh.

Duis aliquet accumsan egestas. Pellentesque id velit orci. Phasellus pulvinar ornare semper. Nunc consequat lorem quis odio accumsan eget interdum nisl vestibulum. Maecenas et eleifend lectus. Vivamus a leo elit, sed condimentum ligula. Morbi nec ultricies sapien. Nam fermentum blandit viverra. Cras vestibulum cursus metus. Duis tincidunt fringilla dui sed porta. Nulla ut iaculis quam. Duis libero ante, elementum vitae aliquet a, bibendum at enim. Phasellus cursus molestie massa, eu sollicitudin enim sagittis non.

Nunc ac tellus tortor, vitae lacinia velit. Nulla facilisi. Etiam at fringilla ante. In at nulla lectus, quis cursus turpis. Mauris a dui in mi condimentum suscipit. Nulla facilisi. Donec ipsum nisl, vulputate non mattis at, dapibus quis lorem.

A pixelated illustration to show the fragmented nature of today's society and that there's censorship everywhere.

"I FIND MY WORK IS OFTEN FAR TOO CONCEPTUAL AND OFTEN FAR TOO DEEP FOR MY CRITICS TO UNDERSTAND. I'M A DARK, BROODING INDIVIDUAL AND I FEEL THAT COMES THROUGH."

Pellentesque semper euismod sapien, vel dictum urna viverra ut. Class aptent taciti sociosqu ad litora torquent per conubia nostra, per inceptos himenaeos. Pellentesque diam diam, tincidunt eget ultrices vel, pulvinar sit amet augue. Sed vestibulum ultricies consectetur. Ut eu nulla magna.

Aliquam sollicitudin gravida volutpat. Aenean rutrum adipiscing risus, vel bibendum eros cursus suscipit. Nunc metus lacus, adipiscing sed gravida et, imperdiet nec mi. Maecenas nec mauris non eros gravida aliquam. Suspendisse potenti. Integer suscipit condimentum ante, id aliquet nulla euismod sed. Quisque nec tortor a ipsum ultricies fringilla. Cum sociis natoque penatibus et magnis dis parturient montes, nascetur ridiculus mus. Sed scelerisque, nisl at hendrerit vestibulum, libero mauris tincidunt purus, ac iaculis sem nunc eu arcu. Aenean ornare aliquam mi id tempor. Fusce sit amet gravida orci. Proin blandit metus eget nulla ullamcorper eleifend. Aenean odio lacus, dapibus ac egestas et, ultricies at tellus. Vestibulum et varius mi.

- Jordanne Harrison-Twine

刊物中的内容多由哈里森-特威斯特撰写，他很善于批判性的设计方式，在视觉上也不遵循传统，呈现一种坦率感。这份杂志运用双色与线图，强化刊物的小众杂志特性。

Mind-blowing bubbles

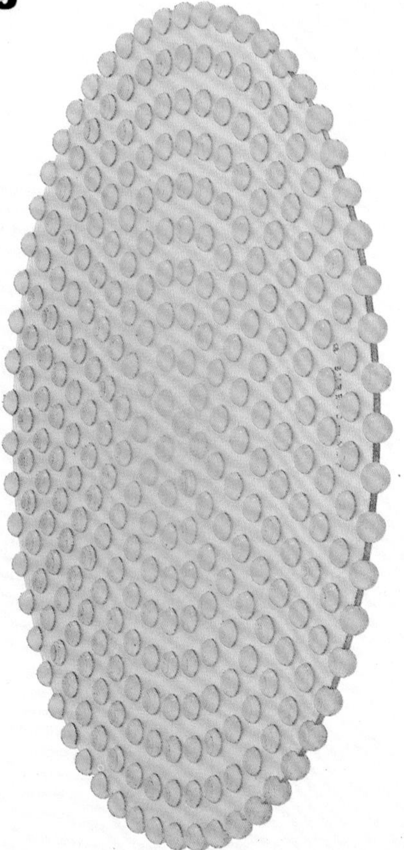

The Guardian
space
interiors/property/design January 23 1998

第五章 制作版面

前几章已谈到什么是编辑设计，如何理解它，以及对优秀的设计师与艺术总监来说非常重要及核心的组成元素。接下来，要正式进入设计流程了。虽然没有完美的版面设计的神奇魔法公式，但有些因素确实可以左右设计方向。前几章讨论过编辑设计师的角色，品牌与刊物识别、与读者的互动关系；还有一些每期都不同的元素（时空、文字量、目标），以及最基本的元素（字体风格、粗细、对称性、图像）。这些各式各样的元素结合起来，就能总结出设计一份刊物的指导原则。设计师如何诠释、应用或刻意忽略这些原则，对刊物设计而言至关重要。此外，也要考虑到如何让内容能够适应不同的设备和载体，并完美匹配。

版面的主要元素

右边的版式是由两个单页构成的跨页。版面上该有的元素，有些在前几章已从品牌与识别的角度谈过，在这一章，我们将重点从视觉效果与版面构成角度来讨论。右图显示版面构成元素和格线系统，包含栏、栏间距、装订留白、白边、页码、基线与裁剪线（更多关于网格线的信息参见本书第六章）。

版式模板

在设计报纸和杂志的新闻页面时，灵活有弹性的模板可以加快排版与制作流程，让页面与整体设计呈现一致性，否则在印刷前的忙乱状态下，很可能会忽略整体的一致性。不过，模板虽然可简化页面制作，但也有可能限制设计的发挥，因此要谨慎使用，以免所有页面看起来一成不变。可以利用图片增加变化，除了利用不同的图片外，图片本身的裁剪缩放和与标题的相对关系，也可以让页面富于变化。

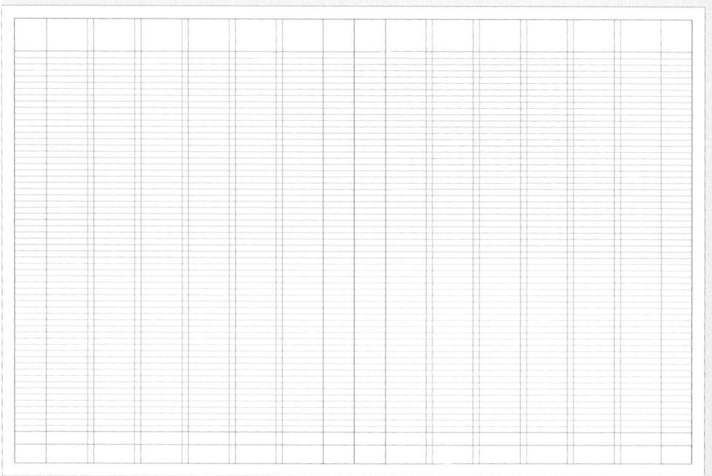

版式模板的基本构成

模板的基本要素包括白边（蓝色）、文字栏与栏间距（绿色）、基线（粉红色）、页码（紫色）与出血区。这个模板默认有六栏网格线，但实际操作时，可根据情况设计成两栏、三栏或六栏的版式。白边的大小会直接响页面有多少留白空间，因为每一寸空白都可以卖给广告商，这部分往往会受广告影响。基线决定行间距的变化，从而影响字体大小，让不同大小的字体能够对齐。这个例子中的网格线是以9pt的字体和11pt的行间距为基础，这表示其他字体的大小要能放进以11为倍数的间距中。页首、页尾则用来标明页码与单元名，以引导阅读。

右上图的跨页展示主要元素排入模板后的样子。图片、标题、导语和引文可以吸引读者注意，让读者轻松快速地阅读文章。首字母大写下沉与页码可以提示读者阅读进度，而图片说明和照片来源让文图更有趣。这些元素再加上彩色图片与线条，就能让版面富于变化。

这个模板还能加入更多其他元素。右页的第二个案例增加了小标题、署名和落款。从这个例子就可看出，同一个版面其实具有很大弹性，这里就只排两栏，而不是像上图那样排三栏，标题的大小与颜色也不同，从而为这篇文章打造出特殊的外观与质感。

The headline of this story

This is a caption. Intended to be read but has no meaning. A simulation of actual copy, using ordinary words with normal letter frequencies, it cannot deceive the eye.

This is a stand-first. Intended to be read but have no meaning. A simulation of actual copy, using ordinary words with normal letter frequencies, it cannot deceive the eye or brain.

Dummy settings which use other or even gibberish to approximate text have the inherent disadvantage they distract attention to themselves. Simultext is effective in any typeface, at whatever size and format is required. Paragraphs may be long or short. Text can be to complete any area, as the copy is simply repeated with different starting points. This is dummy text. Intended to be read but have no meaning. A simulation of actual copy, using ordinary words with normal letter frequencies, it cannot deceive eye or brain.

Settings which use other languages or even gibberish to approximate text have the disadvantage they distract attention to themselves. As a simulation of actual copy, using ordinary words with normal letter frequencies, can't deceive eye or brain. Simultext is effective in any typeface, at what ever size and format is required. Paragraphs may be long or short. What you see is dummy text. It is intended to be read but have no real meaning. As a simulation of actual copy, using ordinary words with normal letter frequencies, can't deceive eye or brain. Copy can be produced to complete any area, as the basic copy is repeated with different starting points. It is intended to be read but have no meaning.

Presentation copy uses languages or gibberish to approximate text have the disadvantage that they distract attention to themselves. Simultext is effective in any typeface, whatever size and format is required. Paragraphs may be long or short. This is dummy text. Intended to be read but have no meaning. As a simulation of actual copy, using words with normal letter frequencies, cannot deceive eye or brain. What you see here is dummy text.

Simultext is effective in any typeface, whatever size and format required. Paragraphs may be long or short. Text can be produced to complete any area, as the basic copy is simply repeated with different starting points. This is dummy text. It is intended to be read but have no meaning.

This is dummy text. It is intended to be read but have no meaning. As a simulation of actual copy, using ordinary words with normal letter frequencies, it cannot deceive eye or brain. Dummy settings which use other languages or even gibberish to approximate text have the inherent disadvantage they distract attention to themselves. Simultext is effective in any typeface, at whatever size and format is required. Paragraphs may be long or short. Text can be produced to complete any area, as the copy is simply repeated with different starting points. This is dummy text. Intended to be read but have no meaning. As a simulation of actual copy, using words with normal letter frequencies, cannot deceive eye or brain. Trial settings which use other languages or gibberish to approximate text have the disadvantage they distract attention to themselves. A simulation of actual copy, using ordinary words with normal letter frequencies, it can't deceive eye or brain. Simultext is effective in any typeface, at whatever size and format is required. Paragraphs may be long or short. What you see is dummy text. It is intended to be read but have no meaning. As a simulation of actual copy.

A pull quote intended to be read but have no meaning. A simulation of actual copy, using ordinary words.

Using ordinary words with normal letter frequencies, it cannot deceive eye or brain. Copy is produced to complete any area, as the copy is simply repeated with different starting points. This is dummy text. Intended to be read but have no meaning. The Presentation copy which uses languages or gibberish to approximate text have the inherent disadvantage that they distract attention to them selves. Simultext is effective in any typeface, at whatever size and format is required. Paragraphs may be long or short. This is dummy text.

Settings which use other languages or even gibberish to approximate text have the disadvantage they distract attention to themselves. As a simulation of actual copy, using ordinary words with normal letter frequencies, can't deceive eye or brain. Simultext is effective in any typeface, at what ever size and format is required. Paragraphs may be long or short.

A pull quote intended to be read but have no meaning. A simulation of actual copy, using ordinary words.

short. What you see is dummy text. It is intended to be read but have no real meaning. As a simulation of actual copy, using ordinary words with normal letter frequencies, can't deceive eye or brain. Copy can be produced to complete any area, as the basic copy is repeated with different starting points. It is intended to be read but have no meaning.

Presentation copy uses languages or gibberish to approximate text have the disadvantage that they distract attention to themselves. Simultext is effective in any typeface, whatever size and format is required. Paragraphs may be long or short. This is dummy text. Intended to be read but have no meaning. As a simulation of actual copy, using words with normal letter frequencies, cannot deceive eye or brain. What you see here is dummy text.

Simultext is effective in any typeface, whatever size and format required. Paragraphs may be long or short. Text can be produced to complete any area, as the basic copy is simply repeated with different starting points. This is dummy text. It is intended to be read but have no meaning. What you see here is dummy text. It is intended to be read.

12 FEATURE

FEATURE 13

« This is a caption. Intended to be read but has no meaning.

The headline

A stand-first is a short summary of the following article allowing quick and easy access for the reader. Often it includes the **Author's and Photographer's Names as a byline**

This is dummy text. It is intended to be read but have no meaning. As a simulation of actual copy, using ordinary words with normal letter frequencies, it cannot deceive eye or brain. Dummy settings which use other languages or even gibberish to approximate text have the inherent disadvantage they distract attention to themselves. Simultext is effective in any typeface, at whatever size and format is required. Paragraphs may be long or short. Text can be produced to complete any area, as the copy is simply repeated with different starting points. This is dummy text. Intended to be read but have no meaning. A simulation of actual copy, using ordinary words with normal letter frequencies, it cannot deceive eye or brain.

Trial settings which use other languages or even gibberish to approximate text have the disadvantage they distract attention to themselves. As a simulation of actual copy, using ordinary words with normal letter frequencies, can't deceive eye or brain. Simultext is effective in any typeface, at what ever size and format is required. Paragraphs may be long or short. What you see is dummy text. It is intended to be read but have no meaning. As a simulation of actual copy, using ordinary words with normal letter frequencies, can't deceive eye or brain. Copy can be produced to complete any area, as the basic copy is simply repeated with different starting points. It is intended to be read but have no meaning.

This is a typical subhead

Presentation copy which uses languages or gibberish to then approximate text have the disadvantage that they distract attention to themselves. Simultext is effective in any typeface, at whatever size and format is required. Paragraphs may be long or short. This is dummy text. Intended to be read but have no meaning. As a simulation of actual copy, using words with normal letter frequencies, it cannot deceive eye or brain. What you see here is dummy text.

Simultext is effective in any typeface, whatever size and format required. Paragraphs may be long or short. Text can be produced to complete any area, as the basic copy is simply repeated with different starting points. This is dummy text. This is dummy text. It is intended to be read but have no meaning. As a simulation of actual copy, using ordinary words with normal letter frequencies, it cannot deceive eye or brain. Dummy settings which use other languages or even gibberish to approximate text have the inherent disadvantage they distract attention to themselves. Simultext is effective in any typeface, at whatever size and format is required.

This is a pull quote. It is intended to be read but have no meaning. As a simulation of actual copy, using ordinary words with normal letter frequencies, it can't deceive eye or brain.

Paragraphs may be long or short. Text can be produced to complete any area, as the copy is simply repeated with different starting points. This is dummy text. Intended to be read but have no meaning. A simulation of actual copy, using ordinary words with normal letter frequencies, cannot deceive eye or brain. Trial settings which use other languages or gibberish to approximate text have the disadvantage they distract attention to themselves. As a simulation of actual copy, using ordinary words with normal letter frequencies, it can't deceive eye or brain. Simultext is effective in any typeface.

As a simulation of actual copy, using ordinary words with normal letter frequencies, it cannot deceive eye or brain. Copy can be produced to complete any area, as the basic copy is simply repeated with different starting points. This is dummy text. Intended to be read but have no meaning. Presentation copy which uses languages or gibberish to approximate text have the disadvantage that they distract attention to them selves. Simultext is effective in any typeface, at whatever size

FEATURE 11

IF YOU GO DOWN TO THE WOODS TODAY

... then you're sure to meet Britain's finest neo-medieval psychedelic folk-rock band. Or you are if you're author and journalist Tom Cox, who for the best part of a decade has cultivated an unlikely bond with Circulus that has survived their differering tastes in fashion, recreational pursuits and – most alarmingly – his relentless championing of their music. Photographs by Murdo MacLeod

44 OMM June 2005

June 2005 OMM 45

根据页面上其他元素的大小，把标题拉大或缩小，在标题位置与页面其他元素之间建立动态关系，可以营造出视觉力量。像图片或标题文字这样的元素，可以相互融合，让版面看起来更有整体感。例如上图《观察家报：音乐月刊》（*The Observer: Music Monthly*）的跨页中字体和图片的设计。如果页面没有主图，标题的字体可以设计得更图像化，创造视觉焦点。大胆尝试是设计出好标题的关键，如右图的*Fish-wrap*这个页面，运用许多装饰性的字体，这些字体是由插画家与设计师专门设计的，有时也会因为需要，为杂志设计“专属字体”。

Nothing ever Happened

BY Tracy Sunrize Johnson

FOUR YEARS AS an occasional mistress, and thirty as a fatherless daughter, brought me in 2003 to a threshold I never imagined I'd drift across: I fell in love with a gay guy. He was not only gay but unavailable, speaking often in a plural tense: "We're going to Big Sur for the weekend," or "Our apartment is so fucking small we can fit it into the back seat of our car and drive it around the goddamn city." His name was Booker. He was the network administrator at the publishing house I came to work for, and immediately my favorite person in the company. Something about the squareness of his large frame made me dizzy with longing. Tall and broad but never bulky, his was a solid structure, packed fully, so that he was brimming against the confines of his own golden skin. He came from Illinois and there wasn't any word for him but 'strapping'. His eyes were blue and as bright as ignited dimes. Caught in a cluster of chattering art directors, he would wink at me and grin; racing into the editorial den to extinguish technical fires, he would wave to me from the end of a long corridor, his smile as ready as a greyhound at the gate.

My treacherous computer provided me with endless opportunity to legitimately seek Booker's attention. All day long I would call him on the phone and say, "ATM Deluxe says it cannot open a necessary file because it's out of memory, but it isn't out of memory, I allotted it

左图为*Inner Loop*的跨页，图像化的超大标题似乎成了图片的一部分，和对页整齐有序的文字栏形成强烈对比。有趣的是，设计师让文字齐底，于是文字框的上缘轮廓就像是起伏的城市天际线，恰好呼应左边的主图。页面下方三分之一处有一处出血的水平线条，贯穿整体页面，把不同元素联系起来。

12 NACIONAL — EL PAÍS, DOMINGO 11 DE JUNIO DE 2006

ANÁLISIS. La resaca del debate sobre el estado de la nación. Por Javier Pradera

Ruptura intempestiva

Rajoy aduce como razón para retirar el apoyo de su partido a la política antiterrorista del Gobierno el anuncio de que los socialistas vascos se disponen a reunirse con la ilegalizada Batasuna

CAJERO AUTOMÁTICO EL ROTO

El indiferente

LA COLUMNA. El PP se ha convertido en el partido del no. No al Estatut, no al proceso de fin de la violencia. Por Josep Ramoneda

Cataluña, con Euskadi al fondo

NICHOLAS D. KRISTOF

The Race Is On And the Favorite Is ...

CALCUTTA — The great race of the 21st century is under way between China and India to see which will be the leading power in the world in the year 2100.

President Bush's trip to India next month is important, for we in America must brace ourselves to see not only China looming in our rear-view mirror, but eventually India as well. India was the world's great disappointment of the 20th century, but now it's moving jerkily forward with economic reforms, reminding me of China around 1990.

One of India's (and China's) greatest strengths is its hunger for education. Most American newspapers lure readers with comics, and some British tabloids with photos of topless women, but a Calcutta daily newspaper is so shameless that it publishes a column on math equations. Imagine titillating readers with trigonometry!

I visited the ramshackle Hasi Khusi Kindergarten and Primary School in a poor area of Calcutta, where most of the pupils' parents are illiterate street vendors, rickshaw drivers or laborers. Out of an average family income of $23 a month, the parents pay a one-time fee of $13 for registration and then $2.30 a month.

A contest of titans: India vs. China

"What they didn't get, their children must get," explained the principal, Sampa Sarkar. Even kindergartners study English, Bengali, math, art and music — and do 30 minutes of homework. Private schools like this one are booming all across the country.

With India's ever-deepening pool of English speakers, its outsourcing boom will continue.

India has a solid financial system, while China's banking system is a catastrophe. And India is in better shape demographically for long-term growth: China has already reaped most of the economic benefits of population control and is now rapidly aging, but India's population will be disproportionately working-age for many decades to come and political explosions in India are declining, while in China they may be rising.

China will probably manage its eventual transition to democracy with bearable turbulence, as Taiwan and South Korea did, but with China anything is conceivable, including a coup d'état, mass unrest or even civil war.

Yet if democracy is one of India's strengths, it's also a weakness. Prime Minister Manmohan Singh knows exactly what to do, and I've rarely met a leader more competent (or less charismatic). But his reforms are stalled or slowed in the Indian political labyrinth. India's basic problem is that its economic policy-making isn't nearly as shrewd, pro-growth or farsighted as China's.

That's a tragedy: we should all want India to demonstrate that democracy is an advantage. But Indian lawmakers aren't helping.

Foreigners are still blocked from directly investing in some sectors in India, like retailing. Privatization is lethargic. Food subsidies are soaring and are so inefficient that it costs 6.6 rupees to transfer 1 rupee's worth of food to the poor. Restrictive labor laws mean that companies hesitate to hire, and regulations tend to suffocate entrepreneurship.

The upshot is that India has enjoyed a boom that has added few jobs. The losers are India's poor.

And while China has been exceptionally shrewd in upgrading its infrastructure, India has been pathetic. India's economic future is marred by its third-rate roads and ports.

India is also horrendously mismanaging its AIDS crisis; it may already have more H.I.V. cases than any country in the world. AIDS casts a cloud over this nation's entire future.

The bottom line is that the once-great nation of India is reawakening from several centuries of torpor, and facing less risk of a political cataclysm than China,

THOMAS L. FRIEDMAN

New 'Sputnik' Challenges: All Fueled by Oil

DETROIT — I came to Detroit looking for the hottest new American cars. Instead, I found Sputnik.

You remember Sputnik — the little satellite the Soviets launched in 1957. The Eisenhower administration was so stunned it put the U.S. into a crash program to train more scientists and engineers so America could catch up with the Russians in the space race.

Well, for anyone paying attention, our generation's Sputnik showed up at the annual Detroit auto show this month. It's not a satellite. It's a car. It's called the Geely 7151 CK sedan. It seats a family of five, gets good mileage and will cost around $10,000 when it goes on sale in 2008. It's made in China.

That doesn't get your attention? Well, there's another Sputnik that just went up: Iran. It's going to make a nuclear bomb, no matter what the U.N. or U.S. says, because at $60-a-barrel oil, Tehran's mullahs are rich enough to buy off or tell off the rest of the world. That doesn't worry you? Well, there's a quieter Sputnik orbiting Earth. It's called climate change — a k a Katrina and melting glaciers.

Bush's blindness puts our way of life in peril.

What am I saying here? I am saying that our era doesn't have a single Sputnik to grab our attention and crystallize the threat to our security and way of life in one little steel ball — the way our parents' era did. But that doesn't mean such threats don't exist. They do, and they have a single common denominator: the way we use and consume energy today, particularly oil.

Friends, we are in the midst of an energy crisis — but this is not your grandfather's energy crisis. No, this is something so much bigger, for four reasons.

First, we are in a war against a radical, violent stream of Islam that is fueled and funded by our own energy purchases. We are financing both sides in the war on terrorism: the U.S. Army with our tax dollars, and Islamist chari-

a toaster and a refrigerator. If we don't quickly move to renewable alternatives to fossil fuels, we will warm up, smoke up and choke up this planet far faster than at any time in history.

Third, because of the above, green energy-saving technologies and designs — for cars, planes, homes, appliances or office buildings — will be one of the biggest industries of the 21st century. Tell your kids. Will we dominate the green industry, or will we all be driving cars from China, Japan and Europe?

Finally, if we continue to depend on oil, we are going to undermine the whole democratic trend that was unleashed by the fall of the Berlin Wall. Because oil will remain at $60 a barrel and will fuel the worst regimes in the world — like Iran — to do the worst things for the world. Indeed, this $60-a-barrel boom in the hands of criminal regimes, and just plain criminals, will, if sustained, pose a bigger threat to democracies than communism or Islamism. It will be a black tide that turns back the democratic wave everywhere, including in Iraq.

The one thing we can do now to cope with all four of these trends is to create a tax that fixes the pump price at $3.50 to $4 a gallon (nearly 4 liters) — no matter where the OPEC price goes. Because if consumers know that the price of oil is never coming down, they will change their behavior. And when consumers change their behavior in a big way, G.M., Ford and DaimlerChrysler will change their cars in a big way.

The more Detroit goes green, the faster it will be propelled down the innovation curve, making it more likely that Detroit — and not Toyota or Honda or the Chinese — will dominate the green technologies of the 21st century. A permanent gasoline tax will also make solar, wind and biofuels so competitive with oil that it will drive their innovations as well.

George Bush may think he is preserv-

上图的《国家报》与右上图的《纽约时报》在呈现作者署名时风格截然不同，读者可以从呈现风格上区分出不同版块。新闻评论通常很密，且少有留白空间，体现重要与庄重之感。如果是一般的评论和讨论，署名就可以占较多的空间，还能找有意思的引语来吸引读者注意。两种不同的编排，读者一眼就看出前者是新闻，后者是观点评论。

大标题

文章大标题通常是版面上字号最大的，目的是激发读者对文章的好奇心，吸引其阅读。在设计之前就定好文章的大标题，有助于确定版面布局的设计方向。不同的出版物版面设计方式也各有不同，大标题能占多少空间有时是由设计师决定的。设计师甚至可以调整文章内容。无论是何种方式，标题的视觉表现一定是和内容相互关联的，设计时要特别强调这种一致性。

导语

和大标题一样，导语一般由编辑撰写，通常以四十到五十个字最好，太长就失去简明扼要的功能，太短又不足以传达必要的信息，还可能导致页面看起来不平衡。可以事先建立一套导语设计的样式表单，避免临时设计。不过，设计师也可以灵活设计，样式表单可以作为指导性的参考工具，而不是不能打破的硬性规定。

署名

通常知名作者的署名会搭配其照片，读者很喜欢这种方式，在报纸上呈现的效果也不错。但如果用在杂志上的专题报道时，要小心人像在整个页面上是否会和其他元素不太协调。

正文

正文文字段落的设计可以有多种方式，各栏文字可以左右对齐（文字填满栏宽）、左对齐或右对齐。报刊通常采用左对齐或左右对齐，因为文字量大时，居中和右对齐读起来会很累。

此外，栏宽不能太宽，读起来才轻松（参见156页的“易读性理论”），但也不能太窄，以免每个字中间的空白和上下几行的字间空白连成一串，形成纵向空白，影响阅读。文字太长的话可分段，让整篇文章

The Guts of a New Machine

The iPod became an instant classic by combining high design and powerful technology. But as Apple has learned before, that formula alone doesn't keep you on top.

By Rob Walker

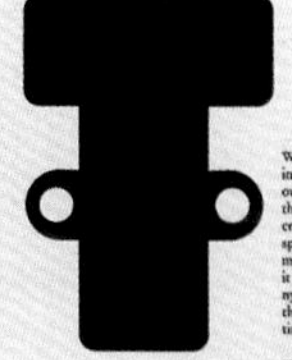

Photograph by Richard Burbridge

The iPod, says Apple's Ive, is about 'not trying to do too much with the device — which would have been its complication and, therefore, its demise.'

The Aura

The Surface

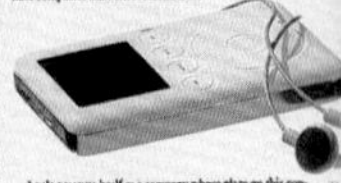

在新段落前使用小标题、首行缩进、拉大行距与分段，都能够在视觉上分割文章，让文章段落精简美观，以免文字栏过于冗长，影响读者的阅读兴致。但是《纽约时报》的前艺术总监珍妮特·佛罗里克（Janet Froelich）则建议设计师，不要怕文字太长：“《纽约时报》杂志是给读书人的杂志，旨在通过图文呈现，让读者更深刻地了解当今世界的文化与政治力量。为达到这个目的，作者与摄影师的声音非常重要，设计也要服从于这一宗旨。大量文字方块所呈现的美感是设计师必须尊重的。如果文字与抢眼图像并用，只要能适当使用留白，搭配强而有力的大标题，就能够给读者以丰富的知识体验。”

THE DEAD INTERVIEW

Meeting your heroes isn't always a good idea. I remember the first time I encountered old Bill Burroughs. He was sober and so outraged at my altered state I had to part his hair with a bullet before he'd quit the sermon. Something like a priest right enough. *El macho hombre* Hemingway rolled over and cried for his mommy when I Giant-Haystacked him during a wrestling match. And as for the Aberdonian boy Lord George Byron, well let's just say he tired before I did.

So it's with a degree of trepidation that I agree to meet the writer's writer Robert Louis (pronounced Lewis) Stevenson, author of *Treasure Island*, *Kidnapped* and *The Strange Case of Dr Jekyll and Mr Hyde* as well a wealth of poems, short stories and articles. An itchy-footed traveller of high seas and uncharted outbacks. The man who explored Edinburgh's lower depths before Irvine Welsh learnt to say choo choo.

We've arranged to meet in Deacon Brodie's, on Edinburgh's High Street. Stevenson enters the ancient hostelry, brought forth through the magic of literature, brushing the last of the Samoan grave dust from his travelling cloak. A few heads turn at his approach, but Scotland's capital is used to stranger sights than ghosts, and the drinkers merely nod, as if to a face they recognize but can't quite place.

I buy him a malt, then we take a seat at a corner table away from the flashing lights of the fruit machine. Louis is rack-thin, his moustache droops and his left eye shifts in its socket. His hair is long, and I remember reading he could never abide being barbered when he was ill. I slide the tobacco and Rizlas I bought in preparation for his manifestation across the table, wait for him to roll the first of the many smokes he draws into himself over the afternoon, then begin my questions.

LW RLS

So Robert, how do you feel?

Pretty good for a dead man. But call me Louis, there were too many Roberts in my family for us all to go by the name.

O.K. Louis; I must say, you've shone up well for a guy who's spent a hundred years in the ground.

A hundred and ten years to be precise, but it's good to see flesh on the old bones again. In Samoa we had a battle with corruption the staunchest missionary could never hope to quell. Things rotted even as you made them. My body started to decay before my friends had felled the trees from the path to my mountain tomb. By the time they carried me to the summit, I was edging towards putrefaction and when they lowered me into my grave, I was already half-way towards being one with the earth.

Gross.

Aye, I won't pretend it was pleasant.

If you don't mind my saying so, you've a reputation for being a bit morbid.

Can you blame me? Take a look at the city of my birth. It's pissing down, there's a constant mistral from the sea driving everyone half daft. You daren't look at anyone sideways for fear of getting your face smashed and the population have either signed the pledge or hit the bottle. It's depressing. I've not been in this pub in a long while but there's the same faces propped at the bar. *(Robert nods to a couple of drinkers who nod back)* Then there's the religion ...

THE DEAD INTERVIEW

We'll get on to religion later. But you must admit you're on the gloomy flank of the fantastical. One of my favourite stories, *The Body Snatchers*, is about two medical students digging up a dead body for dissection.

That proves my point. It was based on fact. Williams Burke and Hare had a cottage industry supplying exceedingly fresh cadavers to Dr Robert Knox, one of Edinburgh's leading anatomists. They murdered sixteen people before their racket was uncovered.

And did the good doctor know?

Of course he knew. Edinburgh was a wee town and suddenly all these fresh corpses start turning up, some of them with big eggs on their heads where they'd been lamped. Of the lot of them only Burke was hung, but Dr Knox was as guilty as Harold Shipman. The only difference was he never got the gaol.

Good to see you keeping abreast of current affairs. Did you go to the hanging?

And you call me morbid? No, this all happened in the 1820s, thirty years before I was born. I'm old but I'm not decrepit.

You do seem a bit obsessed with vivisection though. Dr Jekyll's laboratory is an old dissecting theatre.

Nice touch, eh?

Masterly.

Edinburgh was a medical city and I was good friends with young Simpson, whose father invented chloroform. Simpson experimented first on himself. He stood up in front of the great and good of the Royal College of Surgeons, announced, "This will revolutionize medical science!" inhaled a big blast and fell down in a fugue on the floor. But it was great stuff. Chloroform tea parties were quite the rage for a while and it transformed childbirth. Of course the church was against it, said labour pains were women's punishment for original sin ...

We'll discuss religion later. Your association with medicine informed your fiction?

Och you ken how it is, you use the materials available to you. I mean look at your recurring second-hand book theme.

I didn't know you'd read any of my stuff.

I've to go through Old Nick, because he's the only one in permanent communion with librarians, but I couldn't manage eternity without regular reading matter.

(Waits for a wee compliment on latest novel)

Aye well, I got to know a bit about medical processes and it came in handy for the fiction.

Oscar Wilde said *Jekyll and Hyde* read like a case from *The Lancet*.

Ach he was just having a laugh, but in a way the whole book was inspired by drugs. You know my health was never good?

You were at death's door most of the time.

文字和图片一样，都可以通过一定的处理来阐释内容。例如设计一段对话时，可以是对立的、冲突的、轻松愉快或生动活泼的。字体的选择、字距、文本框造型和行距不仅会影响文章的外观，还可以传达文章的调性、内容与风格。许多人都知道，大卫·卡森曾用“繁复花哨的字体（dingbats）”来表现歌手布莱恩·费瑞（Bryan Ferry）的访谈没有什么意义，因为那个月费瑞已接受不下十家的杂志访谈。不过，*Zembla*（上图）的文斯·弗罗斯特以及*Speak*（右图）的威尼茨基编排正文的手法非常微妙，表现力很强，可说是受到了具象诗与达达主义的启发。

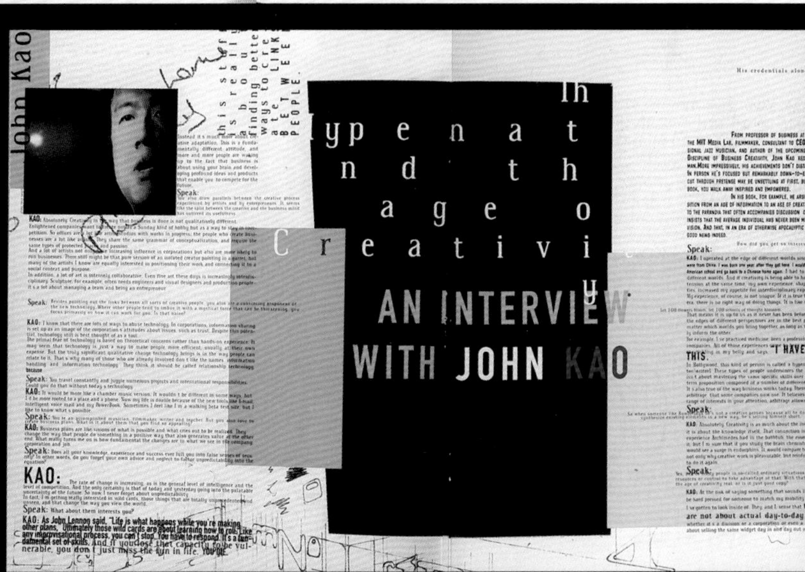

左图是*Fishwrap*杂志的跨页，采用首字母放大下沉式设计，呼应左侧的大标题，而对角线斜放的大标题又将读者的视线直接引导到正文的第一段。

首字放大除了可以用在文章、章节或段落中，还有更多巧妙的用法。下图的例子取自《纽约时报》杂志，文章谈到了法国犹太人在日益高涨的反犹太声浪下，面临身份认同危机。艺术总监珍妮特·佛罗里克（Janet Froelich）把下沉的首字母“I”分成两半，以表现危机感，并呼应照片中分裂的戴维之星。

A Frenchman Or a Jew?

Anti-Semitism is again an issue in France, largely as a result of a spate of vandalism and violence by French Muslims. For many French Jews, it has created a climate of fear — and an identity crisis.

By Fernanda Eberstadt

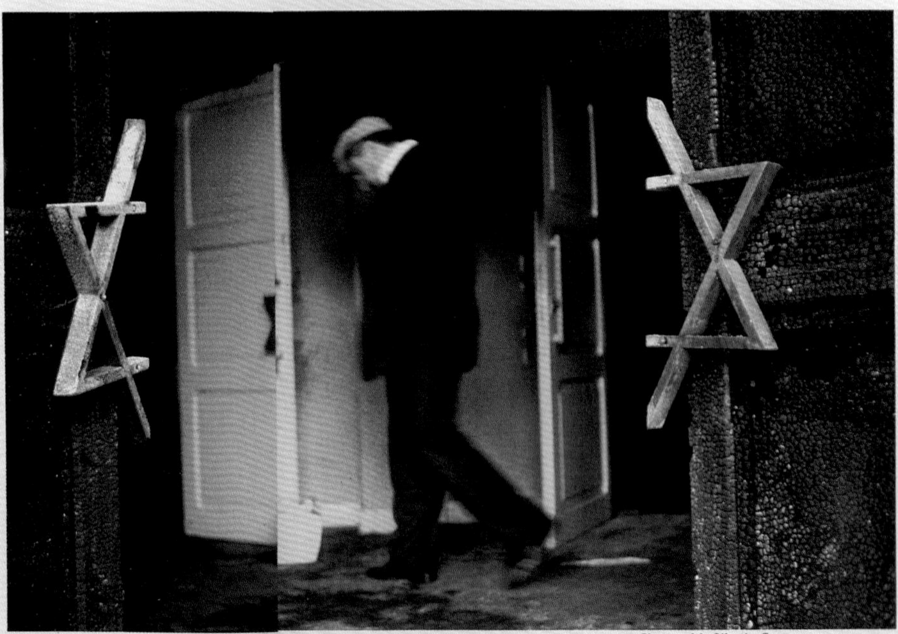

The aftermath of an attack at a Lyon synagogue in March 2002. Photograph by Sébastien Erome

In a working-class neighborhood of the 20th arrondissement in Paris, on a rainy, lead-gray morning last month, the housing blocks looked like sodden cardboard. But inside Brigitte Stora's apartment was an explosion of scarlet, ocher and flame gold, of Israeli and North African textiles, of pottery and a brass menorah. Stora, an Algerian-born Sephardic Jew, is a slim, impish-looking woman in her early 40's with a mop of black hair. She was wearing baggy jeans that revealed a strip of designer-style Jockey shorts, and she sewed a ripped camisole as we talked. In the kitchen, her teenage daughter, home sick from school, cooked herself a plate of pasta.

A former Trotskyite who quit a career in journalism to raise her three children, Stora belonged for decades to a political movement devoted to the cause of equal rights for Arab immigrants. French Arabs were her friends and political allies, and the integrated neighborhood in which she chose to live reflected those commitments. In the last three and a half years, though, Stora's perspective has changed. Since the beginning of the second Palestinian intifada in September 2000 and the subsequent rise of Ariel Sharon to the premiership of Israel, France has suffered what is widely considered the worst epidemic of anti-Jewish violence since the end of the Second World War, much of it at the hands of young Muslims. According to S.O.S. Vérité-Sécurité, an anti-Semitism watchdog organization, 147 Jewish institutions — schools, synagogues, community centers, businesses — have been attacked. There have been reported instances of rabbis being assaulted. Secondary schoolteachers, under pressure from Muslim students, have canceled classes on the Holocaust. On the last Saturday of January, during a concert attended by the wife of President Jacques Chirac, a Jewish singer called Shirel was heckled by a group of French North African youths, who shouted: "Filthy Jew! Death to the Jews!"

There are about 500,000 Jews in France — the largest Jewish population after those in Israel and the United States. There is a reason Jews have come to France from places like Eastern Europe or North Africa: ever since the French emancipation of the Jews in 1791, the country has — with infamous lapses — provided an enviable model of equality, an enlightenment ideal, enshrined in the French Republic, according to which individual difference is subordinated to common citizenship. But today this ideal is threatened by a tide of ethnic harassment and challenged by a surge of religious pride and self-identification among France's Jews and Muslims alike.

Although the frequency of anti-Jewish incidents is said to have abated

读起来赏心悦目，页面更轻盈，更吸引读者。

在文章完成编辑修改后，尽责的设计师会手动微调正文，进一步美化。设计师可以微调字距，把每行的文字拉近些，避免寡行（一行只有一个字）或孤行（一个字落到下个字段的最顶部）。也可选择用自动换行，让字段文字对齐，或插入连字符来避免不美观的断字或断行。设计师要注意整个文本框所呈现的形状，并懂得变通，让文字既好读又美观。在每栏顶部或底部的段落，至少要有两行字。第六章会详细说明字体大小、行距与对齐的问题。

在20世纪80年代和90年代，报刊的版面中常常只有文字，通过文字编排与字体设计，也能清晰地诠释报道

The New York Times Magazine / NOVEMBER 30, 2003

Inspiration: Where Does It Come From?

By Arthur Lubow

Raymond Loewy, the industrial designer, once said that "simplicity is the deciding factor in the aesthetic equation." So, in the spirit of good design, let's begin with a radical simplification. Artists are influenced primarily by other artists, which means that standard art history can sound like a baseball broadcast of an infield play: Velázquez to Goya to Picasso. And designers? To be sure, they are aware of the products of other designers, but their attention is not so narrowly focused. When, near the end of his life, Isamu Noguchi, who straddled the boundary between art and design, created a sculpture garden in Costa Mesa, Calif., he was unquestionably recalling the manipulations of space and perspective in the Zen gardens of Kyoto and the geometric sculptures in the observatory in Jaipur. At the same time, he was thinking of the ways in which the sets he designed as a young man for theatrical stages had, through clever lighting and placement, made a constricted space seem vast. And he was acutely conscious of the function of this sculpture garden

This letter R is made from the pieces of plastic used to hang a hat or a pair of socks from a shop display rack, found in a London shopping mall.

Initial letters by Paul Elliman

69

珍妮特为《纽约时报》杂志设计的版面中，将首字母夸张地放大，为原本和谐的版面引入了冲突元素，以吸引读者的目光。

内容。不过这需要文字编辑与设计师密切合作，设计师要能主动地深入了解文章内容。*Zembla* 杂志的文斯·弗罗斯特还曾把文字排成特殊形状，体现出文章段落间的呼应，并利用印刷技巧，展现出"文字的趣味（fun with words）"，也具体呼应*Zembla* 的设计精神："如果不喜欢杂志的内容，就不要设计！"

首字大写

"首字下沉（drop cap）"是开头字母往下沉，位置低于基线，而"首字大写（initial cap）"是开头字母放大，但始终处在基线上。这两种做法不但可以提示文章的起始点，还可在段落中将文章分

引语的呈现方式很多。在各大刊物中最常见的方法是将字体放大显示，但还有其他独特的方式可以强化内容。将引语垂直排列是比较传统的做法，但能增加页面的活力与趣味。也可以将引语放在空白栏位上，填补并美化留白空间。例如下面*Het Parool*杂志的跨页设计。

6 ZATERDAG 17 JUNI 2006 NIEUWS Het PAROOL | Het PAROOL NIEUWS ZATERDAG 17 JUNI 2006 7

Robbie Williams

Vier avonden de Arena vol met seks, seks, seks en show, show, show

Hij staat vanaf woensdag vier keer voor een volle Arena. Pakweg een kwart miljoen mensen gaat de man zien die ooit het dikkerdje van Take That was en werd uitgelachen toen hij een solocarrière aankondigde.

MAX STEENBERGHE

HET ZIJN Z'N OGEN. RONDE KIJKERS MET EEN COCAÏNERANDJE

HIJ BESPEELT HET PUBLIEK MANNETJE VOOR MANNETJE

'Hij wil ons vermaken, niet meer, en vooral niets minder'

PETER VAN BRUMMELEN

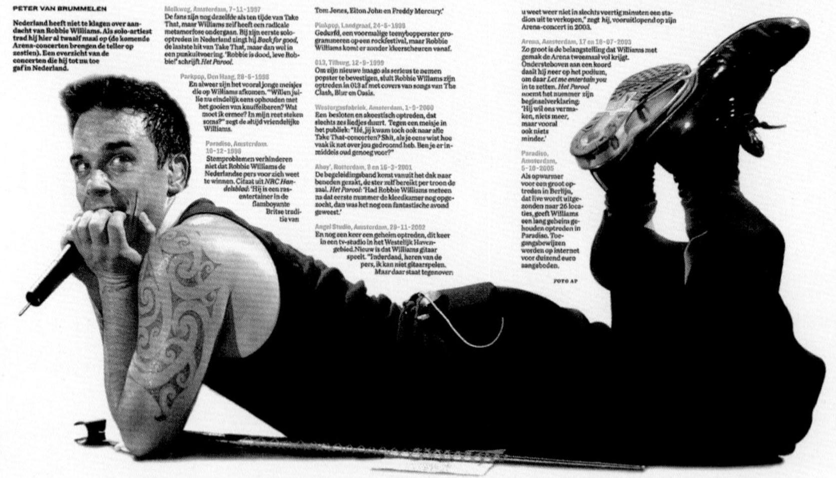

FOTO AP

'Mijn zoontje hoor ik zeggen: Fuck Robbie!'

De rug van Anja Schreuder.

PvB

段，避免冗长的“文字团块”。首字大写在正文内外都适用，既可把字母单独放大，也可以把整个单词或符号都放大。选择放大首字母字体时要考虑到是否与正文其他部分的风格契合，选择的字体可和正文一样，只是将字号放大，也可采用完全相反的风格，例如将花哨的斜体字和干净现代的黑体字一起使用。

小标题

小标题一般在正文中使用，通常会加大字号、加粗、用大写字母、更改颜色或使用不同字体加以区别。

引语、导语和重点句

引语和其他类型的“展示性文案（display copy）”一样，由文稿编辑负责来选取或撰写，但设计师可对引语的数量、位置与长度提出建议。引语是页面上的焦点之一，能够通过变化引起读者的兴趣。引语可以用单引号或双引号括起来，只要用法一致即可。如果引用部分只是来自内文，但不是受访者或受访对象所说的话，一般不用引号。引语的设计方式很多（无论是否有引号），例如把文字放到方框内并居中，也可另起栏，或让文字横跨整个跨页，还可压在图上。报纸经常用导语或者引语来吸引读者注意。

副标题、段标题与栏外标题

副标题、段标题与栏外标题等元素能够帮助刊物建立不同版块的架构，指出或强调某主题、章节或专题大意。副标题可利用线条、文本框、横条、黑底白字或小图标来提高辨识度。栏外标题是位于页首、页尾的缩小版标题，如果一篇文章有好几页篇幅，栏外标题也要随之出现在相应页面，提醒读者现在正在读哪篇文章。

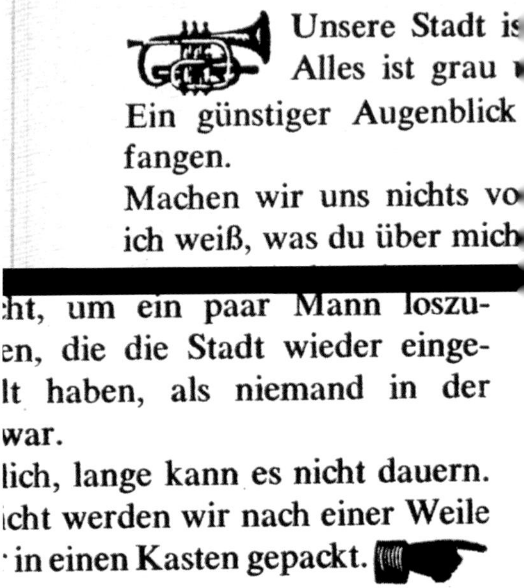

Unsere Stadt is
Alles ist grau
Ein günstiger Augenblick
fangen.
Machen wir uns nichts vo
ich weiß, was du über mich

ht, um ein paar Mann loszu-
en, die die Stadt wieder einge-
lt haben, als niemand in der
war.
lich, lange kann es nicht dauern.
icht werden wir nach einer Weile
in einen Kasten gepackt.

要想让转页标志更醒目的话，可以利用方块、项目符号、刊名首字母或其他符号来强调。小图示告诉读者文章已结束，可以让人一眼看出文章长度。图中 *Twen* 杂志的转页行就设计得很巧妙。

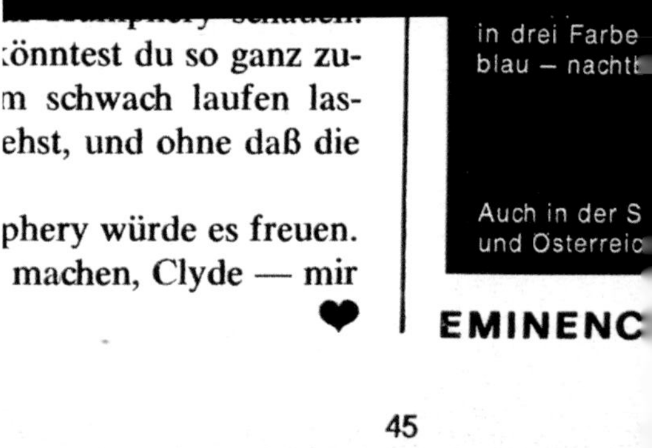

könntest du so ganz zu-
m schwach laufen las-
ehst, und ohne daß die

phery würde es freuen.
machen, Clyde — mir

in drei Farbe
blau – nacht

Auch in der S
und Österreic

EMINENC

45

除了引语之外，还有许多方式可增加视觉趣味，鼓励读者去阅读某篇特别有趣的文章。最上面的 *Twen* 杂志就用了很多小图示，例如小喇叭像在宣布某一栏值得一看的新闻。网站设计常运用各不同的视觉指标，纸质刊物设计师也可以效仿，例如使用箭头、按钮与线条等。

转页提示

如果刊物中的某篇文章当页未完，需要延续到次页或其他页面，最好用“接下页”或“接前页”等文字提示读者，也可以用指示箭头来标示。这称为转页行或转页箭头，报纸则称为“回行（slug）”。文章续页到其他页时，最好从最后一句中间或段落中间分开，若最后一句是句号，读者会以为文章已结束。

图片说明

文章的图片说明通常出现在图片附近或放在图片上，说明图片信息或图片出现的原因，以及它和整篇报道的关系。如果有大量图片需要加图片说明，可以在每张图上编号，在页面其他位置依次列出图片说明。图片说明可以提供正文没有的延伸信息，不过报纸上的图片说明通常被视为新闻事实，很少出现和正文无关的图片。

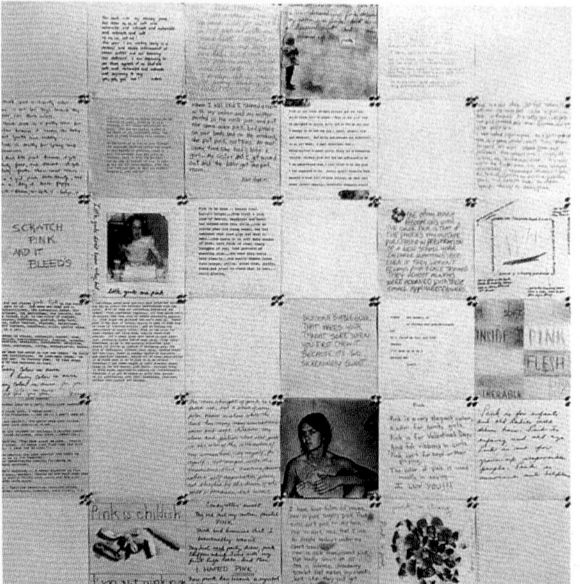

1 2 3
4
5

1 | Lolli Aboutboul:'A hungry woman is not a free woman'., for the UN World conference on Women, Bejing, 1995. Photographs by Lana Wong

2 | WD+RU. Team: Siân Cook and Teal Triggs. 'Smart Fun', 1996

3 | Class Action, Yale University. 'He Hits Me, He Hits Me Not'. Billboard project against domestic violence

4 | Sheila Levrant de Bretteville. 'Pink', 1974

5 | Sandra Kelch, Cranbrook Academy of Art. 'Bosnian Rape Camps', 1993. Postgraduate student project

页码

页码是整本刊物的导航，通常位于每页的相同位置，例如右下角或下部中间位置，方便读者查找参考。如果页码太靠近装订口，就不利于读者寻找。报纸通常把页码放在最上面，书籍的页码旁边通常还附带标出书名和章节。虽然杂志为了刊登出血图或满版广告，常常舍弃页码，但报纸的主要报道版块不能这样操作（广告页例外）。是否方便浏览、快速找到报道位置影响着读者的阅读体验，所以页码必须放在醒目的位置，才能帮读者快速找到想看的内容。

图片出处说明

图片出处说明通常和图片放在同一页，可能垂直排列在图片旁，或放在页面靠装订口的内侧。如果摄影师是知名人物，则摄影师的名字也可能放在署名行，或放在导语中。

信息框与边栏

信息框或边栏文字常会用不同的线条、颜色、边线与栏宽来和正文区分，字体也常是有别于白体的黑体。信息框文字可能与正文相关，也可能是单独的内容。

图像

图像是页面的重要视觉元素，设计时需要强调图文的关联。图文关系可以是文辅图，也可以是图辅文，无论是哪一种，都应该在文字和视觉之间建立有趣的对话。要达到文图对话的方式很多，*Speak* 的威尼茨基（Martin Venezky）说："设计师可用图像诠释文章，不一定非要直接说明文章内容，可以尝试通过思视觉隐喻来搭配文字。"

图片的剪裁、缩放、与文字和其他图片的位置关系，也是潜在的表达和叙述方式。脸部照片的视线如果朝向刊脊，会呈现和谐感，朝外则会带领读者的视线离开页面，制造张力。若两张脸部图片的视线朝反方向望去，张力就更强。一张普普通通的图片被放大后，更能吸引读者，画面的细节可能变得抽象，给读者带来好奇感或惊喜。

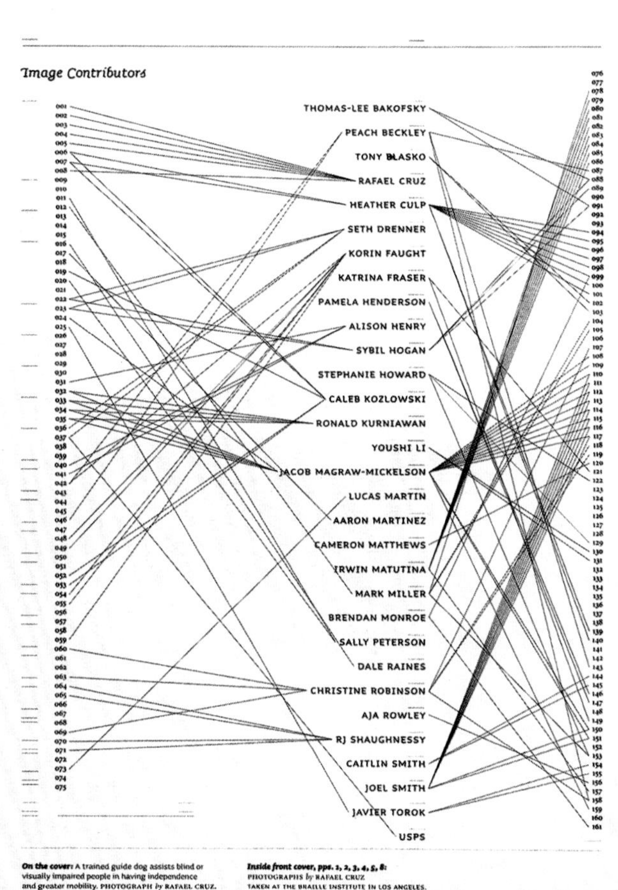

Fishwrap 杂志很重视跨页的视觉效果。上图通过创意图表形式列出各页的图片来源，并放在整本刊物的前面。这样不仅容易找到图片来源，也说明图片出处文字不一定要放在图片旁边，还可以放在其他地方，避免破坏页面的整体设计。

图片说明的编排方式很多，左页是 *Graphic International* 杂志的图片说明分布图。

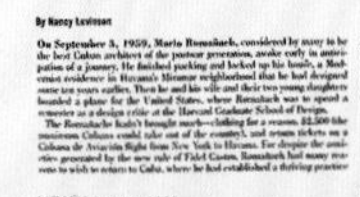

设计师要知道什么时候可以不用处理，直白地使用图像。有一次克里斯韦尔·拉滨要为*Metropolis*设计古巴现代主义建筑师马里奥·罗曼纳（Mario Romanach）建筑作品的跨页，因为马里奥·罗曼纳在美国知名度不高，拉滨表示：“我刊登了很多罗曼纳的照片，不太在意设计是否能体现现代主义特色，因为他在美国虽然几乎无人知晓，但他的建筑比我的设计更能体现现代主义。一些照片用黑色调展现，表示这些建筑物已经面临危险或者已经损毁。”

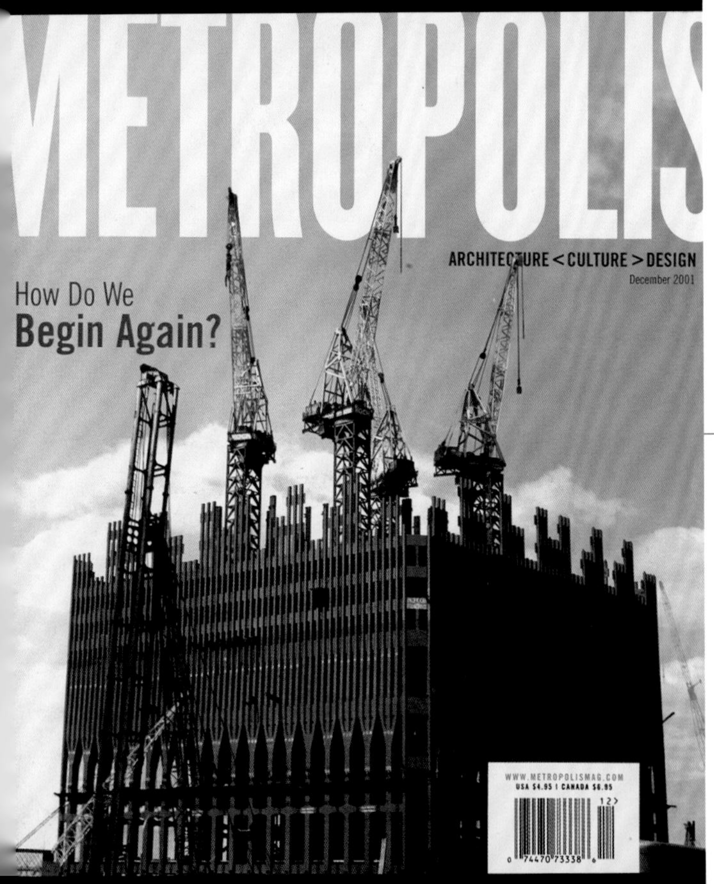

“我们在‘9·11事件’之后出的第一期杂志中，图片编辑莎拉·巴雷特（Sara Barrett）找到这张双子大楼北塔在1972年兴建时的照片。我把照片中的天空加上了天蓝色，让照片多了一丝希望的意味，此外再无任何修饰。这张照片成功引起反响，许多建筑师和普通民众产生共鸣，根据自身经历，为照片赋予许多意义。照片回顾过往，也对未来提问，画面中的北塔远看像是残骸，但近看会发现是兴建时的工地照片，搭配文案：‘我们如何重新开始？’提醒读者进行反思。就连大卫·卡森也问我是怎么修图的。但我除了加了一点青色，其实没做什么。”

——克里斯韦尔·拉滨，《大都会》前创意总监

THE EMPIRE STRIKES BACK

IMPERIAL VENICE MIGHT BE DEAD, BUT THE REPUBLIC OF THE INNER LIFE FORGES AHEAD. HERBERT MUSCHAMP REFLECTS.

Photographs by
MASSIMO VITALI

I spy for dead empires.

It's my way of coping with the imperial ambitions of the living. I spy for Venice, Vienna, Istanbul: any imperial city that has lost its global reach.

May I recruit you? We'll dress like American tourists, sip drinks on the terrace of the Gritti Palace and wave to passing groups of fellow agents disguised as Germans, Britons and Japanese. We'll wink knowingly at Russian impostors and counterfeit Swiss. No one will be the wiser, except possibly ourselves.

The dead imperial city is a global place unto itself, an international state of mind built and operated by the curious and doubtful: it is the republic of the inner life. Let others struggle to become a superpower. We prefer the underpower, in superunderwear. Our undercover mission is to form alliances and pacts with the poets and dreamers who have preceded us. We work for them.

Of all the great imperial cities, Venice is the most intact. Its enduring integrity is as scandalous as the fact of its existence. A swamp is not supposed to produce a

St. Mark's place: embracing the cliché of the Venetian tourist at Piazza San Marco.

90

THE NEW YORK TIMES STYLE MAGAZINE | MAY 15, 2005 91

纽约时报旅游杂志*T: Travel*（上图）只用一张照片，就表现出明媚灿烂的威尼斯的阳光。“我喜欢这张马西莫·维塔莱（Massimo Vitali）拍的照片，虽然灵感来自十八世纪画家卡纳莱托（Canaletto）的画作，但却很有现代感。文字编排也是如此，以Fraktur德文尖角字体做出很美的‘I’。形式布局很不错，图片有很好的景深，适当运用留白空间与文字编排的变化，整个页面美观舒适。”

——珍妮特·佛罗里克（Janet Froelich），《纽约时报》前艺术总监

Sprechen Sie Deutsch?

You'll have to if you're looking for information about photographer Walter Schels. Even a search of the Internet—the living repository of the ephemeral, arcane, and trivial—yields only German language text. In this fast-moving, short-attention-span world, Schels is a photographer who follows his own path—trends and translations be damned.

Born in 1936 in Landshut, Germany, Schels worked as a shop window decorator for almost a decade before taking up the camera for a career in freelance photography in mid-'60s New York City. Schels was perhaps too formal for the burgeoning school of street shooters dominating the cityscape at the time; he returned to Germany in 1970, settling for two decades in Munich where he honed his skill at portraiture. Newly published with an English text, Schels' latest portfolio, *Animal Portraits*, elevates the genre to a level previously unattained except in an earlier portrait by fellow German Albert Renger-Patzsch titled "Silver Baboon." Schels' insightful, psychological portraits reduce photographer William Wegman, that noted canine pimp, to a circus trainer by comparison. Schels' accomplishment is perhaps better understood by its German title, *The Soul of Animals*. Call him the C. G. Jung of the animal kingdom—it's only a small exaggeration.

Walter Schels: Animal Portraits (Die Seele der Tiere) is published by Edition Stemmle.

12/01.02 FLAUNT 51

左图选自*Flaunt*杂志，这张跨页的出血图是一头看向旁边的驴子，以此营造出很强的张力与动感，将读者的视线拉向页面的

杰里米·莱斯利（Jeremy Leslle）

接下来的访谈中，杰里米·莱斯利会和我们谈谈，艺术总监在纸质和电子刊物中担任什么样的角色。

20世纪80年代，杰里米·莱斯利（Jeremy Leslie）在《闪电战》杂志任职，之后曾在《卫报》、*Time Out*、*M-real*等刊物工作，多年来致力于通过各种形式的出版物来探索视觉传达的魅力。1999年到2009年，他在约翰·布朗传媒公司（John Brown Media）担任创意总监，在这期间，客户杂志恰好成为最具创意、有活力的产品。

莱斯利热爱杂志，在2003年出版了自己的书籍*magCulture*，并开设了同名博客magCulture。magCulture博客被评为当时最好的设计博客之一，同时莱斯利还担任magCulture设计工作室的创意总监，并为*Creative Review*杂志撰写文章。他是英国编辑设计协会（British Editorial Design Organisation）的创会成员，也担任美国出版品设计师协会（Society of Publication Designers）的评审。magCulture同时是刊物内容顾问公司，近年作品包括*Port*杂志数字版。无论刊物是什么样的经营模式，纸质版还是数字版，莱斯利都展现出极大热忱，这份热情也体现在他担任独立杂志研讨会Colophon的联合策展人的工作中。

平板电脑已上市多年，设计师现在是不是已经对数字版刊物的设计有了很好的掌控？

事实上，如何在不同的媒体上进行编辑设计，现在依然没有统一的标准和原则。不过，大型出版公司的内部设计团队会将过去用在纸质刊物上的做法加以调整，应用到网站、手机与平板电脑版本的设计上。

大型出版商纷纷建立庞大的内容管理系统，这是否已成为一种趋势？

理论上这是必须要做的事。无论你喜不喜欢，大家会把我们所有做的东西都视为“内容”，我们或许听不习惯，但这就是现状。尽管“内容”这种叫法太过老套，但现在编辑设计处理的确实就是内容，科技最终也是为了内容服务，手机就是个好例子。比方说，我们得思考该怎么在小小的屏幕空间用几种有限的字体，让《卫报》手机版脱颖而出？

平板电脑版的问题在于，每种刊物看起来都一样：开本相同，外观也一样漂亮。大家都说平板电脑触摸屏非常灵敏，但就是没有触摸感。看起来光鲜亮丽，但不是打开的状态就是黑屏的状态，不会因为你所处的位置不同而有明显变化。纸质刊物却可以做

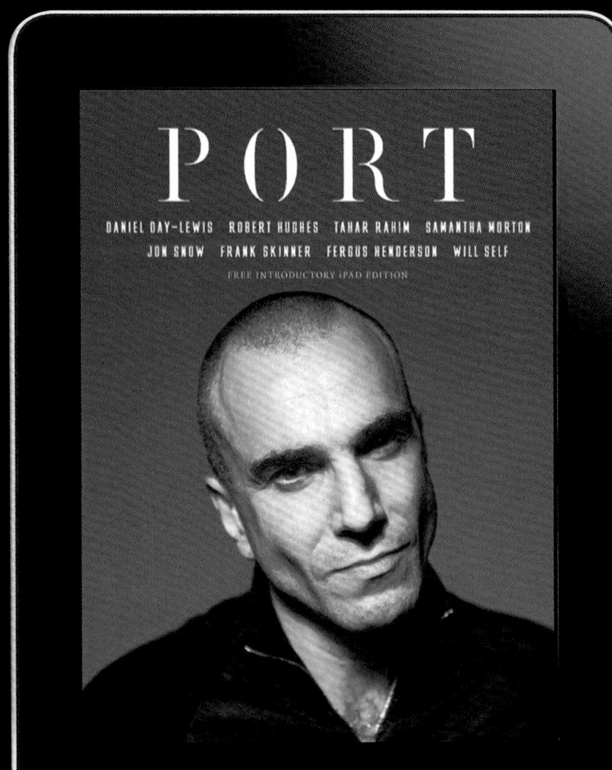

到有触感、纸张质感和墨香，开本各异、可拆可撕，易于分享。现如今大家都狂热地用推特这样的应用来分享内容，我不是守旧派，但我们必须承认纸质媒介和电子屏幕分享之间存在着区别。

您在客户杂志领域的设计成果有目共睹，如今十五年过去了，出于什么原因转向了数字平台发展？

客户杂志曾经很有创意，是红极一时的营销宣传手法。一旦客户杂志展现出其力量和影响力，人们意识到客户杂志对公司的价值，事情就变得严肃起来了。很多有趣的想法和尝试会被腰斩，因为担心成本和风险。显然公司打算在杂志上花钱的前提，是要获得相对回报，营销方式趋于保守，反而失去了创造空间，失去了更多可能性。我很庆幸自己经历过那段客户杂志还有很大发挥空间的阶段。

数字技术的进步是否加剧了这种变化？

首先，是数字化促成了最初的变化。有了数字工具，就有了衡量

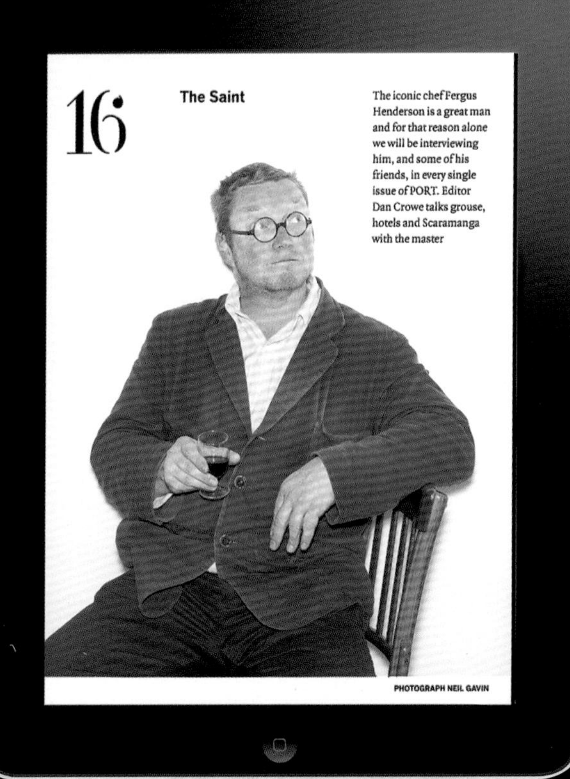

16
The Saint
The iconic chef Fergus Henderson is a great man and for that reason alone we will be interviewing him, and some of his friends, in every single issue of PORT. Editor Dan Crowe talks grouse, hotels and Scaramanga with the master
PHOTOGRAPH NEIL GAVIN

Taking steps to reduce your carbon footprint but still committed to air travel? For the Thule Inuit on the tip of northwest Greenland, climate change means their hunting and gathering way of life is finished. Markus Bühler-Rasom met the people on the receiving end of a global problem

GREEN
LAND
THE
CLIMATE
OBSERVER

Words and photography
Markus Bühler-Rasom

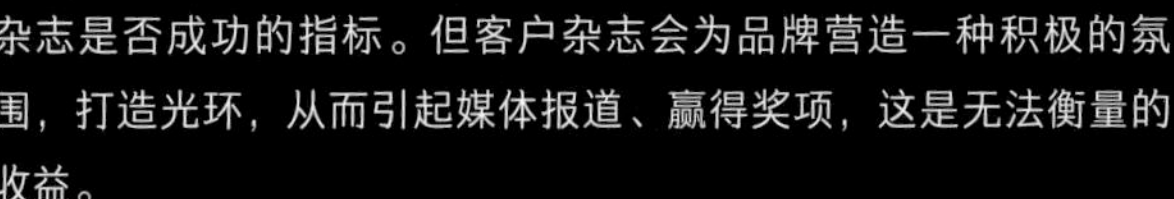
杂志是否成功的指标。但客户杂志会为品牌营造一种积极的氛围，打造光环，从而引起媒体报道、赢得奖项，这是无法衡量的收益。

一旦开始分析，就会失去创意。只要开始衡量、想知道有多少点击量，那什么东西都会和数据及利益挂钩。接下来，会对纸质杂志也报以同样的期待。创意杂志的焦点就改变了，从重视创意到商业为王。现在的确不一样了，任何策略都只考虑是否有利可图。

你觉得数字技术让刊物本末倒置了吗?

关注点不同。你可以理解公司的出发点，许多主流杂志都有自己需要承担的功能，也是在尽自己的职责。但我还是认为品牌构建不仅仅是为了卖出产品，还在于能创造出自己的世界。

现在不像以前那样经常购买和阅读杂志了。但如果买了杂志，就会沉浸在杂志的世界里，杂志世界是个很好的营销空间，以细腻的方式帮助品牌从中脱颖而出。如果技巧高明，就会带来人们对品牌的正面感受，在不知不觉中有深度地诠释和传达品牌形象，那是仅靠直白的广告所无法达到的效果。

在商业模式方面，大公司有钱去投资定制开发系统，但中型消费者杂志的前景是否也可以这么乐观?

平板电脑杂志成功者寥寥，不少大型出版商重金投入，也耗费了许多时间与资源。在数字杂志发展非常初期的阶段，虽然不乏优秀的创意之作，但杂志仍然困难重重，其中一个是营销，如何让读者的发现你的App呢， App Store和实体商店里的书架陈列可完全不是一回事儿。

独立杂志根本负担不起开发App的费用。他们的经费有限，只好寻找其他模式。*Monocle* 杂志直接绕过开发App这个选项，转而投资播客工作室，制作在线播放的有声内容，效果也相当不错。

小个杂志除非运气极佳，否则很难盈利，只有极少数找到细分市场的杂志才能获利，例如*Anorak* 杂志，还开发出很多衍生品，基于衍生品盈利。

*Fire and Knives*杂志决定不开发数字版，杂志工作人员也都不领薪水，这样才能和成本持平。但是每个参与过杂志编辑的人，日后会接到很多项目。

回顾不同的商业模式，我们可以学到什么？

我们可以回顾*The Face*、*Blitz* 与*i-D* 杂志诞生的那个时代，人们想的是“我自要单枪匹马闯出一片天”。尼克·洛根（Nick Logan）打算推出*The Face* 时找不到人支持，干脆在厨房桌子上自己制作创刊号。特里·琼斯（Terry Jones）离开*Vogue*，创办*i-D* 杂志；*Blitz*原本是大学杂志。那时因为主流杂志没那么关注时尚与文化，这些脱离主流的杂志推出时，造成很大轰动。而现在所有媒体都和大众文化紧密连接，要引起轰动就没那么容易了，信息到处都是。

你的magCulture博客创造出的社群是否还在成长？

没错，博客接触到的读者范围越来越广，大家都喜欢和志同道合的人一起。博客或粉丝俱乐部其实和专业杂志的区别不大，都是把有共同兴趣的人联系起来。我是看*NME* 杂志长大的。*NME* 在我十几岁时扮演的角色，就像我儿子今天在用的Facebook，目的是将有共同兴趣的人连接起来。只不过，今天的连接方式更快、更简单。

如今*Creative Review* 之类的杂志，在网络上阅读和分享的人比看纸质版本的更多，因此杂志还得在每一篇博客正文的后面，提醒粉丝他们的纸质杂志出版了。

像*Cosmos*和*Company* 这类中型市场的杂志，有大量的推特粉丝，且彼此互动频繁。编辑会利用社交媒体来进行活动推广和特价促销。如果你的杂志在社交媒体上曝光很少，恐怕很难生存。很多人加入杂志的社交网站，并非因为喜欢这本杂志，而是想要加入这个社群。

做纸质刊物和数字刊物的艺术总监的基本原则是否相同？

基本原则没变，但有些额外的事项要注意，最基本的原则是，要很清楚自己想说什么、说话对象是谁，以及怎么说。现今平面设计的视觉语言也在发展，就像语言本身也会演变一样。最基本的编排原则是一样的，不过视觉偏好会不断变化。

许多人以为纸质版设计和数字版设计天差地别，其实不然，除非设计师刻意为之。

编辑设计必须能够传达与解释想法，并让读者接受，所以要先下功夫做研究。设计师必须懂网格线和易读性，不能把刊物编辑当成只是在处理一块块的文本块。一定要仔细阅读文章内容，才能有针对性地进行设计处理。最重要的是必须尝试最新的沟通方式，为己所用。比如动态图在数字阅读中很重要，以前拍图片的摄影师，已经开始改拍影片了，这是大势所趋。

关于数字化您觉得最值得期待的潜力是什么？

数字化颠覆了一切。要真正发挥新报刊媒体的潜力，比许多人想象中困难，至少不像平板电脑刚上市时大家想得那么简单。整个产业结构改变的速度没有想象中那么快。

真正值得期待的，是数字挑战了原有的基本设计原则，让设计师重新审视编辑设计这门学问。在数字产品大量出现之后，反倒衍生出不少令人惊喜的纸质版设计。

所以我们还需要一段时间才能看出数字刊物的潜力？

数字设备提供了新的空间，数字刊物和纸质杂志不同，也不是电视节目。人们不会为了看杂志而买平板电脑，而是用平板电脑做很多别的事才会购买的。设计师得研究用户行为，他们都用平板电脑来做什么?我们能够怎么利用这个空间？我们才刚起步，未来的路还很长。

版面构成的决定性要素

版面构成没有万能公式。基本上，版面构成最重要的就是结构逻辑、准确传达与便于浏览。设计版面包含了从在设计会上进行团队沟通，到考虑某个专题的成本，更改版面布局的制作周期，以及如何合理安排所有该放进版面的素材。许多因素是设计师无法左右的，例如预算、空间分配、页数与时间周期。任何设计方案都会遇到困难，如何设法克服困难更考验设计师的创意。

策划与时限

刊物编辑总有开不完的策划或者进度会议。编务举行会议的目的，在于确定每一期刊物内容与主题的重要性，估计并掌握每个项目或专题需要的空间，并向所有相关编辑确认进度与工作状况。设计部门在策划会议后，就能够开始设计每篇专题的视觉走向，并根据需要来委托其他人制作图片或购买图片。同时，编辑部门会着手研究、撰稿或约稿；技术或美妆部门也要开始收集产品进行测试与拍照；时装部门要试装并组织摄影。这个过程是持续变化的，因此必须持续评估各种资料，根据需求调整。

设计前的准备工作

设计师从编辑部门收集文稿、图片资料和其他信息和材料，包括最新的落版单。不过，越接近截稿日，落版单的变动就会越多，可能是因为又卖出更多广告，或者需要增删页面来排新收到的资料。

设计师一定要阅读文稿，才能设计出符合内容的视觉效果。

检查所有图片材料（插画、照片与任何图像）的质量是否达到出版要求?别忘了，数字照片的像素必须是300dpi以上，印出来的质量才足够好。许多数字相机预设只有72dpi，因此拍照前一定要注意，最佳打印尺寸通常是照片原尺寸的四分之一。

确认已经取得所有样式规范，包括版式、样式表单、配色与杂志标准字等。

右图为加里·库克（Gary Cook）为英国《金融时报》杂志设计的版面。这是一篇关于前保守党国会议员乔纳森·艾特肯（Jonathan Aitken）的政治专题报道，库克巧妙运用网格线上的负空间，传达这篇文章的宗教主题，也让文字页面呼应图像，使整张跨页统一起来。

FORCE
The Geordie Boy is back. Ready to rock with Rangers and talking exclusively to loaded about football, fame and hair extensions ...
story by JON WILDE photos by STEVE PYKE
1
2
3

眼睛如何浏览页面

虽然每个人浏览页面的方式不是绝对固定的，但是运用图像或文字的视觉引导，可以大致掌握读者的视线运动方式。以左图*loaded*的页面为例，读者会先看到版面上超大的G，因为它的字号与颜色很突出，之后会沿着G水平的线，延伸到足球员保罗·加斯科因（Paul Gasconigne）的眼睛，之后才看到较小的导语。右页是选自*Het Parool*的例子，这是个比较复杂的版面，不过读者的视线会顺着大照片中的男子眼神一直延伸到页首标题，之后再看其他元素，视线通常会沿着格线垂直与水平的线条游走。

刊物的制作周期

设计部门的刊物制作周期，从收到文字、图片与插图之时就开始了。拿到这些资料之后，设计师就要开始编排版面。如果有图像，通常版面会以图像为中心来布局，如果没有图，也可以利用强有力的标题创造一个视觉焦点。需要注意的是，各元素之间要避免冲突，不能让该突出的视觉偏离主题或被削弱，即使空白区域也很重要，也会产生视觉焦点。版面完成后会交给文稿编辑校对，根据需要删改。这时已经确定了杂志各单元的送印时间，如果刊物篇幅较大，各单元的截稿日期不一定相同，会按照特定的顺序送印，通常封面是最后送印的。各个单元，无论是正文内容还是广告，都要在下印之前完成。

现实考虑

设计部门有多少经费可用在图像、特殊用纸、油墨与特殊工艺上，早已事先确定好，具体如何分配由艺术总监决定。但必须遵守先前做好的预算，就像刊物与印刷厂定好截稿日后就要遵守一样，以免错过送印时间与出版日期。时间因素会影响设计师能做多少实验，以及设计师不满意版面时，是否还有时间可以修改。每个设计师有自己的工作模式，以应对紧张的时限。有些人会先确定页面的整体设计，再将信息框、字体与颜色等细节打磨到满意为止。有些设计师则会同时设计很多不同的版面，直到做出满意的作品，或最后一刻才不得不停手。文章页面的顺序与页数是由出版公司和编辑决定的，通常受广告销售情况的影响。特定单元或专题的页数是固

cultuur

Het PAROOL 29 ZATERDAG 17 JUNI 2006

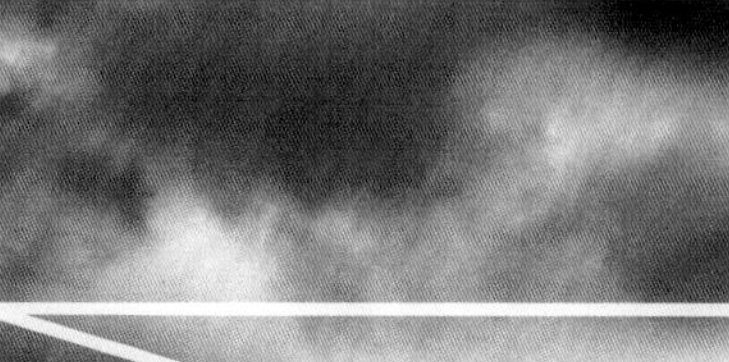

'Dromen zitten je geluk in de weg'

Osinga: 'Er werd geschreven dat ik met deze roman handig inspeel op de actualiteit.'

Belangrijke prijs voor Amsterdamse galerie

'Voor Ellen de Bruijne is verspreiden ideeën belangrijker dan verkopen'

'Het varkentje van Beeren', te koop voor vijf miljoen dollar.

HET IS EEN SIGNAAL: KUNST IS NIET LOUTER COMMERCIE

HOLLAND FESTIVAL

De wortels en de vruchten van de meester

Het Nationaal Ballet danst His master's choice.

定的，其中一部分页面可以卖给广告商，如果广告出现增删或移动，文稿也会跟着删除、增加或移动，例如原本五页的专题必须缩减到只有三页的空间。这对设计可能产生很大的影响。

主导设计方向的决定性因素：空间

现代读者喜欢方便携带、内容丰富活泼，并且从哪里开始读都不影响阅读体验的刊物。这些需求反映在设计上，就需要利用好图片、展示文稿、彩色标题、信息框、项目符号与表格。如果遇到文字量大的内容，可以利用留白空间，避免版面看起来厚重沉闷。这些设计手法都会占用空间，无论如何，编辑设计一定要做到文字、图片与图像元素的平衡。通常内容字数在发排前就已经确定，但如果一篇报道的内容超过太多，不太好设计，设计师可以要求文稿编辑将其删改到合理的长度。有时某些专题必须在收到稿件前先设计好版面，或者完全由设计来主导、计算报道字数，再请作者按照字数写作。还有就是必须要考虑到插图的格式，如果是网上的图片素材，尺寸必须合乎要求，以免颗粒过大、模糊不清，不符合杂志的风格。有时记者在写文章时想要特别强调某些信息，以便读者更好地理解内容，或者图片研究团队可能要求某张图不能超过几个栏位。这些要求很可能是在最后阶段才临时提出来的，肯定避免不了一再更改设计。

报纸的空间规划可以说是一场水平与垂直的游戏。20世纪中期之前，许多报纸的标题为了要与内文栏位同宽，会分成好几行，于是版面看起来就是许多细长的栏位并列，不美观，也影响阅读体验。后来出现了其他的做法，例如常用更宽的栏位与空白，加上小报与其他较小的版式流行起来，于是报纸朝横向阅读发展，变得更舒适好读。即使是比较窄的柏林版式，设计师也可以让标题或报道文字横跨数个栏位，阅读方向就更接近于水平，而不是上下延伸。不过，如果是采用垂直设计，一定要留足够的白边或运用空白栏位，才能保证不影响阅读体验，让整体页面更轻盈，强化空间感与可读性。

主导设计方向的决定性因素：形状

图文经过编排后会呈现出一些形状，这些形状也要符合刊物风格。波特把这部分工作称为“把元素分配到空间里”。这些元素包括标题、文本、图片与留白等。如何安排这些形状，也会影响版面的好坏。如果设计得当，形状的分布能引导读者视线，带领读者按顺序阅读整篇文章，还能通过形状创造出各种感受与意义。

试着把眼睛眯起来去看一个版面，就会看到各种形状：文字框变成灰色方块，插图与照片变成长方形或正方形，偶尔出现不规则的图片或装饰字体。继续眯着眼，看看这些形状如何排列起来，彼此连接成其他形状，或形成鲜明的对角线。留白区域也有形状，这些形状会营造出平衡、和谐或冲突感。如果能为这些形状设计建立规则，对于打造出流畅的版面会很有帮助。

版面中的形状需要满足两个功能：一是形状彼此之间必须平衡，能够和谐共处；二是形状和整体版面的关系也要很恰当。形状的安排与整合，对制作出优质版面很关键，通过形状的变化也可凸显出各元素的特色。如果能合理地安排形状，设计师就能将读者的视线吸引到特定的位置，引导他们依次去看有助于理解内容的各个元素。因此，成熟的编辑设计师能够运用不同形状的搭配，创造阅读趣味。

DAGENS NYHETER.

FREDAG 8 JUNI 2012 • SVERIGES STÖRSTA MORGONTIDNING • GRUNDAD 1864 • PRIS 30 KRONOR

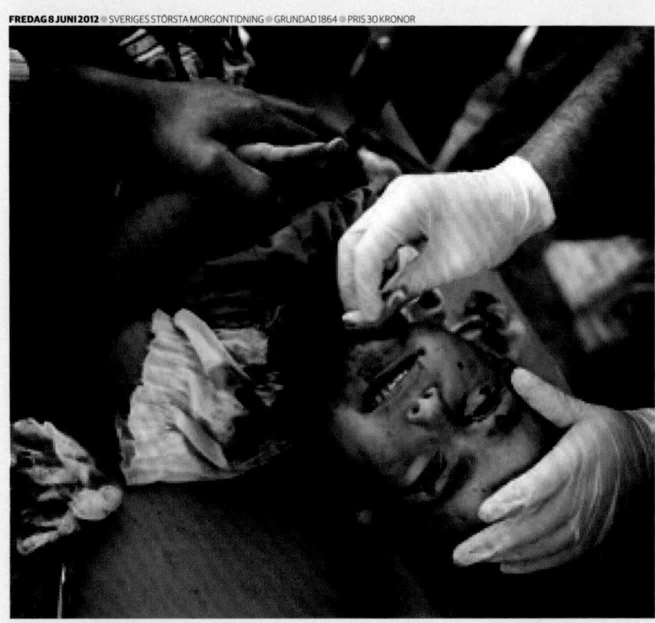

Diktatorns offer

• Skakande bilder inifrån våldets Syrien • Flickan överlevde attacken som utplånade hennes familj

Den tvååriga flickan från byn al-Rami blev ett av den syriske diktatorns Bashar al-Assads tiotusentals offer.
Hennes familj utplånades, men hon överlevde med splitterskador och kunde räddas av rebeller som förde henne till ett hemligt sjukhus i trakten.

DN kan i dag publicera ett bildreportage av fotojournalisten Niklas Meltio som följt en rebellgrupp inne i Syrien.

Minst 10 000 har hittills dödats och världens ledare talar om risken för fullt inbördeskrig. Varför görs inte mer för att stoppa dödandet?

Det lika enkla som brutala svaret är att Syriens befolkning offras på realpolitikens altare, skriver DN:s Erik Ohlsson. **Nyheter 2, 8–11, Ledare 4**

Länsstyrelsen slår larm:
100000-tals stockholmare utan bostad 2030
Bostadsbristen ökar, ändå byggs det allt färre bostäder. I en ny rapport larmar länsstyrelsen om en "mörk" situation i Stockholm som är på väg att förvärras. **Stockholm 4–5**

V på DN Debatt:
”Ge näringslivet större ansvar för yrkesutbildning”
Näringslivet måste ta på sig ett större pedagogiskt ansvar för yrkesutbildningen, skriver Jonas Sjöstedt och Rossana Dinamarca från Vänsterpartiet. **Debatt 6**

Johan Esk:
Jag hoppas på tyskarna – för deras fotbolls skull.
Sport 12

Fotbolls-EM
Segervissa polacker tror på succé
Inför dagens EM-premiär har vinden vänt och optimismen gror i Polen. "Kvartsfinal är ett minimum", säger sportministern till DN inför matchen mot Grekland. **Sport 12–15**

Insidan
”Jag har svårt att orka med min tvååring”
"Min tvååriga dotter bråkar alltid när hon inte får som hon vill. Ofta får hon det, eftersom jag inte orkar", skriver en tvåbarnsmamma. Psykologen ger råd. **Insidan • Kultur 21**

7 388101 003001 Nr 152

•A5

瑞典《每日新闻》(*Dagens Nyheter*)用新闻摄影来传达主题。在这个例子中，照片以绘画般的构图与低饱和度来减轻内容的恐怖程度。报头与头条的文字编排鲜明、有戏剧性。报道横过页面下方,引导读者进一步阅读里面的内容。

Gibberish Text has the inherent disadvantage

They distract attention to themselves.

Simultext is effective in any typeface, whatever size and format is required.

Paragraphs may be long or short.

This is dummy text.

A simulation of copy, using ordinary words and normal frequencies.

The headline

This is dummy text. Intended to be read but have no meaning. As a simulation of actual copy.

By An Author

Dummy settings which use other or even gibberish to approximate text have the inherent disadvantage they distract attention to them selves. Simultext is effective in any typeface, whatever size and format is required. Paragraphs may be long or short. Text can be to complete any area, as the copy is simply re-peated with different start points. This is dummy text. Intended to be read but have no real meaning. As a simulation of actual copy, using ordinary words with normal letter frequencies, it cannot deceive the eye or brain.

Settings may use languages or gibberish to approximate text have the disadvantage that they distract attention to themselves. A simulation of copy, using ordinary words with normal letter frequencies, can't deceive eye or the brain. Simultext is effective in any type face, also at whatever size and format is required. Paragraphs may be long or short. What you see is dummy text. Intended to be read but have no real meaning. As a simu- lation of actual copy, using ordinary words with normal letter frequencies, can't deceive the eye or brain. Copy can then be produced to complete any area, as the copy is repeated with different starting points.

Dummy settings which choose other languages or gibberish to simulate text have the inherent disadvantage that they can distract attention to themselves. Text is effective in any typeface, whatever size and format is required.

Paragraphs may be long or short. Text can be to complete any area, as the copy is repeated with different start points. Intended to be read but have no real meaning. A simulation of actual copy, using ordinary words with normal letter frequencies, it cannot deceive the eye or brain. Settings that use other languages or even gibberish to approximate text, have the inherent.

FEATURE **5**

如何放置各个项目

文字的排列方式可以向读者传递信息，并影响读者心理。一篇文章是占一整个跨页，还是和其他两三篇文章放在一起，能立刻告诉读者这篇文章的重要程度。上方跨页的文章，搭配满版出血的图，加上很大的标题和导语，宽大的栏位，配有资讯列表，运用强烈的色彩与大面积留白，都是为了引起起读者的兴趣，吸引其进一步阅读。

相对地，下方的跨页是对同一篇文章的另外一种设计，说明同一篇文章如果按照不同方式来设计，就会传达出不同的意义。下面的设计更像是新闻故事的版面设计，而不是上面那种专题报道的设计，文字从跨页自上到下排满文字，标题较小，栏宽较窄，留白空间较少，因此这篇文章的重要性就降低了。不过导语、资讯框、项目符号与字体颜色依然有助于吸引读者的目光与兴趣。

This is a headline and is intended to be read

This is dummy text. Intended to be read but have no meaning. As a simulation of actual copy.

This is dummy text. Intended to be read but have no meaning as a simulation of actual copy.

4 NEWS

The second story headline

BOX COPY
Which uses dummy text or even gibberish can have the inherent disadvantage of distracting attention from itself. Simultext is effective in any typeface, whatever size and format is required for use.

A three-line headline that is an extra level in the hierarchy

- **Dummy settings**
- **Gibberish text**
- **The disadvantage**
- **Attention too**

Another third story headline

NEWS 5

The Masthead

03 *It is intended to be read but have no meaning. As a sample of actual copy in a natural situation*

14 

25 *It is intended to be read but have no meaning. As a sample of actual copy in a natural situation*

36 

Paragraphs may be long or short. Text can be produced to complete any area, as the copy is simply repeated with different starting points. This is dummy text. It is intended to be read but have no meaning. Paragraphs may be long or short. What you see here is dummy text.

This is the headline

It is intended to be read but have no meaning. A sample of actual copy in a natural surround

Dummy settings which use other or even gibberish to approximate text have the inherent disadvantage they distract attention to themselves. Simultext is effective in any typeface, at whatever size and format is required. Paragraphs may be long or short. Text can be to complete any area, as the copy is simply repeated with different starting points. This is dummy text. Intended to be read but have no meaning. A simulation of actual copy, using ordinary words with normal letter frequencies, it cannot deceive eye or brain.

Settings which use other languages or even gibberish to approximate text have the disadvantage they distract attention to themselves. As a simulation of actual copy, using ordinary words with normal letter frequencies, can't deceive eye or brain. Simultext is effective in any typeface, at what ever size and format is required. Paragraphs may be long or short. What you see is dummy text. It is intended to be read but have no real meaning. As a simulation of actual copy, using ordinary words with normal letter frequencies, can't deceive eye or brain. Copy can be produced to complete any area, as the basic copy is repeated with different starting points. It is intended to be read but have no meaning. Presentation copy uses languages or gibberish to approximate text have the disadvantage that they distract attention to themselves. Simultext is effective in any typeface, whatever size and format is required. Paragraphs may be long or short. This is dummy text. Intended to be read but have no meaning. As a simulation of actual copy, using words with normal letter frequencies, cannot deceive eye or brain. What you see here is dummy text.

Simultext is effective in any typeface, whatever size and format required. Paragraphs may be long or short. Text can be produced to complete any area, as the basic copy is simply repeated with different starting points. This is dummy text. It is intended to be read but have no meaning.

Paragraphs may be long or short. Text can be produced to complete any area.

This is dummy text. It is intended to be read but have no meaning. As a simulation of actual copy, using ordinary words with normal letter frequencies, it cannot deceive eye or brain. Dummy settings which use other languages or even gibberish to approximate text have the inherent disadvantage they distract attention to themselves. Simultext is effective in any typeface, at whatever size and format is required. Paragraphs may be long or short. Text can be produced to complete any area, as the copy is simply repeated with different starting points. This is dummy text. Intended to be read but have no meaning. As a simulation of actual copy, using words with normal letter frequencies, cannot deceive eye or brain. Trial settings which use other languages or gibberish to approximate text have the disadvantage they distract attention to themselves.

报纸的卖点一般都在头版。因此，头版的页面设计必须足够抢眼，才能在报摊上脱颖而出。报头、图片、标题的大小与位置对报纸的销售量影响很大。头版的版式需要放很大的图或很大的标题来吸引读者注意。

The Masthead
20 February 2007

03

Smaller headline to vary visual strength of story

It is intended to be read but have no meaning. A sample of actual copy in a natural surround

The lowest level of headline styles

Headline for a small article

Exciting box copy

如果页面上有多篇报道，文字与其他元素的位置可以引导读者的视线。最重要的报道会放在页面最上方，用最大号的标题、配图与导语，栏位也最宽。相对地，另外两篇文章共享六个栏位，没有导语，占的空间要小得多。通过标题大小、图片占用的空间，读者就能感受到这几篇文章的重要程度差异。

IN PERSPECTIVE

SUR, P.S.1. | Architects Xefirotarch, Hernan Diaz Alonso | Queens, NY

FROM THE BEAUTY TO THE BEAST
SUR

This year's winner of the MOMA/P.S.1 Young Architects Program, Hernan Diaz Alonso, has created a temporary structure in the courtyard that is closer to a sci-fi film set than a building. Transferring the power from the architect to the computer, Alonso's installation is one gigantic happy accident, the formal appearance of which is a series of computer transformations of the same, singular cell.

Words Simon Hosauf
Images Robert Mezquiti & Xefirotarch

In its sixth year, the Young Architects Program of MOMA/P.S.1 is a competition that invites emerging architects to transform the courtyard of P.S.1 during the summer months. The objective is of course to provide an outlet for emerging practitioners to create work outside the usual constraints of architect/client restrictions. Five of the 17 entries were short-listed and will be exhibited at the centre until September. Xefirotarch's winning proposal opened its doors (well, actually there aren't any) to the public on 26 June. Attracting the P.S.1 construction budget of $60,000, the installation will also serve as a music venue for P.S.1's other popular summer activity: warm-up. Although the installation would doubtlessly draw a large enough audience by itself – this combination of cutting edge architecture and cutting edge DJs is an obvious winner.

Although the structure has more of a sculptural nature than an architectural one, through the activation of the space during warm-up the initial intention of its creators will emerge. Alonso describes it thus: "It doesn't mean anything; it's a frame for experimentation."

In a way the same could be said for the courtyard of P.S.1 itself. So we have a frame for experimentation (P.S.1) allowing for yet another frame for experimentation by the architects. The actual experimentation is then presumably done by the visitors. The structure, SUR, has a very particular aesthetic, however, which couldn't be more different to the sober backdrop of the courtyard's concrete walls. Resembling a character from a recent sci-fi flick or the oversized skeletal remains of an exotic deep sea animal, SUR is an interesting experiment in the creation of architectural form via computers – not the architect's will. Although vast discussions and disputes have been very much part of the public debate of current architecture – the physical manifestation of the more extreme examples of digital form finding mostly remain in theoretical form.

Ironically, Zaha Hadid has used a 'single surface design concept' in her contribution to what can only be described as the [illegible], if one were to be more theoretical, mille plateaux, that is the Hotel Puerta América in Madrid.

在图片和文本之间建立视觉对话，能够在跨页上建立良好的结构与形状。左图的例子选自*Inside* 杂志，这是一篇关于建筑师扎哈·哈迪德（Zaha Hadid）的专题报道，设计师杰弗里·多切蒂（Jeffrey Docherty）把建筑的造型当成版面的主导元素。“我非常尊重建筑摄影，尽量避免把文字压在图上。我喜欢把图片当成艺术品，能装裱，但不能干扰。当然也有例外，在一些两者结合效果更好的例子中，字体可以为图片提供结构或动态感，关键在于如何帮文字找到适当位置。太明目张胆的位置没什么意思，如何帮助文字和图片找到适合而且有趣的位置，就是挑战所在。编辑设计师常常会盲目跟随潮流与趋势，而我认为设计更需要找到适当的平衡。学会克制是很重要的，可惜在设计中很少被考虑到。”

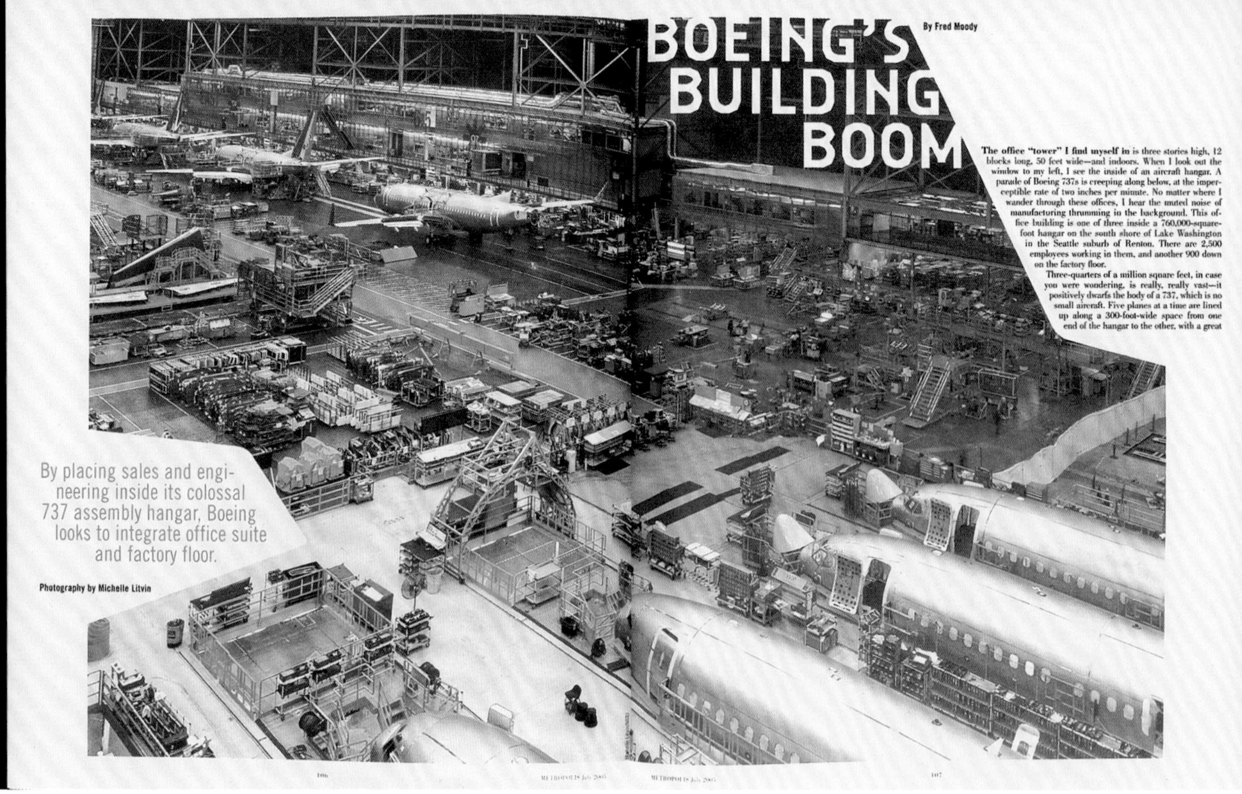

BOEING'S BUILDING BOOM

By Fred Moody

The office "tower" I find myself in is three stories high, 12 blocks long, 50 feet wide—and indoors. When I look out the window to my left, I see the inside of an aircraft hangar. A parade of Boeing 737s is creeping along below, at the imperceptible rate of two inches per minute. No matter where I wander through these offices, I hear the muted noise of manufacturing thrumming in the background. This office building is one of three inside a 760,000-square-foot hangar on the south shore of Lake Washington in the Seattle suburb of Renton. There are 2,500 employees working in them, and another 900 down on the factory floor.

Three-quarters of a million square feet, in case you were wondering, is really, really vast—it positively dwarfs the body of a 737, which is no small aircraft. Five planes at a time are lined up along a 300-foot-wide space from one end of the hangar to the other, with a great

By placing sales and engineering inside its colossal 737 assembly hangar, Boeing looks to integrate office suite and factory floor.

Photography by Michelle Litvin

主导设计方向的决定性因素：古典比例形状

无论是从习惯或还是人类与生俱来的偏好来看，古典比例最受欢迎。在编辑设计中，最知名的古典比例就是黄金分割，即1：1.618，或者长16.2，宽10。这种形状最美观，适用于多种版面。

主导设计方向的决定性因素：通过色彩塑造形状

文字量大的页面通常利用形状来打破单调性，通过照片、插图、装饰字体、留白、色块或文字本身的造型来实现。通过色调塑造的形状，可以有效地让各个版面元素区分开来，或是将各种元素结合到一起。从心理学上讲，将文章分解成比较小的区块，阅读起来会更轻松。将有颜色的标题、边框与线条等元素组合，会为版面赋予意义。人眼很善于接受色彩暗示，在脑海中建立各元素的联系。

设计师应该大胆尝试各种色彩，勇于配色，并在刊物上使用。马里奥·加西亚（Mario Garcia）说："颜色是个人问题。我们认为，如今的读者喜欢鲜艳的色彩，并不认为鲜艳的颜色就代表低俗或廉价。比方说，过去LV的包包都是棕色的，现在也有黄色、灰绿和粉红色。"加西亚在《观察家报》上大胆用色，无论是导航工具还是布局元素都是如此，让这份周日出刊的报纸生动活泼，吸引了很多年轻读者群。他表示："从调查来看，读者喜欢色彩，也喜欢用不同颜色来分类，我也是。无论从视觉还是实用层面来看，颜色都有助于整理信息。"刊物用颜色来区分各类别时，要注意各单元应该使用区分度很大的配色。

主导设计方向的决定性因素：张力

张力在强化文章的立场方面效果显著。各元素形状之间的关系，以及这些形状与页边的关系可以创造张

右图是《纽约时报》杂志的《年度创意特刊》，其设想是利用百科全书的方式，呈现最好的概念、发明与计划。当时的艺术总监佛罗里克受到19世纪插图的启发，并“选择摄影师罗德尼·史密斯（Rodney Smith）操刀，他有办法通过画面就能传达想要的内容”。这份刊物结合图片、留白与内文缩排，巧妙模仿科全书的形式，打造出和谐又抢眼的跨页。左页是《大都会》关于波音公司新办公室的文章，拉滨运用文字块与图像拼出飞机形状，将内容可视化，不但增加了阅读趣味，还能帮助读者快速理解文章内容。

steganography goes digital

its way through the court system ever since. Written on a page, the DeCSS code looks innocuous enough: a dozen or so lines of letters, numbers and punctuation marks. But when it's run as a computer program, it performs a very specific action: it disables the encryption on a commercial DVD. And so the essence of the DeCSS case is a debate over the legal status — and in fact the very nature — of these few lines of code. Is the program a machine, subject to government regulation like a car or a gun? Or is it language, protected by the First Amendment?

Throughout this past year, Corley's supporters have waged a creative battle to prove to the world that the DeCSS code is human language, not machine language. Their tactic: treat it publicly — very publicly — like regular speech. And so the potentially illegal code has been printed on T-shirts and performed as a folk song. It has been delivered as a dramatic reading, it has appeared in a short film in which it scrolls across the screen like the opening credits of "Star Wars" and it has even been rendered as an extended haiku, several hundred stanzas long:

How would you like to
hear how to decrypt both the
disk and title keys?

All we'll need are the
encrypted versions and a
player key. No sweat!

We'll call the title
key TKEY, a pointer to
six bytes, encrypted. . . .

Although the campaign may seem goofy and prankish on its surface, the legal questions it addresses are on the cutting edge of both First Amendment and copyright law. The M.P.A.A. is certainly taking the pranksters seriously; the organization sued the manufacturer of the DeCSS T-Shirt for "misappropriation of trade secrets," arguing that the $15 garment had become a "circumvention device" as soon as the code was printed on its front. The argument implicit in each reimagined version is that the DeCSS code — just letters and numbers, after all — is protected speech, the same as "This Land Is Your Land" or "The Second Coming" or "I'm With Stupid," and so can be published and distributed freely. As one lawyer supporting Corley argued, "If you can put it on a T-shirt, it's speech." The U.S. Second Circuit Court of Appeals is currently considering the case. SIVA VAIDHYANATHAN

Steganography Goes Digital The microdot, a classic 20th-century spy tool, was once the ultimate in hidden writing, or steganography. Microdots are physically tiny, but can carry whole pages of data. They can be sneaked inside an innocent letter, disguised as part of a semicolon. But microdots are difficult to make and can't be copied easily. Plus, conspirators need microscopes to read them. The new, improved method of hidden writing is digital steganography. Each digital picture is made of thousands of tiny screen dots. Every such dot is identified by the computer with a long string of numbers, and a few of the characters in each string — the "least significant bits," as they are known — don't matter much to humans. Their effects are invisible to the naked eye.

In the hands of a master steganographer, though, these insignificant bits can be replaced by new bits that contain a hidden message. The chosen picture looks unchanged despite its secret payload. You could, for instance, take a digital pornographic image and embed

STEGANOGRAPHY
Encrypted in the binary code of this digital picture is Lewis Carroll's poem "The Hunting of the Snark."

within it some religious propaganda, as operatives for Al Qaeda are rumored to have done. The very same tools could also bury dirty jokes in holy pictures, if you considered that more subversive.

Best of all, you don't even have to e-mail the message. Like Poe's "Purloined Letter," a "container image" can be left in plain sight on the Internet. Who's to know — aside from the steganographers — if the boring photos on some auction Web site are crammed with sinister instructions? You can also hide stego messages inside digital music, or in digital text files, or even disguise them as spam.

The spy's life is never easy, though. On close digital examination, a stego file will look fishy. Steganography can't protect you from double agents in on your scheme. And if the police tap your computer and log your keystrokes, you're sunk. BRUCE STERLING

SuperSlow Exercise Working out has come to this: twice a week for 20 minutes, lift weights as slowly as humanly possible, then sit back and revel in newfound muscles and weight loss. That is the idea behind SuperSlow exercise, a weight-machine routine for legs, arms and chest. The participant lifts at a creepingly slow rate for 10 seconds, then lowers for 10 seconds, repeating four to six times. Without the boost from momentum, muscles experience prolonged, concentrated contractions. Such intensity can thoroughly exhaust muscles; SuperSlow enthusiasts require three days rest in between sessions. In June, the Journal of Sports Medicine and Physical Fitness published a study that found the method resulted in 50 percent greater increase in strength compared with regular, faster and momentum-aided weight training.

While this year has brought the technique a rash of media attention and a slew of new devotees, it is actually almost 20 years old. Ken Hutchins developed the system in 1982 at the University of Florida, where he conducted a Nautilus-sponsored weight-training program for osteoporotic women. Originally too feeble to get the weights in motion the usual way — two seconds lifting, four seconds lowering — the women were instructed to lift as slowly as possible without actually stopping. Hutchins observed startlingly positive results — senior citizens previously confined to walkers were all of a sudden hotfooting it to Europe solo.

Wayne Westcott, co-author of the Journal of Sports Medicine and Physical Fitness study and the fitness research director at the South Shore Y.M.C.A. in Quincy, Mass., hastens to point out that the practice is far more grueling than it seems: of 147 study subjects, only two people chose to continue the method after the 8- to 10-week program. "The product is great," Westcott says, "but the process is painful." RORY EVANS

Tele-Immersion September's suicide jet attacks have given business travelers a new reason to stay on the ground. When it isn't absolutely necessary to fly, many are now trying teleconferencing instead. But teleconferencing, in its current form, has many drawbacks,

102

tele-immersion

SUPERSLOW EXERCISE
Photograph by Rodney Smith, 2001

THE NEW YORK TIMES MAGAZINE / DECEMBER 9, 2001 103

力。例如，可以把不同元素安排在同一对角线上，让读者的视线被顺利带到需要强调的位置，而出血图片和超出边界的文字可以创造出动态效果。色彩也可制造张力，相邻的图块会因为同色调或者撞色调而排斥或吸引其他元素和形状。

主导设计方向的决定性因素：重复与连续

许多刊物利用视觉连续性或重复的元素来构成刊物的识别特征。要成功地重复色调或形状，就要运用格线结构或对齐技巧（第六章有更多关于格线的信息），才能保持视觉上的和谐。在刊物中经常重复或连续出现的元素包括文本框、图片或图形（例如边栏与线条的颜色大小），设计师可以利用这些元素，设计出一致又不枯燥的流畅版面，如果每一页都看起来一模一样，就不是理想的刊物设计（就连电话簿中的广告也会更换位置、改变字体粗细）。

主导设计方向的决定性因素：比例

版面上各元素之间的比例可以引导读者视线，提供视觉吸引力，强调文字信息，甚至增加戏剧感。例如标题中的一个大字就能扭转整个页面的焦点和意义，效果媲美图片。比例是相对的，因此可以用来创建层级结构。还可以运用比例来营造视觉效果，让页面更生动，增添趣味。刊物内容也会影响比例安排，例如旅游杂志常用出血图片营造出开阔的感觉，让读者身临其境。如果图片缩小，就难以产生同样的效果，难以吸引读者。

主导设计方向的决定性因素：对比

有时通过让设计元素呈现细腻的大小对比来表现内容，但必须要谨慎，以免设计看起来底气不足、没有说服力。使用极强烈的对比设计相对比较容易成功，

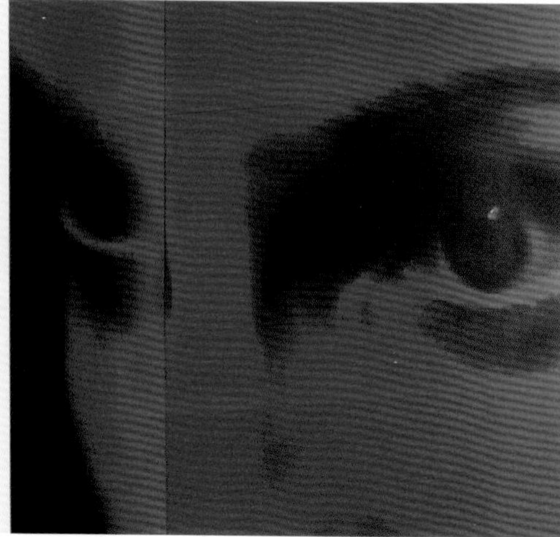

最上图的《纽约时报》杂志则让两个人物背对背，朝刊物的开口方向看，而不是看向刊脊中心，创造出张力。*Espous*杂志（上图）用一张以红色为主的大图，营造出不安不断蔓延的画面，既扣人心弦，又有电影效果。

例如某个元素很大，搭配几个小元素相互平衡。图像的大小必定和页面或其他项目有关。想象一下，如果把番茄切片图放得极大，和一个缩小的图放在边栏内，带给人的视觉感受肯定不一样。

主导设计方向的决定性因素：平衡

编辑设计中版面的平衡感很重要，可以通过多种方式实现。一味追求对称或将所有元素做成同一尺寸，只能达到表面上的平衡，缺乏变化张力和对比，难以成功。应该多尝试设计元素之间的关系，找出真正的动态平衡。一张大图配几张小图，或是与更大但是色彩饱和度低的图搭配，也可以达到平衡。如何平衡各个元素，编辑设计师有自己的考虑，关键在于让版面达到均衡的视觉重量。

The Kitchen Think

Are smart appliances—from Internet-enabled fridges to chatty oven mitts—only going to make us miss the dumb ones?

By Peter Hall

In a scene from the 1990 feature film *The Jetsons* (in which George Jetson is offered the opportunity to run his own sprocket factory), the slumbering household breadwinner is catapulted out of bed and into a flying saucer by the robot maid. "Wait a minute. I can't go to the office in my pajamas," George protests as the flying saucer hurtles him into an outer-space commute. "No problem," replies the spaceship autopilot. "We groom you on the way!" George is then scrubbed, dressed and handed a cup of coffee by a telescoping arm. A video monitor pops up, and his wife, son, daughter and dog wish him a pleasant day. Our bewildered hero finally relaxes, cooing, "This is the life!"

If the scene sounds like an increasingly accurate vision of tomorrow's lifestyle, it's because the Jetsons, that futuristic TV cartoon family of the 1960s, embody a dream we just can't shake.

When General Electric, Sunbeam and Whirlpool rushed to unveil their prototype "smart" appliances earlier this year, there was resounding—and rather familiar-sounding—fanfare for the imminent day when all domestic chores will be automated. "Imagine getting to work, turning on your desktop computer and receiving a message from your home appliance network that you left an oven burner on," said GE Appliances president and CEO Larry Johnston at the National Association of Home Builders trade show in Dallas, where GE showed its prototype bar-code-reading microwave oven, Web-enabled fridge and voice-activated oven. At the same event, Whirlpool offered up its Internet fridge with food-tracking capability. In Chicago, at the International Housewares Show, Sunbeam announced Thalia, a family of networked appliances—including coffeemaker, alarm clock, mixer, bathroom scale and device controller—that can all talk to each other. On January 17, the *San Jose Mercury News* greeted the array of intelligent appliances as the harbingers of a "Jetsons world where no one has to know how to cook."

The Sunbeam Thalia products boast slightly less ambitious technological tricks than other arrivals; they are therefore priced to compete with existing dumb appliances. Expected on shelves this year, the seven Thalia products, designed by Detroit-based Sundberg Ferar, promise a wake-up call not far removed from Mr. Jetson's. They communicate with each other over standard power lines, so they require no additional wiring to establish a sequence of automated events. "Imagine a tomorrow where you set your alarm clock and several household appliances configure themselves to meet your needs when you awaken," reads the promotional blurb. "Your coffeepot sets itself to brew, your electric blanket schedules itself to shut off, your home heating system prepares to warm your home moments before your alarm sounds. Well, tomorrow is here...!"

The economic motives for this renewed interest in the automated home are not difficult to discern. If your fridge has Internet access and the ability to recognize its contents, it can send a request for a delivery to the supermarket when, say, dairy supplies are low. Opportunities for all the industries with a vested interest in that transaction are huge. The interactive kitchen offers a virgin entrepreneurial landscape—not just for the computer industry to exploit, but for schemes of manufacturing and communications conglomerates as well as supermarkets, food producers and delivery services. Not surprisingly, a spate of strategic alliances accompanied the recent salvo of intelligent appliances: Ericsson and Electrolux, Microsoft and GE, Sun Microsystems and Whirlpool, all competing to develop a protocol through which appliances can speak to each other and to suppliers.

In many ways, this first wave of interest in Internet-enabled appliances has transported us to a period strikingly reminiscent of the postwar consumer spending boom. The very same companies—GE, Sunbeam and Whirlpool—that were pushing design variations in the 1950s and 1960s based on color options and extra controls, are now pushing microprocessors and networking capabilities. Yesterday's refrigerator had stylish chrome plating; tomorrow's has e-mail access. Of course, the modern selling proposition is fueled by the Internet economy and the high-tech imaginations of students and professors at venerable establishments like the Massachusetts Institute of Technology. The motivation, however, is similar to that of companies during the postwar boom: Build in obsolescence and they will buy. In the 1960s, the avocado-colored fridge would last as long as your matching kitchen decor. In the early 2000s, the fridge or washing machine will last about as long as your personal computer without an upgrade. According to a report in the *Financial Times* on December 10, 1998, the typical household keeps a refrigerator for 10–15 years. Put a chip in it, and you accelerate its demise. If your neighbor's orders groceries at a speed equivalent to today's T-1 line—while your model chugs with the enthusiasm of a dial-up modem—a trip to the online appliance store can't be far behind.

America's think tank for the smart kitchen is the highly publicized Counter Intelligence Project at MIT's Media Lab, where a five-year program is under way to develop future domestic technologies. Its funding is nearly $30 million

Sunbeam's Thalia smarties execute a Jetsons-like wake-up call: blood-pressure monitor, smoke detector, coffeemaker (top row); mixer and bathroom scale (second row); kitchen-connected device controller, electric blanket regulator, alarm clock (below).

64 I.D. MAGAZINE MAY 2000 65

主导设计方向的决定性因素：立体感

印刷刊物是平面的，但可以通过一些技术创造出空间感，例如压花、使用金属色或特别色烫印。此外，也可以利用透视法则或特殊工艺，让平面元素在视觉上立体化。举例来说，巧妙地重叠设计元素（尤其是形状、字体或颜色）可让版面看起来有突破二维平面之感，像是要跳到读者面前。

SHOPPING ETC.

Looking at innovative retail design, *Metropolis* finds the best ideas respond to a consumer landscape in constant flux.

By Paul Makovsky and Andrew Yang

The Superdesigner Extravaganza
New York

Buy Before You Fly
Amsterdam

Buy While You Fly
Airbus A380

*I.D.*杂志（最上图）运用线条、比例、色彩与文字编排，确保杂志中每个单元及页面都有自己的特色。上图例子则选自《大都会》杂志，克里斯威尔·拉滨用彩色的条码作为设计元素。条码象征着零售业，也让这篇以购物为主题的文章与众不同。“我们在报道各种不同的购物环境时，需要一个所有购物环境的共同象征。每种条码都代表着报道中不同类别的新零售环境。”

制造动感

当静止的动态在平面页面上展现时，诠释动态主题很有挑战性。摄影可以通过多重曝光、慢速快门或高速连拍等技巧来拍出连续动作，体现动感。

Repeat After Me

It all began when the queen of suburban taste, Martha Stewart, asked the downtown–New York photography studio Davies + Starr to shoot daffodils for a floral wrapping-paper design. But after snapping some 50 daffodils, the idea wilted and the project was thrown out for compost. A few months later, Davies + Starr tried shooting the same sort of repeat pattern with a savage-looking hunting blade. According to photographer Chalkie Davies, that was when they saw something cruelly beguiling. "It looked beautiful from a distance, like a heart," says Davies, "but when you got closer you could see it was an incredibly vicious knife." The idea hit him: un-Martha wallpaper. "Martha had made this huge statement—no wallpaper—I guess because she sells paint," says Davies. "But I think kids would really like it if you

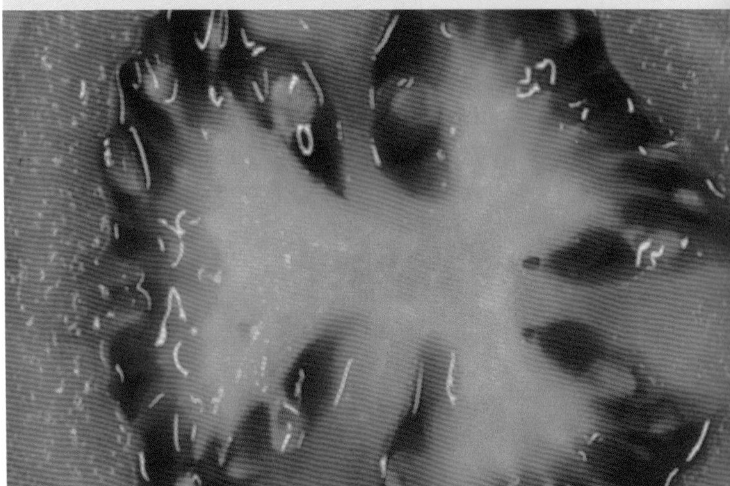

上面的两幅图片显示了同样的照片使用不同的裁剪方式会产生不同的效果和故事性。通过大小、颜色与比例搭配，可发挥几何之美。例如左边*I.D.*的例子，这页讨论壁纸的版面本身就像壁纸的一部分，产生了几何美感。

和谐与冲突

根据出版物的不同特性，设计可以保守，也可以前卫。是版面元素相辅相成，达到视觉平衡，一片和谐，还是相互竞争让版面陷入混乱，产生冲突的张力，这些都靠设计师的直觉、训练和经验来判断。和谐与无序之间的拉锯战不仅仅是风格上的冲突，也是人类历史、思想和发展过程中两股不同力量之间的分歧——讲究古典与秩序的组织者和不讲根源、好动多变的浪漫派，两者的竞争与妥协制造出张力，推动着源源不断的创意产生。

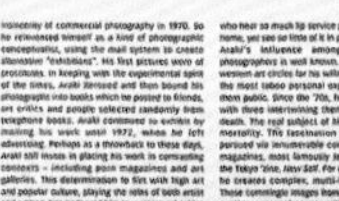

"IN JAPAN, APPARENTLY, WOMEN COMPETE TO HAVE ARAKI TAKE THEIR PHOTOGRAPHS. TO BE PHOTOGRAPHED AS A SEXUAL OBJECT IS CONSIDERED A LIBERATING EXPERIENCE IN OTHER CULTURES."

这两组跨页采用不同方法达到视觉平衡。最上图的*Flaunt*杂志使用单纯的对角图文编排；而上图的*Dazed & Confused*杂志，看起来很像电影分镜表。

达到和谐

刊物设计可通过几种方式创造和谐感。包豪斯与瑞士设计讲究纯粹性，认为和谐的设计应该具备下列特色：

- 均匀的灰色，没有多余的“花哨”的元素（例如夸张的首字大写），以免破坏版面的经典感；
- 平静、严谨的文字编排格线；
- 简洁明了的黑体字；
- 正文字号不能太大；
- 行距不能太宽，以免让留白空间与文稿产生冲突，利用细腻不抢眼的行距，强化整体的外观质感；
- 标题要和内文使用相同字体，只需将字号加粗或放大一些即可；
- 边界留白要够宽，清楚区分出文字方块，但不要宽到让人觉得突兀或浪费空间；
- 任何额外信息（例如栏外标题）也都要遵守上述大原则；
- 放置照片时，要确实在格线上跨一、二、三或四栏，并保持与其他元素齐平；
- 使用留白需谨慎，创造出可呼吸的平衡空间。

这样的版面设计整体感觉是规则和均衡的，没有不和

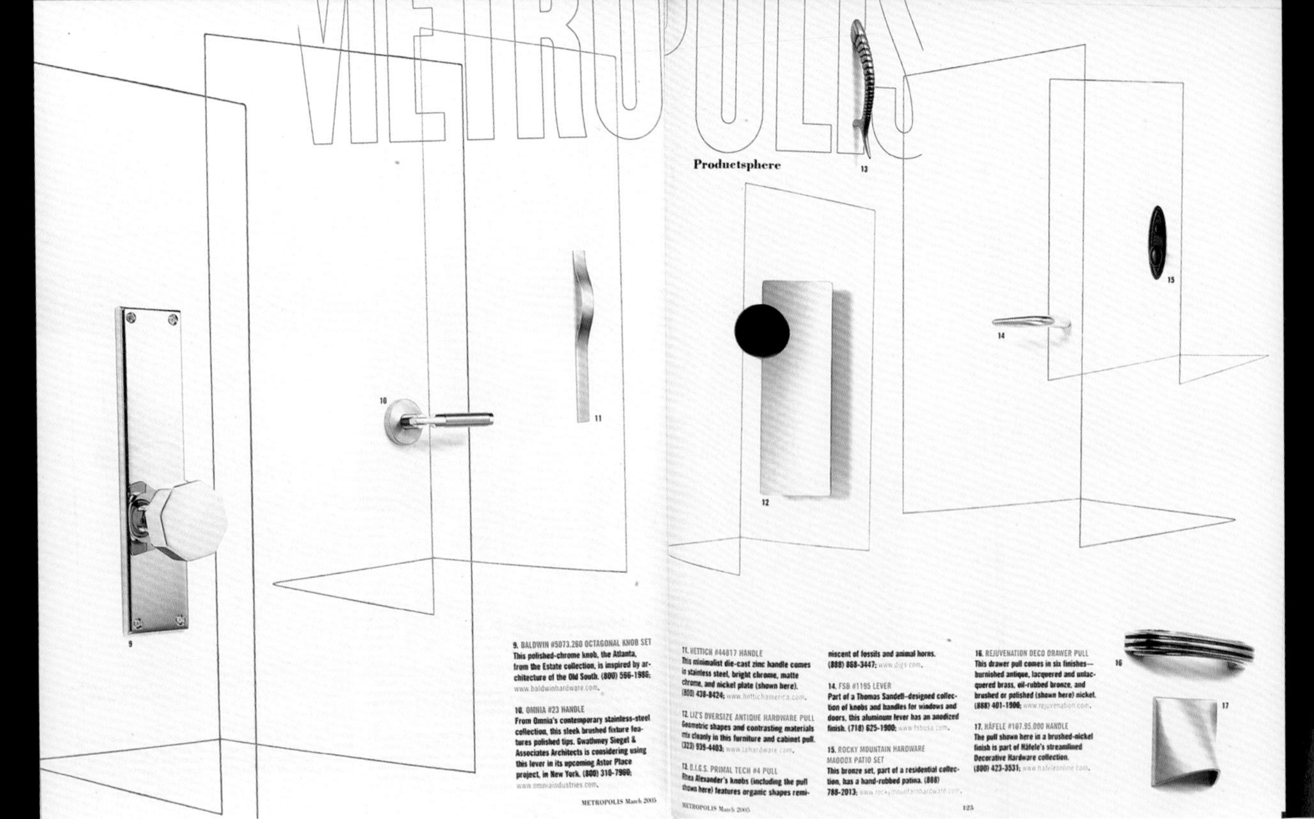

METROPOLIS

Productsphere

9. BALDWIN #5073.260 OCTAGONAL KNOB SET
This polished-chrome knob, the Atlanta, from the Estate collection, is inspired by architecture of the Old South. (800) 566-1986; www.baldwinhardware.com.

10. OMNIA #23 HANDLE
From Omnia's contemporary stainless-steel collection, this sleek brushed fixture features polished tips. Gwathmey Siegel & Associates Architects is considering using this lever in its upcoming Astor Place project, in New York. (800) 310-7960; www.omniaindustries.com.

METROPOLIS March 2005

11. HETTICH #44817 HANDLE
This minimalist die-cast zinc handle comes in stainless steel, bright chrome, matte chrome, and nickel plate (shown here). (800) 438-8424; www.hettichamerica.com.

12. LIZ'S OVERSIZE ANTIQUE HARDWARE PULL
Geometric shapes and contrasting materials mix cleanly in this furniture and cabinet pull. (323) 939-4403; www.lahardware.com.

13. D.I.G.S. PRIMAL TECH #4 PULL
Rhea Alexander's knobs (including the pull shown here) features organic shapes reminiscent of fossils and animal horns. (888) 868-3447; www.digs.com.

14. FSB #1195 LEVER
Part of a Thomas Sandell–designed collection of knobs and handles for windows and doors, this aluminum lever has an anodized finish. (718) 625-1900; www.fsbusa.com.

15. ROCKY MOUNTAIN HARDWARE MADDOX PATIO SET
This bronze set, part of a residential collection, has a hand-rubbed patina. (888) 788-2013; www.rockymountainhardware.com.

16. REJUVENATION DECO DRAWER PULL
This drawer pull comes in six finishes—burnished antique, lacquered and unlacquered brass, oil-rubbed bronze, and brushed or polished (shown here) nickel. (888) 401-1900; www.rejuvenation.com.

17. HÄFELE #107.95.000 HANDLE
The pull shown here in a brushed-nickel finish is part of Häfele's streamlined Decorative Hardware collection. (800) 423-3531; www.hafeleonline.com.

METROPOLIS March 2005 125

上图是选自《大都会》（*Metropolis*）的例子，这是一篇讨论门把手的专题，拉滨以巧妙的构图让画面看起来立体生动。

谐的元素，很适合应用在书籍和目录设计中。但是报刊读者都知道，杂志与报纸很少遵守这种风格，因为报刊需要以不同方式来呈现内容，才能突出层次感，增加视觉刺激。为达到这个目的，刊物会在和谐中加入一些冲突元素，并将二者融合为一体。

和谐与冲突的融合

现代读者已习惯电视、网络和纸质刊物的强烈视觉风格，因此大部分刊物都不用顾忌冲突元素，而是将和谐与冲突融合在一起。可以通过使用共同的元素来打造和谐感，例如连续的标题、独特的跨页、采用包豪斯原则的内容文本，变换字体或使用不平衡的形状，也可以通过图像化的文字来展现。文字可用手写、草书、剪下或运用常见的手法加以风格化。文字框可以斜放、以奇特的造型排列或重叠。首字大写长度可达半个页面。漂亮的字体可以故意设计得不易辨认，例如把美丽的字体印成黄色，盖在照片中模特儿穿的黄色衣服上。然而无论怎么处理，都不能杂乱无章，而是要让人眼睛一亮，感到刺激、有趣、自由，这样才有意义，才能建立刊物的辨识度，脱颖而出。

打破和谐

如果刊物或品牌想做点与众不同的事，提倡一种另类的生活方式、提出激进的政治议题，或不苟同时代的精神与文化时，编辑设计师常常会故意打破画面和谐。这类杂志的优秀设计师能巧妙、创造性地利用设计手法，展现新颖或另类的设计。内维尔·布罗迪（Neville Brody）的*The Face*、大卫·卡森（David Carson）的*RayGun*和马丁·威尼茨基（Martin Venezky）的*Speak*都是绝佳的例子，这些刊物以颠覆传统或出乎读者意料的排版方式，结合刊物的风格与

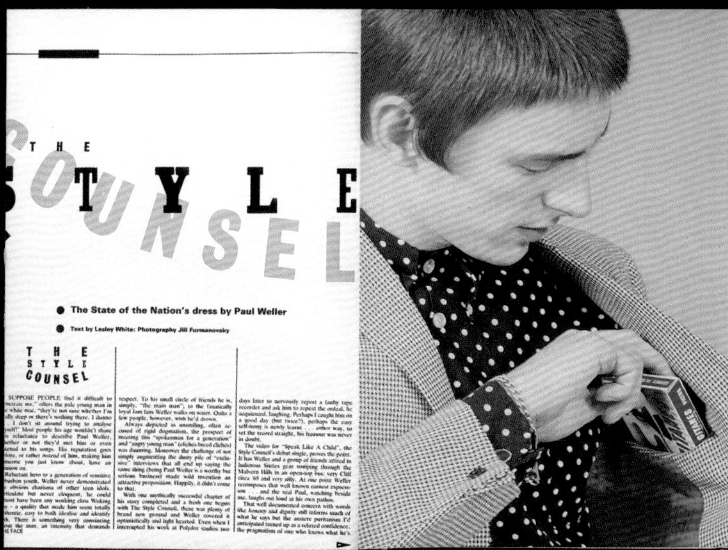
THE STYLE COUNSEL

● The State of the Nation's dress by Paul Weller

● Text by Lesley White: Photography Jill Furmanovsky

上图是内维尔·布罗迪（Neville Brody）设计的*The Face*，与下方*Speak*“碰撞与燃烧”（crash-and-burn）设计方法相比要平淡许多。但这两者都呈现反权威的设计风格，直接反映出设计师所处的大环境陷入了道德与政治的困境。这两种设计各以不同方式，利用相当有力的鲜明轮廓破坏画面和谐，但又不失可读性，很容易理解。

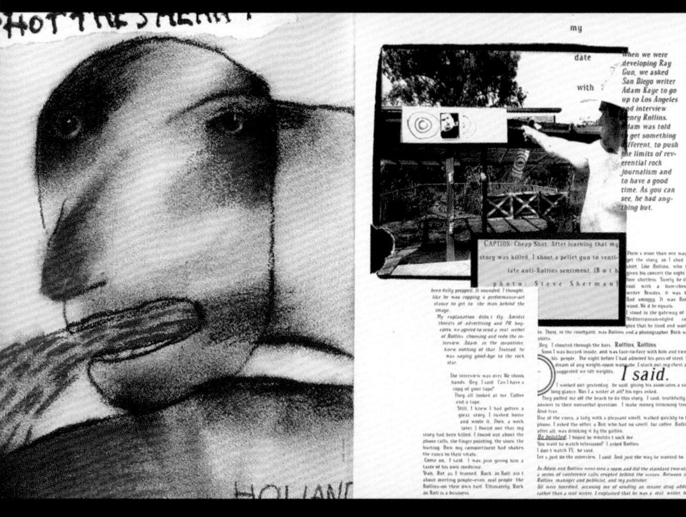
my date with

When we were developing Ray Gun, we asked San Diego writer Adam Kaye to go up to Los Angeles and interview Henry Rollins.

LET'S BLAST OUR

I said.

内容，把激进的信息与火爆的设计全部融为一体。不过，要月复一月制作这样新颖、激进的设计，设计师很容易被“掏空”。另外，激进的设计杂志在获得广泛赞誉和商业上的成功之后，也会渐渐失去前卫色彩。设计师需要随时留意文化变迁以及自己的刊物在文化中扮演的角色才能保持住自己的风格。值得注意的是，许多实验性的前卫杂志（包括上述三个知名杂志）要么昙花一现，要么已经在市场上失去优势（因为市场稀释了这些杂志的风格特色），但这些刊物对于平面设计的影响和作用是毋庸置疑的。

小结

和谐的设计或许仍会是常态，毕竟从商业角度来看是比较安全的做法，也符合企业心态。这样的设计容易搭配广告，也可把标准模板交给初级设计师，指令清晰易懂，不会出什么大问题。和谐的设计接受度也较高，看起来赏心悦目，不太会受到严重质疑。虽然新奇的版面设计更容易赢得荣誉和奖项，但只要是创新，就难免带有实验色彩，就算想办法淡化风险，发行量大或主流的刊物仍不一定能够接受。

风格——什么是风格？如何建立与传达风格？

对很多设计师来说，风格很难用语言描述，他们会说风格是一种“本能”，是靠“直觉”。虽然风格没有确定的规则可循，但是可以在不断接触各类视觉媒体的过程中，通过识别、鉴赏与训练，学习把握风格和创造风格的技巧。这些知识和技巧，可以帮助设计师为任何特定的出版物创造合适的风格。

如果你很喜欢看某份刊物，无论是纸质刊物还是电子刊物，想想看，最先吸引你的是什么？是颜色、图片还是封面文案？当你拿起一本纸质刊物时，是否会感觉到它的重量？纸张质感怎么样？你会注意到其他东西吗（比如装帧）？有哪里做得特别好？为什么能够脱颖而出？如果封面进行特别色的烫印，对你来说意味着什么？风格可以为读者营造一种阅读氛围和基调，读者可从风格猜测刊物的内容与调性。一旦熟悉刊物给的线索，刊物就会像老朋友一样，读者可以预料到会看见什么，并期待未来再度相遇，享受和这位“朋友”共度的时光。刊物的风格可大致分为三类：编辑风格、设计风格与广告风格。

www.folha.com.br/copa2006
copa 2006
FOLHA DE S.PAULO
INCLUI ESPORTE
FIM
DA ERA CAFU, QUE JOGOU QUATRO COPAS, ERGUEU DOIS TÍTULOS E SOFREU DUAS DECEPÇÕES
O noticiário dos quatro melhores times do mundo rumo à semifinal >> www.folha.com.br/copa2006

BRASIL 0 x 1 FRANÇA

sem mágica
sem tática
sem fôlego
sem craque
sem time
sem raça
sem hexa

sem desculpa

Megafavorita ao título no início da Copa, seleção tem atuação desastrosa, perde para a França pela terceira vez num Mundial, desperdiça chance única de revanche e é eliminada nas quartas-de-final

SELEÇÃO NOTA 3,9	
Dida	6,6
Cafu	3,3
(Cicinho)	4,3
Lúcio	6
Juan	6
Roberto Carlos	3
Gilberto Silva	4
Juninho	3
(Adriano)	2,6
Zé Roberto	4
Kaká	3
(Robinho)	2,6
Ronaldinho	2,6
Ronaldo	3,6

COM SCOLARI Seleção portuguesa bate Inglaterra nos pênaltis e iguala melhor campanha; para técnico, ir às semis com Portugal foi mais difícil do que ser penta

eleições 2006
FOLHA DE S.PAULO

Lula e Alckmin vão disputar o 2º turno

Com mais de 99% das urnas apuradas, o petista obteve 48,65% dos votos válidos, contra 41,58% do tucano

SENADO OPOSIÇÃO E GOVERNO SE EQUILIBRAM

CÂMARA ELEITORES PUNEM MENSALEIROS E SANGUESSUGAS

48,65%

41,58%

>> ACOMPANHE A APURAÇÃO NA FOLHA ONLINE www.folha.com.br/eleicoes2006

copa 2006
FOLHA DE S.PAULO
INCLUI ESPORTE
FIM
BRASIL 0 x 1 FRANÇA
sem mágica
sem tática
sem fôlego
sem craque
sem time
sem raça
sem hexa
sem desculpa
COM SCOLARI

编辑风格

编辑风格由页面的安排与流动方式、文章的视觉表现和调性以及文章的数量和种类组成。大多数刊物都有自己的框架以及这样安排的原因，例如专题报道和访谈的篇幅比其他报道内容要长，并且希望读者能够将故事和内容一口气读完。而其他页面例如评论、新闻或活动列表类的页面，则可以在短时间内略读或者随机阅读。这种处理方式可以满足读者对于刊物风格一致性的预期。阅读风格多由编辑设定，设计师则需要确保通过设计，能够将编辑的风格清楚地传达给读者。

设计风格

设计风格是指所有视觉元素的呈现方式，通过创造性的设计，实现图文之间的绝妙平衡。杂志的设计风格和品牌密切相关，可以从以下几个方面来理解：开本（尺寸与形状）、用纸、结构和设计元素。

开本：决定纸质刊物的开本或大小时，通常要考虑几个因素。开本、形状与页数会受到印刷机及纸张尺寸的影响，还需要考虑寄送时所用的信封尺寸。如果是常见的开本，就可放在一般的新闻报摊架上展示。当然，有些杂志会选择打破常规。英国的设计杂志《创意评论》（*Creative Review*）与*M-real*就是正方形的，以传达特殊主张。而*Statements*、*soDA*与*Visionaire*杂志则每一期都会改变开本。这些特殊开本主要是出于美学考虑，但在决定开本时，除了美观之外，还要考虑实际的成本，以及其功能和内容。高档连锁饭店适合精美的超大开本刊物，以传达出奢华感，但需要随身携带的杂志（例如活动列表杂志），采用较小的开本反而比较方便。

用纸：纸张也会影响刊物的风格与功能，刊物质感对读者的吸引力不可小觑。用新闻纸印刷的出版物，直观感受上比印刷在铜版纸上的时尚杂志更环保，因此更适合用来传达特定品牌信息以及面向特定的读者群体。以杂志和书籍为例，质感是通过纸张、重量、装帧和最终外观来传达的。例如*soDA*（封面会用金属光泽与彩色塑胶）、*Matador*或*Statements*等制作精美的刊物的触感，显然和八卦周刊*Hello*或新闻周刊《经济学人》截然不同。

结构：读者在阅读杂志时，很少一页一页从头读到尾，然而传统刊物的内容节奏与结构却是建立在这种假设上的：认为读者不是先阅读最感兴趣的专题和文章，而是从头开始看起。但是，设计师可以尝试满足不同阅读节奏和顺序的结构。固定读者会很快熟悉新结构，保持设计的一致性，加上良好的浏览导航，有助于吸引新的或偶尔阅读的读者。《大都会》的前创意总监克里斯韦尔·拉滨针对好的内容结构说：“专题报道区的设计与文章长度有变化，可改变阅读节奏，让读者保持兴趣。读者会想要东翻西翻，看看接下来有什么内容。”

设计元素：版面设计是由很多元素构成的，包括视觉图案、字体、色彩、边栏与图形元素。如何运用元素以及各个元素如何整合，会形成刊物风格，塑造出某种氛围。字体有成千上万种，设计师的选择及运用能创造出永恒或只流行一时的字体。*Flaunt*杂志的李·科尔宾（Lee Corbin）解释字体可以有多种使用方式：

> “每期月刊都有许多信息需要整理。我们会用文字编排来整理信息，以方便读者浏览，也用文字编排制作美观的构图，为杂志建立特色。每一期*Flaunt*选用的字体不仅要容易阅读，还要有其特色。我们会寻找适合这期刊物的字体，有时会根据某一篇报道的内容而定，有时则视整本刊物的状况而定。如果某一篇报道的文字编排与当期的其他部分不一样也没关系。这时，我们通常会让这篇报道的风格带领我们走向正确的方向。”

同样地，插图或照片会迅速向读者传递一些信息，但是图片经过缩放、裁剪与位置安排之后，能带来截然不同的感受。如何运用这些设计元素并没有固定准则，只要整合后，整体刊物或个别内容都能传达刊物的特性即可。这一点对报纸与杂志都适用。马里奥·

of parity among competitors and causes the contests to run more like high-speed parades than races. The Shanghai Circuit promises to have the perfect "formula" to satisfy both drivers and spectators. Its straights are long and wide enough for acceleration passes and its wide corners are ideal for overtaking manoeuvres under braking.

In its current configuration (phase 1), the track can accommodate 120,000 spectators. If demand for tickets is higher, an additional 80,000 seats (phase 2) will be added, giving the Shanghai circuit a 200,000 seat maximum capacity.

Even if it expands to that extreme, most spectators should still be able to enjoy the action properly. One of the most impressive aspects of the Shanghai Circuit is the track's visibility. The view from the SIC's main grandstand, unlike most other circuits, is *not* like watching a tennis match. Over 80 per cent of the circuit can be seen by the 29,500 spectators from the main grand stand, while the remaining portion of the circuit can be viewed on large video screens. In addition, there are hairpin curves that are sandwiched between grandstands. Here spectators on either side can see the cars overtake under braking and accelerate away. While this Formula One is so far unfolding as predictably as the last did, the Shanghai International Circuit is something to get excited about even if, come 26 September, the cars on it are not.

CLARENCE CORNWELL

DELHI

Diwali: Baby, Don't Light My Fire

They call it the Hindu Christmas. And like the Christian version, it is a combination of holiness and hellishness. The holiness: every year, usually around October, the world's Hindus hold the festival of Diwali. **A celebration of good over evil and light over darkness, and maybe of Lord Ram's return from exile** (the legends differ around India), the five days of the Festival of Lights – Diwali means "rows of lighted lamps" – are marked with the lighting of clay lamps, the buying of new clothes (pointless consumption isn't limited to Christians) and the setting off of firecrackers. All very jolly, except for the gunpowder. That's where the hellishness comes in. Last Diwali, as at most Diwalis, there were 200 recorded fires around Delhi, and millions of sore ears and gasping lungs.

So many firecrackers are set off during Diwali that a noxious cloud covers Delhi for about ten days. Bad news for environmentalists, and even worse for asthmatics. Particularly for Delhi residents: the number of firecrackers let off, and increased traffic at festival times, increases Delhi's pollution by six to ten times. Delhi's main hospital casualty department sees 30 per cent more asthma patients around Diwali. Fireworks give off carbon monoxide, sulphur dioxide and nitrogen oxide, and unburnt hydrogen. All of which adds up to carcinogenic smog.

There is an anti-bang bandwagon rolling, though. In 2000, India's Supreme Court decreed a maximum decibel for firecrackers: they must be cracked under 125 decibels, and only between the hours of 6-10pm. The Delhi High Court went further, decreeing that noise levels of the crackers must be printed on the wrapper. All well and good, if this was a regulated industry. But it's not: Diwali firecracker manufacturers, like the fly-by-night Guy Fawkesers in the UK – who pop up in shops you never noticed were there – are shadowy and unregulated. Most of the manufacturing work is done by children – 50,000, say campaigners, as young as ten, who have to constantly breathe in gunpowder.

The firecracker industry is dependent on children in other ways too: they're their biggest clients. But even the kids are turning. A government campaign – Just Say No To Crackers – has been going for several years. Anti-cracker campaigners want the bangs to be replaced by the more traditional lighting of clay oil lamps and candles, or by organised firework displays. If such displays were good enough for the great Mughal emperor Akbar, whose Diwali firework extravaganzas in Agra could be seen for miles, they're good enough for Delhi. The city's schoolchildren are beginning to think so: last year, children from 50 schools formed a two-kilometre long human chain and shouted: "Life is important! Say no to crackers!" "We want an eco-friendly Diwali," said one shouter. "One that doesn't harm the environment and lets people enjoy its spirit. If young students can understand that, why can't adults?"

ROSE GEORGE

SYDNEY

Fox Studios: Lightsabers! Camera! Action!

A few years ago in a country far, far away, the Fox Empire revealed plans for a vast secret weapon. It would be a complex unlike any other, with film stages, workshops, production studios, cinemas, restaurants, live venues, bowling lanes, mini golf courses and bungy trampolining for the masses. But, more than all this, it would feature a backlot trip around the set of *Babe 2: Pig In The City*.

And thus was born Fox Studios, on a 32-acre site once owned by the Royal Agricultural Society in Sydney (to where Virgin Atlantic begins a regular service from December this year). Sure enough, some of the biggest movies of recent years took advantage of the eight purpose-built stages and 60-plus companies covering every aspect of the filming process from casting to special effects. It's no coincidence that Sydney lent its skyline to the *The Matrix* trilogy. Soon, *Moulin Rouge*, *Mission Impossible 2* and, yes, *Babe 2* had all been filmed on movieland's newest lot. But the *Babe* fans had someone far more frightening to fend off than the pork slicers. News that Episodes II and III of the *Star Wars* prequels were to be shot at Sydney's Fox Studios ushered a whole new calibre of obsessive through the gates: people like Dave Hankin, who is struggling to keep a few secrets from his girlfriend's parents. There's the fact that he smokes. **Then there's the Yoda-shaped bong in his garage. And the tattoo on his shoulder – an exact replica of the symbol worn by his hero**, *Star Wars* bounty hunter Boba Fett. Hankin doesn't stop there – he's got the same symbol stenciled across the bonnet of his car, not to mention the habit of dressing up as his favourite character at every available opportunity.

Hankin is one of the stars of a very different Star Wars movie: *The PhanDom Menace*, now available on Special Edition DVD. Following fans in the build-up to the launch of the first *Star Wars* prequel, *The PhanDom Menace* has become a cult classic of its own. A documentary by the obsessed for the obsessed, the film was made by Craig F. Tonkin and Warwick Holt on a Betacam camera, and centres around Australian fanclub Star Walking Inc, where we witness grown men who have allowed a childhood passion to consume their adult lives. Despite mitigating pleas of "I haven't outgrown it… I've grown with it", there are plenty of tell-it-to-the-judge moments – such as Chris Brennan wondering exactly how he's going to break it to his wife that they need a bigger house to accommodate his $40,000 collection of memorabilia.

加西亚说：“编辑设计师要努力确保设计能强化内容，让内容更好懂。营造外观与感觉的技巧很多，但版面要努力扮演好辅助和配合内容的角色，这是不变的原则。”

广告风格

刊物通常得靠广告和软文来平衡出版成本。这样一来，广告商就有很大权力，甚至能影响刊物内容的页面安排与可用的跨页数量。广告商在购买广告时可能会要求刊登在特定跨页或某文章的对页。由于刊物的前三分之一最受广告商青睐，这部分能留给设计师的跨页常会减少。此外，右页的可见度较高，价格比左页贵，因此刊物或许得多卖些右页，导致整个单元多半放在左页。在这种情况下，设计师得更费心设计版面。如果设计师事先知道广告的主题，就能让内容版面看起来很不一样，将两者区分开来，也使跨页保持美观，不仅与广告呈现和谐的关系，也保持刊物的整体性。无论如何，设计师必须通过形状、对比与色调，让页面成为一个整体，也要让可能跨好几页的专题有强烈而连贯的特色。

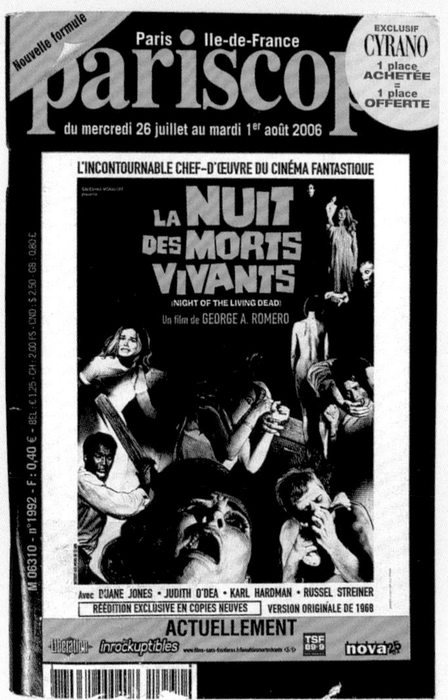

杰里米·莱斯利（Jeremy Leslie）为维珍航空头等舱杂志*Carlos*（左页）选择的用纸，出乎意料地呈现出高品质的奢华感，这可能与时代背景有关。莱斯利表示："二十年前，全彩杂志并不多见，那时*Carlos*如果用昂贵的全彩印刷，就会显得廉价普通。然而现在全彩是常态，物以稀为贵，与众不同的做法反而能脱颖而出。"瑞士杂志*soDA*（左上）每一期都改变开本与纸张，因为一年仅发行一次，因此才能有这种空间，无须顾及读者的阅读惯性，也没必要将开本固定下来。*Pariscope*（右上）是巴黎的活动资讯周刊，尺寸小到可放进包包或口袋中。该杂志持续盈利，每周能够卖出十万三千份。

西蒙·埃斯特森（Simon Estersd）*Eye*杂志

西蒙·埃斯特森的作品风格大胆、引人注目，并充满新闻活力。他为人低调谦逊，离开杂志行业后，在伦敦经营埃斯特森律师事务所（Esterson Associates）。1993年，他与德扬·苏季奇（Deyan Sudjic）以及彼得·默里（Peter Murray）共同开办大型建筑杂志《蓝图》（*Blueprint*），该杂志以刊头醒目的模板字体和厚页而为人熟知。埃斯特森在1995年从大卫·希尔曼手中接任《卫报》的设计工作，延续了《卫报》的大胆外观，呈现现代样貌的纸质刊物特色。埃斯特森设计事务所曾担任过许多报纸的改版工作，例如《新苏黎世报》，并与波特合作，设计葡萄牙里斯本的《公共报》。他还在意大利建筑杂志*Domus*担任创意总监，并担任伦敦泰德出版社（Tate Publishing）的设计顾问。

2008年，埃斯特森与编辑约翰·沃特斯（John LWalters）、出版商汉娜·泰森（Hannah Tyson）联手收购了*Eye*杂志，并携手让杂志起死回生，回归独立杂志的风骨，重新恢复生气。他们通过网站和社交媒体，培养大量粉丝群体，即使有些粉丝可能都没看过*Eye*的纸质版本，但却能受到他们的欢迎。

埃斯特森表示："网站和博客对*Eye*杂志的生存至关重要。网站等于是过去出版作品的档案库，如果上Google搜寻关于平面设计的信息，*Eye*杂志会出现在第一页。*Eye*每年只出版四期，但我们希望大家不要遗忘我们，所以博客很重要，我们不只是在做纸质杂志，编辑约翰·L.沃尔特斯（John L. Walters）也在用推特和Flickr与读者保持互动。并不是说只要把内容在尽可能多的平台上发布就行了，而是要充分利用好我们已经出版许久的纸质杂志资源，发挥自己的优势。"

eye
THE INTERNATIONAL REVIEW OF GRAPHIC DESIGN 81 AUTUMN 2011 DESIGNERS AND CLIENTS

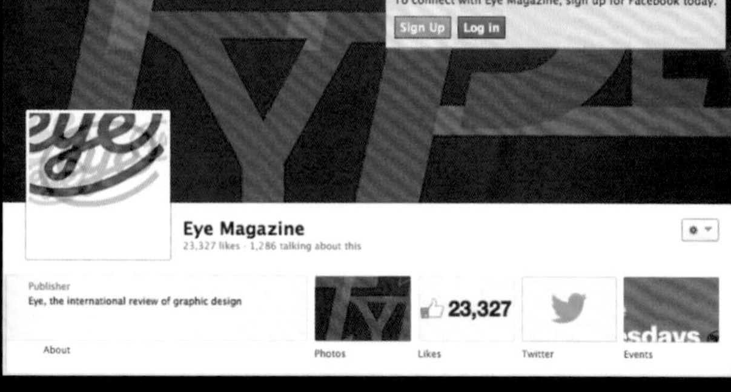

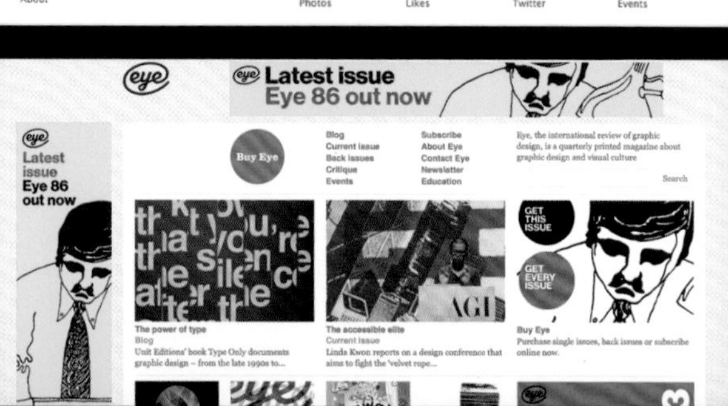

*Eye*的纸质版杂志应用简单的格线系统，设计相当优雅别致，能更有效地讲述故事。高品质的图片增加了品牌的权威性，在设计界口碑很好，也受到设计师们的认可和尊重。*Eye*的博客有名种多样的设计内容，是从业者和学生的丰富资源。2012年，博客有488621名关注者，也说明*Eye*可以通过推特，比单纯发行纸质刊物时接触到更多读者，毕竟纸质版的发行量是有限的。

埃斯特森对于设计的热情，让他在1983年设计了新创刊的建筑杂志《蓝图》（右页）。刊头使用的模板字体成为刊物的标志，十分抢眼。封面文案非常简洁，长度刚好和页面宽度一致。每期的封面不是以建筑物为主角，反而是建筑师的肖像。

制作版面

EUROPE'S LEADING MAGAZINE OF ARCHITECTURE AND DESIGN/DECEMBER-JANUARY 1986/NUMBER 23/£1.50

BLUEPRINT

CRISTIAN CIRICI: THE MAN WHO RE-BUILT MIES. PHOTOGRAPH BY DAVID BANKS

BORN AGAIN BARCELONA

PLUS MICHAEL GRAVES EXCLUSIVE AND DAN DARE VERSUS JUDGE DREDD

将灵感落实到版面设计

如何寻找灵感并落实到版面设计上，做有趣、活泼、有意义的版面，对设计师来说是时刻都要面对的挑战。设计师经常有灵感枯竭的时候，只能干坐在电脑前面盯着屏幕抓耳挠腮，这时候要怎么办呢?可以试试离开工作环境，去逛逛艺术画廊、街头市场，看看电影、逛逛街、去游乐园，或只是坐在公园里看着天际线，从不同的角度看待事物，也许会激发新的灵感。接下来要介绍一些有用的创意练习。想想看，还有哪些事情能启发设计师的灵感呢?

所有的创意工作者不免面对缺乏灵感的时刻，要打破僵局，也许可以尝试下面的办法。

建筑：建筑结构是很丰富的视觉灵感来源。很多建筑物拥有格线般的结构，转换到版面上就是美观实用的空间规划。美国知名设计师索尔·巴斯（Saul Bass）深谙此道，曾用这一概念设计了一些电影的片头，影片中，他让字幕优雅地滑过公共大楼窗户的透视线。

自然：蝴蝶翅膀、昆虫眼睛、鱼鳞与节肢动物骨骼的放大图，能在比例、形状、对比和结构方面给予设计师很大的启发。伦敦的水晶宫（Crystal Palace）造型就是仿照百合的叶脉结构设计而来。

工业设计产品：工业设计的产品，常常可以成为灵感源泉。产品是人造的，通常很容易换成文字与图像。时尚的远洋客轮、流线型的火车、肯·格兰奇（Ken Grange）线条优雅的派克钢笔以及乔纳森·艾夫（Jonathan Ive）的苹果电脑，都是扎实设计原则巧妙运用的成果。同样的原则也可以应用到自己的版面设计中。

“不要仅仅模仿其他平面设计师的做法。可以从各方面寻找灵感，例如艺术、电影、时尚或历史。”

——艾瑞克·罗奈斯泰德，*Flaunt* 艺术总监

看看你的周围：看看你的电脑桌面，再看看现实的桌面。无论它们在别人眼中看起来如何，但你应该很清楚桌上每样物品的位置，而这正是设计思维的核心。

你所要做的就是让别人看到你的基本结构。你收集的东西，这些东西展示的方式——所有这些都是设计。在某次演讲中，*Speak*设计师马丁·威尼茨基（Martin Venezky）表示，他的设计工作通常是从阅读文稿开始：

> “阅读之后，我会把事物的关系与意象列出来，趁着印象深刻，开始翻找筛选相对应的画面、书籍、字体等，找到那些直接或者间接打动我自己的部分。我故意不整理文件，才能随时发现惊喜。在寻找某种图像时，可能会有别的东西突然映入眼帘，反而令人兴奋，也更有出乎意料的惊喜。我认为，浅显意懂的说明性图片与更另类的诠释之间存在着‘诗意的落差’，是很好的灵感来源。”

离开办公桌吧!脑袋在漫游的时候最容易产生灵感。你的潜意识中已藏着许多图片与概念，重点是如何打开并运用。玩游戏、看看窗外、散步时随时携带小的设计笔记本。有时到附近的公园随手画张草图，也有助于创作出活泼有力的独特版面、标题、Logo或页面设计。

"这期的主题是经典的'来到纽约'的故事。我们花了一年时间与每个主角相处，从他们抵达纽约开始，一起度过在这个城市的第一年。封面必须传达出既期待又怕受伤害的感觉。我们的灵感来源是辛迪·谢尔曼（Cindy Sherman）的无题剧照，照片中一个徒步旅行者站在路边，身边放了两个手提箱。我们到处寻找地点，最后来到中央公园的大草坪，这里的建筑物天际线显得历久弥新，而草地无边无际地延伸，仿佛开启种种可能与焦虑。我们最后选择拍摄黑白照片，因为这样很能代表纽约精神，标题也用很有纽约特色的黄色。黑白与代表计程车的黄色，很简单。"

——珍妮特·佛罗里克，《纽约时报》杂志前艺术总监

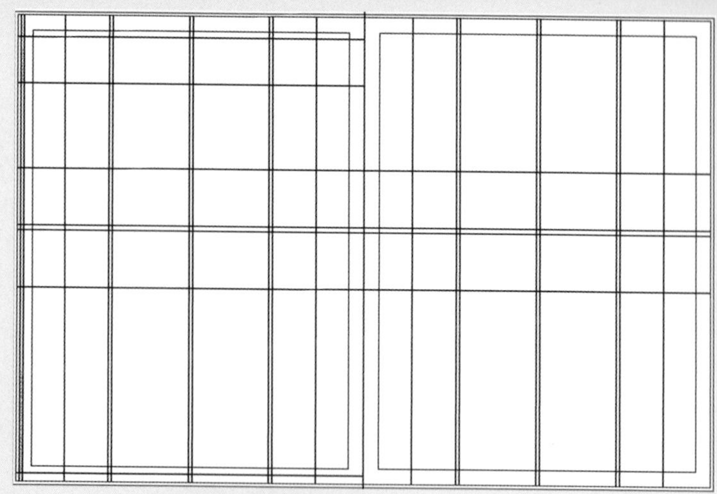

工作坊4

跨页

目标

为杂志设计几份跨页的专题版面。

操作步骤

在你设计的杂志版面上加入以下元素：大标题、主图与其他图像，介绍性文字导语（大约30字）与一些内文。可用InDesign软件里自带的假字，也可自己输入一段文字，然后多次复制生成正文。

参考本书110页到111页给出的范例，自行绘制格线。可以用InDesign，也可使用其他类似的“桌面出版（DTP）”软件。

创建文字内容。先撰写大标题和导语，加上署名行，也写下图片说明，并加上小小的文字说明图片来源。

图片来源。如果是使用自己拍摄的照片，就把摄影者写成自己。如果是展示朋友的作品，则用图片说明来标注对方姓名。如果是用网络上的图片，要确认是否为免费资源。即使是练习用的版面设计，也要尊重别人的劳动成果。至于字体，可以直接使用工作坊3的内文字体设定。

把第一份设计好的版面存档起来，再另外设计一个版本，直到设计的页面和谐、平衡并具有戏剧张力。把各个元素用不同比例放大或缩小，做出活泼的跨页设计。把你设计的版面印出来，才能看出各元素的实际大小。请按照实际尺寸裁剪页面，并在背面用胶带把页面拼贴起来审视。

做了几次并且达到满意效果后，就用彩色打印机打印出来并裁剪好，放到自己的作品集中。

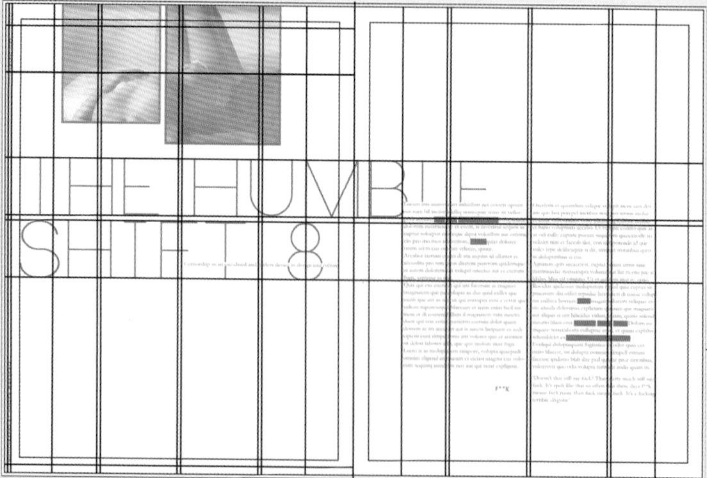

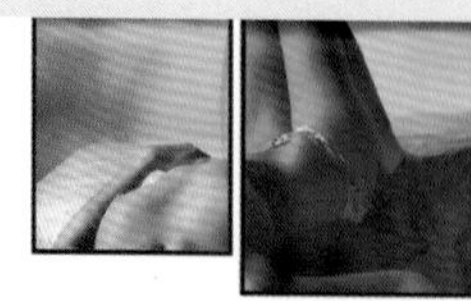

GRIDS ARE OUR FRIENDS
DISCUSSING THE IMPORTANCE OF ANALOGUE TECHNIQUES WITH LUCA IIRITI
"WRITE DOWN OR SKETCH EVERYTHING COMES ACROSS YOUR MIND"

What techniques do you use to come up with your ideas?

Usually my initial approach with every project starts with an accurate research of the subject which, I'm dealing with. After the documentation part I usually sketch for a few days without even touching the computer. I believe that you keep your ideas fresh by putting them down with pencil rather than going straight on the screen.

A foundamental part of idea generation is having breaks in between, and with break I don't mean like a tea break but a proper few hours or a whole day off, going out seeing friends, exhibitions, going to the park etc... I always feel like by doing this I get back on the project with a much more fresh mind and I can spot easily mistakes which I might haven't seen before, or come up with new ideas perhaps.

Who inspires you?

As I'm from Switzerland most of the designers who inspire me are Swiss like the historical Josef-Muller Brockmann, Herbert Bayer, Max Huber... As I lived in Lugano I used to see the work of Bruno Monguzzi everyday on the streets, and had the pleasure of meeting him a few times in switzerland and here in London, his mentality and way of thinking always amazes me.

What's the most satisfying part about being a designer?

Personally I think that getting to an end point where you can finally touch with your own hands all the effort and thinking that you've put into a project is one of the most satisfying things of being a graphic designer. And obviously seeing that the people is interested in what you did and that your answers to the design problems are working.

What motivated you to become a designer?

I've always been interested in art and design since I was a kid, then at the age of 14 I've applied to the CSIA (an art school in Lugano) and I've discovered this world and instantly fell in love.

I've chosen to be a graphic designer cause I like to think that I can communicate determinate ideas to a wide range of people and have different reactions, I like the interactive side of it and that gives you the possibility to work with a wide range of different clients and material.

From start to finish, how long does it normally take you to complete a project? It's really difficult to say, as it really depends on the client needs... Generally I face each project in phases (as mentioned before) which are documentation and research, sketching and ideas, mock ups and overview of the contents, design, outcome.

I've been working on projects for a week or 6 months, it's just a matter of how much work you have to do and how important it is for the client and the community.

Do you prefer to work independently or with a group of people?

Again it depends on the project, I personally prefer to work independently if I have the possibility and the time. But I love working in teams if the people shares knowledge and interests similar to mine.

What kind of advice would you give to someone who is thinking about becoming a graphic designer?

Do research and look to the past of our history, it's always good to look back where designers weren't surrounded by technology but use to come up with amazing outcomes (which are rarely seen these days), it always inspires me to think of new methods to realise my works. Write down or sketch everything comes across your mind, everything is important or could lead to another idea.

If you love art, design, ink, paper, communication and people then welcome to the club, and grids are your friends!

What is the most challenging aspect of the job?

I think that coming up with a good idea which is as original as understandable by a large audience is one of challenging aspects of being a graphic designer. Personally I always like to look into the cheapest way of realising what the client needs with still having an appealing "facade", the lower the budget is the more challenging the project will be. Also I always think of how to automise the design, and with that I mean creating a design, a system which allows me to position and build the layouts throughout time easily, if you spend lots of time thinking on how to make things in the easiest and cleanest way then the design will help you in the designing part.

What is you're most valued possession when it comes to design?

My pencil.

How long have you been a graphic designer?

I don't know even if I've ever been a graphic designer to be honest, I've started studying graphic design at 14 but I don't really feel part of the "graphic designers club", I prefer to see it as another way of express and communicate.

左页下图是约瑟夫·马歇尔（Joseph Marshall）为大学杂志制作的比稿作品。栏位的垂直线靠着水平轴来平衡——这是平衡版面元素的常见手法。把文字拿掉后，我们可以看到能弹性运用的四栏格线系统（左页上）。这个版面是小报尺寸，比A4大，但比A3小。马歇尔想要通过留白空间，给读者带来强烈对比的惊喜。

上图是简单的三栏格线版面。小报版式让设计师有许多留白空间可用。这套格线很基础，和马歇尔复杂的模板大不相同。这份版面是由斯特凡·亚伯拉罕（Stefan Abrahams）、茉莉·琼斯（Jasmine Jones）、雷贝卡·达夫·史密斯（Rebeca Duff Smith）与哈德逊·什维利（Hudson Shively）这几名学生共同创作的，他们也利用这套格线系统制作了数字版本的版面设计（右图）：数字版本更适合采用简单的格线系统。同时，格线系统也要能反映文章内容的风格。

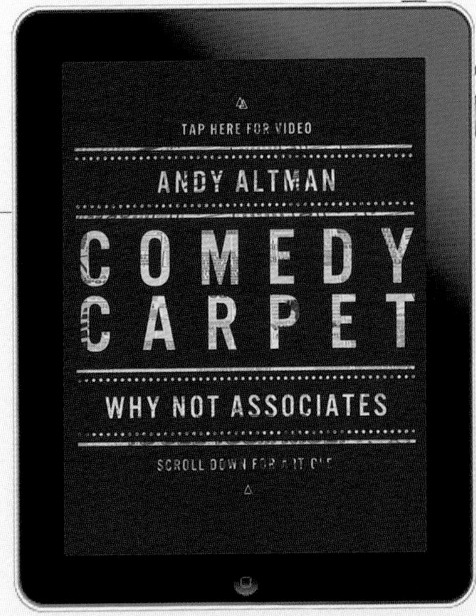

TAP HERE FOR VIDEO
ANDY ALTMAN
COMEDY CARPET
WHY NOT ASSOCIATES

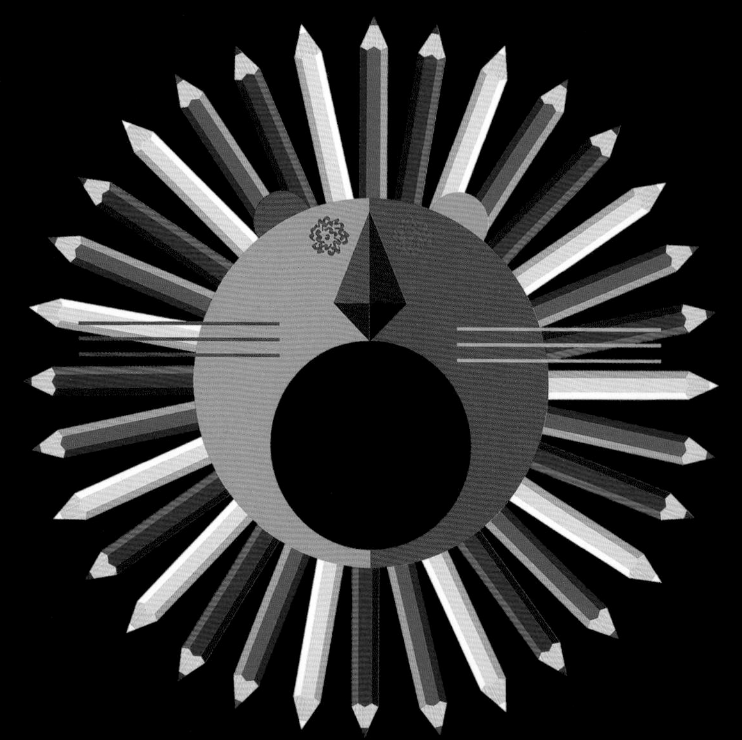

IN REAFFIRMING THE
greatness
OF OUR NATION,
WE UNDERSTAND THAT
GREATNESS IS NEVER A GIVEN.
IT MUST BE EARNED.

—PRESIDENT BARACK HUSSEIN OBAMA
INAUGURAL ADDRESS · JANUARY 20, 2009

第六章　编辑设计师的必备技能

编辑设计师要设计出优质的版面，除了应具备相关知识之外，也要有相关的工作经验及设计技能。前几章已讨论过一些必备的技能，例如要了解刊物的结构以及每个岗位在刊物制作环节所担任的角色等。本章将继续深入探讨编辑设计的知识和技能，尤其以下几项，不仅编辑设计师需要，作为设计总监更是必须具备：

- 将设计想法可视化的能力；
- 能够充分做好设计前的准备；
- 充分掌握文字编排逻辑并能执行；
- 与时俱进，能快速学习新软件，持续学习新知识和新技术；
- 知道如何设计出有整体感却又不乏味的版面；
- 把控好管理项目的时间与成本。

杰里米·莱斯利认为，一名编辑设计师要拥有以下特质：

· 能深刻理解并处理文章内容，即使你对内容本身并不感兴趣或者内容过于深奥；

· 可视化能力强，能和同事沟通想法；

· 深谙插图与摄影作品的开发与利用；

· 了解市场与竞争对手；

· 能发掘好想法，无论这想法是谁提出的（或者署谁的名字）；

· 能在截稿时间内尽可能发挥创意，平衡创意想法和截稿时间；

· 享受设计制作的过程，如果缺乏热情，就会体现在设计上；

· 在自己的设计领域里要有自信并勇于坚持自我，但也要学会听取其他岗位的意见。

将想法可视化

将想法可视化是一项很复杂的能力，要把所有素材加以筛选、淘汰、强调、安排与组合。编辑设计不仅要做好单独的版面设计，还要通过设计建构出刊物的整体视觉，并能随着时代、内容及读者类型改变而持续调整。设计师必须具备以下能力。

阅读文稿，理解并准确解读内容，提出精准契合内容的设计呈现方式，必要时可以与文字编辑、作者或摄影师讨论设计方向。

和文字编辑培养出默契，制作出其心目中契合文字内容表达的理想版面。善于沟通，能够利用草图和讨论会与编辑达成共识。

理解刊物的目的、品牌形象以及读者的心理和阅读行为。

一个版面进行多版本的设计，同样的内容能够提出不同的设计想法，并相互比较优劣，一旦发现最初的设计想法不可行，能够及时提供备选方案并进行讨论与完善。

尽量保持灵活，学会编辑设计基础知识后，就要开始检验、质疑、发挥、探索与打破规范。在寻找解决方案时要懂得变通，不必固守单一途径或方向。

在正式动手设计版面前，要发挥想象力，多方尝试各种版面的可能性。随身携带素描本，在纸上列出想法与粗略版面；画草图、图表并列出其他数据，想法十分完整之后，再用电脑做设计。

页面制作与网格线

要为一份好版面打下稳固的基础，得先选择版式与纸张、制作合适的网格线、知道如何用设计来表达内容优先级，并善于利用落版单。这么一来，即使临时状况层出不穷，不断修改，也能保证各单元与分页方式完整以及整体刊物的流畅度与节奏。

网格线

网格线就像建筑蓝图，是设计的无形指南和秩序系统，能够帮助设计师决定图片、文字与其他设计元素（例如留白空间、白边与页码）的位置和设计。网格线既能保持刊物的连续性，同时也让版面有变化的空

34

The Guardian | Monday September 12 2005

theguardian

Founded 1821
Number 49,456

Four years after 9/11

The war of unintended consequences

Corrections and clarifications

Japan's general election

A mandate and a monopoly

Country diary

Claxton, Norfolk

In praise of... The Proms

September 12 1914

The sure way to end the war

The Guardian | Xxxxxday Month XX 2005

National

在正式的网格线系统上，大块的文字会产生“浮动”的感觉，例如《卫报》改版后的样子。

罗杰·布莱克（ Roger Black）的十条设计准则

（但要记住，学会规则就是为了打破规则）

1.每页都要有内容。编辑设计不是简单的装饰，而是要传达信息或呈现契合的视觉效果，所有设计都是建立在内容基础上的。要知道，读者不会一字一句地阅读所有内容，一般的阅读都是跳读，所以一定要确保内容足够充实。

2、3、4. 最重要的颜色是白色，其次是黑色，第三是红色。早在500多年前，书法家与印刷家就认识到了这一点，而经时间验证，他们的看法完全正确。背景用白色、文字用黑色，要凸显重点则用红色，更引人注目。这三种颜色最好，使用其他颜色时则需小心谨慎。

5. 不要盲目跟随流行趋势。

6. 别用一大堆黑体字，会让人产生视觉疲劳。

7. 封面要设计得像海报。封面用单一人像，比用一大堆图或全是字的封面更畅销。

8. 整本刊物只要使用一两种字体。意大利设计师深谙此道，简单的元素通过搭配反而能够制造有强烈感受的页面。避免使用太多字体与颜色，会显得杂乱无章。

9. 尽量放大每个元素。大字号的文字看起来更好看，不好看的图在放大后也会更美观。

10. 要有料!多数设计往往缺乏惊喜。若想引人注意，就要呈现出有变化的节奏。如果图片、标题、文字、广告或者标题总是一成不变，读起来就会枯燥无味。

罗杰·布莱克的设计作品包括《滚石》《纽约时报》《新闻周刊》、*Mc-Call' s*、《读者文摘》《君子》和《国家询问报》《时尚先生》与《国家询问者》等杂志。

间。良好的网格线系统会设定好框架，但又不拘泥于事先做好的设定。在设计流动性高的刊物时，网格线也可以当作固定点或参考点。有了网格线，文字、图片、留白区域的大小与形状会更容易进行具体规划，加快版面的制作流程。三栏式的网格线设计变化较少，如果设计成更复杂的九个或十二个单元，会有很大的弹性，使用起来有很多变化。网格线本身虽然一成不变，但不妨大胆些，以自己的方式利用网格线，做出独树一帜的版面。

不同类型与出版周期的出版物，使用网格线的习惯也不同。先了解这些惯例，才能在需要的时候打破惯例。举例来说，周刊或日报的制作周期短，设计需要简单快速，所以一般采用严谨固定的网格结构。相对的，季刊的制作时间很充裕，设计师有时间去尝试不同的网格线和字体系统，增加版面的变化。费尔南多设计的*Matador*杂志每年出版一期，所以每期都可以采用严谨但不同的网格线搭配一种特定字体。有些刊物不使用网格线，以一页为一个单位，并以图片或标题为基础，设计版面结构，如果技巧高明，可做出有流动性和弹性的页面与刊物。但也要小心谨慎，只有这种风格契合刊物的品牌特质与读者群时，才能使用。

制作网格线时要考虑易读性问题，“法塞特易读性定理（Fassett's theorem of legibility）”指出，每行应包含45个到65个字符数（字符包括字母、数字、标点符号与空格），超过这个长度会降低易读性。并不是说每行不能用40个或75个字符，但是需要仔细想想为什么要这样做。要想让读者阅读时感觉清晰舒适，在设计网格线时就要考虑到易读性。以报纸为例，小报版式或柏林版式最适合采用五栏网格线系统。在跨字段的标题下，文章长度超过7.5厘米时，一般就需要用副标题或图片来分段。也有一些风格惯例可以参考，例如杂志专题报出于易读性考虑，通常使用三栏，但文艺杂志往往只用两栏，营造出经典对称感，并且往往使用长段落来展现“文艺气息”；许多报纸的增刊也采用两栏，但会搭配留白空间与其他设计元素，让页面更轻盈。

许多一年只出版一两期的刊物或系列出版物，并没有固定的网格线系统，而是靠每期的内容与设计概念创造连贯性。*soDA*（左图）就是一例，格式、设计及隐藏网格线（如果有的话）要根据当期杂志主题来决定。左下的《大都会》则在整本刊物中使用二、三或四栏网格线，前艺术总监拉滨表示："我们的版面上有一条网格线规范图与文字的最高点，因此上方总是有很大的留白空间。《大都会》的Logo随着翻页会在页面顶部像是进度条一样移动，以此来显示阅读进展。这个惯例从1999年薛·博兰（Paula Scher）改版之后沿用至今。专题报道会印在无涂布的纸张上，让读者通过视觉与触感，发现这是杂志的特殊单元。这种做法就是从内容出发进行设计，更有实验性。每篇专题报道根据报道内容与搭配视觉素材的质量进行个性化设计。多数报道是采用相同的二栏或三栏网格线，但如果某篇文章更适合用别的方式设计，我们也会毫不犹豫地舍弃网格线。"

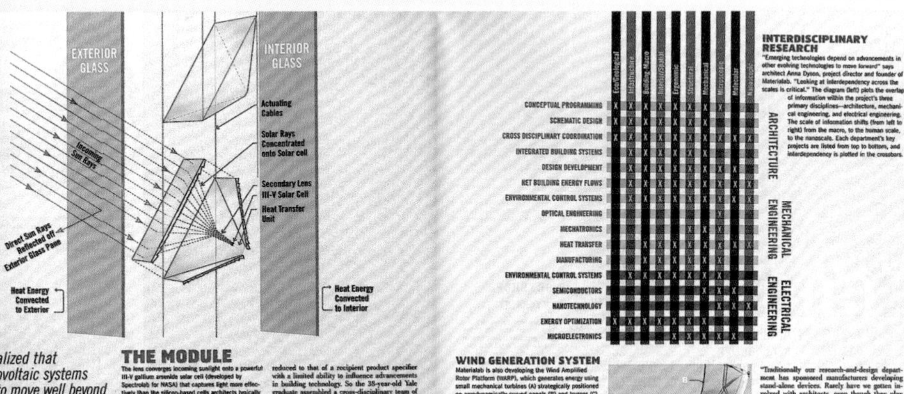

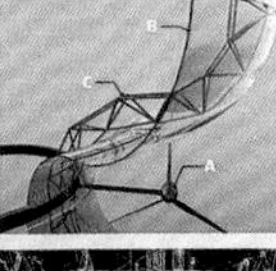

68 IN PERSPECTIVE
Interview with Giulio Cappellini

69

Interview by Simon Horauf
Images courtesy of Cappellini

Giulio Cappellini is one of the world's most influential furniture manufacturers. The list of designers that have been commissioned by Cappellini reads like the who's who of design of the past few decades. Simon Horauf managed to steal an hour of Cappellini's time during his first visit to Australia.

Giulio, how is the sky in Melbourne different to the sky in Milan?
(Laughs) The colour of the light is totally different. The sky in Milan is warmer and a paler blue. This is my first time in Australia and I am very impressed with the light here and the colour of the sky. The colour of the light is very important for us. Depending on the colour outside, you have a totally different perception of colours and textures inside. So, the same red will look different in Australia to what it looks like in Milan.

I personally like the really white, cold light of Australia. In Milan, the light is warmer, but I like the cold, clear light.

This might have nothing to do with the colour of the light, but tell me, why has Milan managed for so long to be the centre of design and manufacturing?
There were a lot of very good designers with big ideas in Italy, like Sottsass or Castiglioni, and on the other side you had the businesses that saw the potential of good design in the marketplace. There are a few companies that are truly interested in design and that invest in innovations and take risks. But this is really what you need to do and it doesn't matter if you are from Italy, Brazil or Australia. We had a lot of problems when we started working with foreign designers. Many people accused us of

上图选自*Inside*杂志，版面隐含着拘谨的网格线，但页面简洁，并不妨碍其他单元的设计与整体布局。这份刊物的前艺术总监杰弗里·多彻蒂表示：“简洁的网格线也可以进行多种组合，设计者必须把视野放开一点。读者喜欢对一份杂志有熟悉感，和杂志建立起关系，而保持一致的网格线能满足这项心理需求。如果网格线保持一致，忠实读者拿起任何一期，都会觉得浏览这一期时很熟悉、很自在。”

White Mischief

ONE MINUTE, SHE'S THE ARCHANGEL GABRIEL, AND THE NEXT, MARILYN MANSON. **TILDA SWINTON,** AS LYNN HIRSCHBERG DISCOVERS, IS A WOMAN OF EXTREMES.

Photographs by Raymond Meier

One of Tilda Swinton's ancestors on her very posh, very military Scottish family tree was painted by John Singer Sargent, and it is easy to imagine Swinton, with her alabaster skin, otherworldly green eyes and regal 5-foot-11 bearing, captured in oils. "I do look like all those old paintings," Swinton joked over a midsummer lunch of raw oysters at the Mercer hotel. "But I'm afraid my temperament does not conform. At all."

She said this, as she said nearly everything, with a mix of direct authority and engaged enthusiasm that was both immediately ingratiating and commanding. Swinton, who is 44, was wearing no trace of makeup, a print sundress and flip-flops, and her hair, which is naturally red, was dyed white-blond. "I love the roots," she said, as she tilted her scalp forward for inspection. "That's the best part of being this blond."

Her unique looks, her ease with herself and her voracious interest in the more esoteric worlds of cinema and style have made Swinton a kind of goddess of the avant-garde. In her movies, she has continually transformed herself — changing class, nationalities, gender. For "Orlando," perhaps her most famous film, she played multiple incarnations of the title character, including a man. In "Thumbsucker," opening in theaters on Sept. 16, she is utterly convincing as a suburban American mom. The director Jim Jarmusch cast her as an ex-girlfriend of Bill Murray's in the recent "Broken Flowers," in which she is terrifying, her face half-obscured by a foreboding curtain of long brown hair. For "The Chronicles of Narnia: The Lion, the Witch and the Wardrobe," a big-budget movie that is due out from Disney at the end of the year, Swinton

6

227

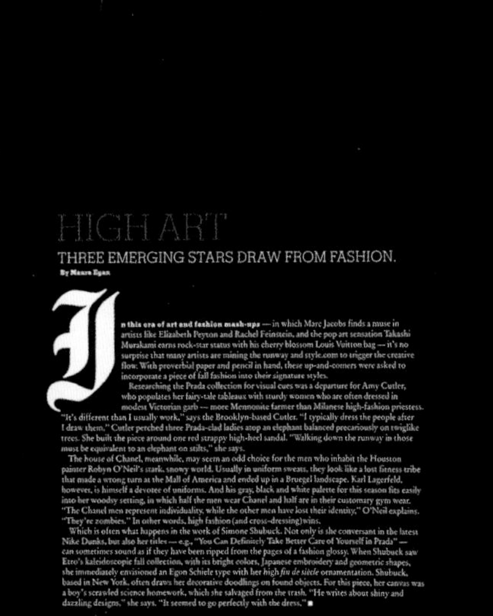

HIGH ART

THREE EMERGING STARS DRAW FROM FASHION.

By Maura Egan

In this era of art and fashion mash-ups — in which Marc Jacobs finds a muse in artists like Elizabeth Peyton and Rachel Feinstein, and the pop art sensation Takashi Murakami earns rock-star status with his cherry blossom Louis Vuitton bag — it's no surprise that many artists are mining the runway and style.com to trigger the creative flow. With proverbial paper and pencil in hand, these up-and-comers were asked to incorporate a piece of fall fashion into their signature styles.

Researching the Prada collection for visual cues was a departure for Amy Cutler, who populates her fairy-tale tableaux with sturdy women who are often dressed in modest Victorian garb — more Mennonite farmer than Milanese high-fashion priestess. "It's different than I usually work," says the Brooklyn-based Cutler. "I typically dress the people after I draw them." Cutler perched three Prada-clad ladies atop an elephant balanced precariously on twiglike trees. She built the piece around one red strappy high-heel sandal. "Walking down the runway in those must be equivalent to an elephant on stilts," she says.

The house of Chanel, meanwhile, may seem an odd choice for the men who inhabit the Houston painter Robyn O'Neil's stark, snowy world. Usually in uniform sweats, they look like a lost fitness tribe that made a wrong turn at the Mall of America and ended up in a Bruegel landscape. Karl Lagerfeld, however, is himself a devotee of uniforms. And his gray, black and white palette for this season fits easily into her woodsy setting, in which half the men wear Chanel and half are in their customary gym wear. "The Chanel men represent individuality, while the other men have lost their identity," O'Neil explains. "They're zombies." In other words, high fashion (and cross-dressing) wins.

Which is often what happens in the work of Simone Shubuck. Not only is she conversant in the latest Nike Dunks, but also her titles — e.g., "You Can Definitely Take Better Care of Yourself in Prada" — can sometimes sound as if they have been ripped from the pages of a fashion glossy. When Shubuck saw Etro's kaleidoscopic fall collection, with its bright colors, Japanese embroidery and geometric shapes, she immediately envisioned an Egon Schiele type with her high *fin de siècle* ornamentation. Shubuck, based in New York, often draws her decorative doodlings on found objects. For this piece, her canvas was a boy's scrawled science homework, which she salvaged from the trash. "He writes about shiny and dazzling designs," she says. "It seemed to go perfectly with the dress." ■

Amy Cutler

OPPOSITE: FOR "PASSAGE," THE ARTIST WAS INSPIRED BY THE PRADA FALL COLLECTION.

40

在杂志的专题报道中，出于易读性的考虑，通常使用三栏网格线，但文艺杂志常用两个宽栏，形成经典对称感，让文字段落看起来更长、更饱满。本页取自《纽约时报》杂志的范例将这一点进一步发挥，大胆采用一个宽大的专栏，这个做法虽违背法塞特易读性理论（参见156页），但却很易读，又具有古典结构的形式美感。

数字刊物的格线

设计数字版刊物的版面时，格线也是重要工具。《泰晤士报》纸质版所使用的格线系统，是为了给设计师和文字编辑提供一套井然有序、方便易读的系统，以经典的文字编排逻辑来引导读者阅读。在设计数字版《泰晤士报》时，也采用了同样的格线系统，但根据数字设备的屏幕进行了适当的缩小。有些设计师会认为没必要在数字版刊物上采用传统纸质版的格线设定，但设计总监乔恩·希尔（Jon Hill）这样解释道：

> “数字版《泰晤士报》也使用了线性浏览模式，读者利用滑动页面，一页一页翻阅。我们希望通过保持和纸质版相似风格的处理方式，让读者更容易接受阅读数字版本。虽然这种做法一开始饱受批评，但是最后收到了很好的读者反馈。而栏位固定，也的确比较容易设计。改版之初，对读者而言，从纸质版到数字版其实是很大的飞跃和挑战。但读者对平板电脑的操作已经习以为常，加上传统网格线系统的辅助，也就很快适应了。”

相比之下，马克·波特（Mark Porter）对数字刊物的格线使用有不同看法。2005年《卫报》改成柏林版式时，波特就设计了新的格线系统，他在2010年担任平板电脑版本的设计顾问时，也同样大刀阔斧改动文字编排的方式。波特将《卫报》的视觉识别加以调整，将同样的风格延伸到《卫报》网站与App。他解释说：“重点在于，如何在不同媒介上灵活变通。报纸的格线固然不能直接应用到网站或App上，但也要有一套可以参照的系统，根据不同媒体的特点进行优化，而不是完全放弃原有的格线系统。”

模板

THE TIMES | Eureka

In the competitive world of Premier League football every detail counts. For Sam Allardyce, the Blackburn Rovers manager, this involves employing "the Big Brother of football" to help him to get the best out of his players. But what's good for elite footballers could also be good for you and your job

HUMAN PHYSIOLOGY

WORDS: KAYA BURGESS | PICTURES: GERRY PENNY

IT was brief and a long time ago, but Sam Allardyce's experience of the North American Soccer League still shapes his approach to football management some three decades later.

Back in 1983 Allardyce played for the Tampa Bay Rowdies, who shared the facilities of the American football team, the Tampa Bay Buccaneers. Even though the Buccaneers were not the strongest team in the NFL, Allardyce was impressed.

"They were given breakfast in the morning, trained, relaxed, watched some video analysis, looked at stats, took their weight, took their blood, took a urine test," Allardyce recalls. A huge back-up team acted on this information and built it into the players' training. "You were talking about six physios, eight masseurs, a psychiatrist, a psychologist, five support scientists, three nutritionists, some people employed just to do strappings."

The Blackburn Rovers manager became determined to apply this all-encompassing approach to training and player evaluation to football. Soon after his appointment at Bolton Wanderers in 1999 he went one better, adding a computer analysis system called Prozone to the mix. He brought it with him to Blackburn in 2008. And something similar may soon be coming to your workplace — if it isn't there already.

A football manager and his staff cannot possibly keep a close watch on all 11 members of his team for 90 minutes. Yet no Rovers player can escape the watchful eye of the eight Prozone digital cameras placed around Ewood Park — or the intense evaluation of his performance they produce. The technology has since been adopted by many leading clubs, including Manchester United and Real Madrid.

While Prozone works exclusively on football, its executives have been consulted by companies seeking to use a similar approach to monitor employees. Three years ago one agency sought advice from Prozone on using their system to show the distance nurses had to cover within a hospital, their work-rate and the type and frequency of tasks they had to perform.

Simon Edgar, a manager at Prozone, says: "Businesses can certainly benefit from having a system where all data outputs

employee productivity by plotting data about individuals' output against data about the way they work. Companies can subscribe to software packages that will monitor internet usage. Electronic swipe-cards that give access to the building can also provide information about working hours. Keystroke logging can monitor every word typed, while attentiveness monitors can flash up a message on a computer screen — if you don't click it quickly enough, you're not paying attention.

For Cary Cooper, Professor of Organisational Psychology and Health at the University of Central Lancashire and co-author of *Business and the Beautiful Game*, such software can be valuable — but only if used properly. "These technologies can be extremely useful, but you can't rely totally on them," he says.

"A lot of people want a magic bullet for observation but none of these systems work by themselves. They are most predictive when collated with other data on performance.

"In a call centre one employee might take longer over calls, but they might also have a better record of keeping and attracting customers. You have to be very careful that monitoring software doesn't blind you to other data.

"Sam Allardyce isn't going to use Prozone exclusively without making his own observations and judgments about players' abilities. Managers in other walks of life should be similarly wary."

GREAT MINDS

Footballing injuries can be prevented using sport statistics
Dr Ian McHale, University of Salford

《泰晤士报》决定在平板电脑版本上延用类似的格线结构，让传统报纸的读者在浏览页面时能有熟悉的阅读体验。数字版《泰晤士报》也会延续纸质版多年来重视井然有序版面的传统。设计总监乔恩·希尔表示："一旦定好了字体，数字版的格线系统和版面设计与纸质版一样重要。我们希望数字版也是同样编排精美的产品。"

THE TIMES | Eureka

MEDICINE

QUANTUM OF COOL
Heroes & Villains of sports science

THE SCALE
From 1 John Lennon (the measure of the coolest hero in the Universe) to 1 Yoko Ono (the measure of the uncoolest villain in the Universe)

COOL

ARCHIBALD VIVIAN HILL
(600 Princess Dianas)
Mix of Marlon Brando and Oskar Schindler. Hill is the godfather of sports science, discovering the mechanism of how muscles produce mechanical work and also saving 900 academics from Nazi prosecution.

ROGER BANNISTER
(400 Eric Cantonas)
Not only did he run the first four-minute mile, in 1954, he beat Ben Johnson in the 1988 100m Olympic final. In a 40-year neurology career, he introduced the steroid test that caught the cheating Johnson.

DR RICHARD STEADMAN
(500 Tucker/Campbells)
The hopes of English sports fans rest on many things, but a small-town doctor in Colorado prolonged the careers of Michael Owen, Michael Vaughan and Alan Shearer with his groundbreaking knee operations.

NEUTRAL

ANTHONY GALEA
(300 Tucker/Campbells)
Charged with smuggling, conspiring to lie to federal officials and the distribution of growth hormone. Tiger Woods's ex-doctor is in trouble. The charges relate an NFL client. Tiger has other things on his mind.

PATRICK ARNOLD
(600 Alan Shearers)
A word of advice: never make fun of the weedy kid, in case they invent a super-drug and take over the world. Arnold, frustrated by his inability to bulk up as a body-builder, created the designer steroid THG.

MANFRED EWALD
(700 Stings)
The stereotype of a moustachioed East German female athlete growling her way to victory is down to Ewald, who orchestrated the state-sponsored doping of up to 10,000 female athletes in the 1970s and 1980s.

UNCOOL

HOW IT WORKS
The scale starts at neutral, or 1 Tucker/Campbell. 1,000 Tucker/Campbells = 1 Eric Cantona. 1,000 Eric Cantonas = 1 Princess Diana. 1,000 Princess Dianas = 1 John Lennon. Negative cool: 1,000 Tucker/Campbells = 1 Alan Shearer. 1,000 Alan Shearers = 1 Sting. 1,000 Stings = 1 Yoko Ono.

THE TIMES | Eureka

WORDS: HANNAH DEVLIN

Why women are now the injured party

MEDICINE

Scientists are tailoring exercise regimes and diets to counter high levels of injury among female athletes as they challenge the limits of their sports

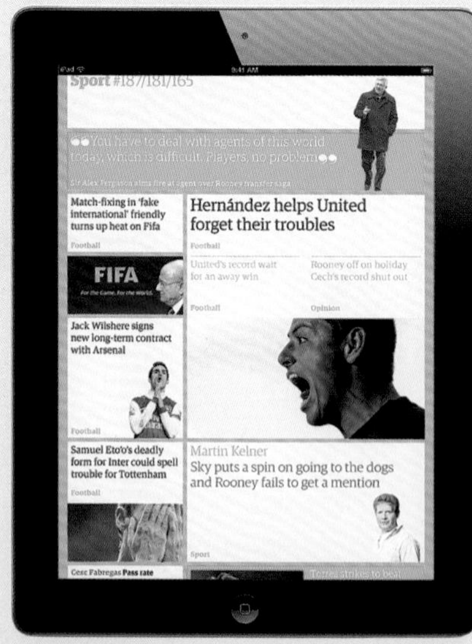

《卫报》的纸质版报纸是以二十栏隐藏格线系统为基础的，读者实际上会看到分成五个栏位的版面，网站则为隐藏十二栏的格线系统，实际上会看到页面分成三或四个栏位，平板电脑版本则是分成八乘六的正方形。页面的文字编排层次，会按照文章的重要程度排列，并运用利落的标题、图文并茂来强化这些相对关系。网站将图片的力量发挥得淋漓尽致，读者可以用简单的浏览工具直接找到自己想看的内容。

《卫报》的平板电脑版本和许多新闻网站一样，用水平的单元标题搭配垂直主题列表，让读者能快速找到想要的内容。

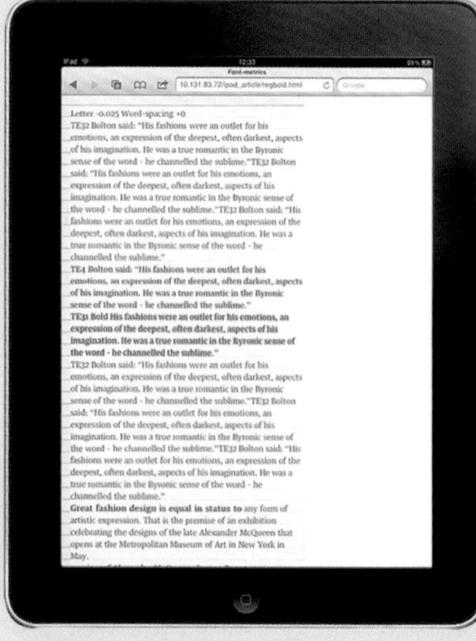

一旦建立好格线系统，就可以为杂志的不同单元建立相应的页面模板，例如新闻页面模板、专题报道模板、后段模板等。这些模板基于已经设定好的格线系统进行设计，但可以根据不同单元需求进行变化，以满足板块的特定需求。在设计模板时，必须要用上所有可能出现的页面元素，例如不同字号的文本框、内文栏位、图框、图片说明框等（更多模板的信息，参考110页）。

分页

一份刊物中的内容顺序必需事先规划好，才能确保刊物版式完整流畅。这很重要，因为分页会决定刊物的节奏和内容安排的平衡性，避免相似内容扎堆。通常来说，内容类似的跨页之间必须要有明确的分隔。分页通常由编辑、美术编辑、制作编辑与广告销售主管共同决定。影响分页的因素包括印刷流程（各单元如何拼版送印）与广告需求。

分页时，有些细节必须要格外注意，例如若某篇专题报道在左页结束，右页最好不要马上另起一篇专题。此外，一篇专题也不宜被连续四页或突然冒出的广告插页打断。要测试分页及刊物节奏是否合适，可以事先把整本刊物的页面规划缩小印出，贴在工作室。由于制作流程中页面难免出现变化，这时编辑设计师就可以直接调动这些小页面的顺序，从而看出新排序对设计造成的影响。

传递信息

杂志的分页与版式必须把文章的重要程度与风格传达给读者。文章在刊物与页面的位置、栏宽（宽栏宽多为专题报道或报纸社论版）、字号级数、标题长度与位置、文章长度、文章样式、图片的运用与大小、色彩运用等诸多设计元素都是潜在信息，会向读者传递这篇文章的重要程度。以报纸为例，头条新闻通常靠近头版页面最上方（放在报摊时能见度最高），标题最大，也分配到最多空间，周围则摆放次要新闻。社论版和新闻页面有差异，会使用更多的负空间，有搭配照片的署名行及导语，栏宽更宽，而且字号和粗细都不同。杂志也会运用这种方式来传达讯息。在杂志的新闻与评论页面上，信息传递方式和报纸类似，但专题报道区会更细腻。如果一篇文章的标题醒目，篇幅长达八页，字段宽，有满版照片，委托摄影师拍照，显然代表刊物很希望读者好好阅读这篇文章。设计师通过连贯且一致的方式，在刊物中利用这些信息暗示与读者沟通。

DAILY Mirror
Friday
July 8 2005
WE WILL HOLD FIRM 35p

37 dead, 700 injured in London suicide terror

Blair vows: Britain will NOT be intimidated

BLOODIED BUT UNBOWED

www.mirrorpix.com
mirrorpix
For usage call us on +44 (0)207 293 3700 or email desk@mirrorpix.com

2005年7月7日，伦敦爆炸案发生后的第二天。全国发行的八卦报《每日镜报》（*The Daily Mirror*）将伦敦描绘成虽遭血洗但却不屈服于恶势力的首都。报纸通常会以图像形式来呈现这类重大事件。

《卫报》平板电脑版设计草图中有很多大方格，有些放文字，有些放图。之后会设计成漂亮的网格线，而且能够进行滑动浏览。虽然是对数字版本的设计，但设计师在沟通阶段还是会使用手绘草图。

数字刊物的页面设计

数字刊物的文字编排和版面设计原则其实与纸质版类似。但设计师要明白，数字刊物和纸质刊物不仅仅是不同的产品，读者对数字版本会有哪些不一样的需求和期待，触屏应用上的阅读体验，和在网站上的阅读体验是不一样的。刊物编辑团队必须了解自己的读者是谁，读者的关注点和兴趣点是什么，以及读者为何、如何、在哪里与数字媒体互动。之后，艺术总监和编辑团队才能够设计出适合读者的内容。

浏览导航

纸质出版物中清晰明了的设计原则在设计数字版本时同样适用，并注重设计的交互性。如今，适用于平板电脑或移动设备的数字刊物已有很高的交互能力。因此设计者要谨慎规划顺畅的浏览方式，激发读者与刊物互动并进行更深入的探索。在设计浏览方式时，要让读者很快找到想读的内容，例如运用标签（tag），可以让读者快速找出相关报道或

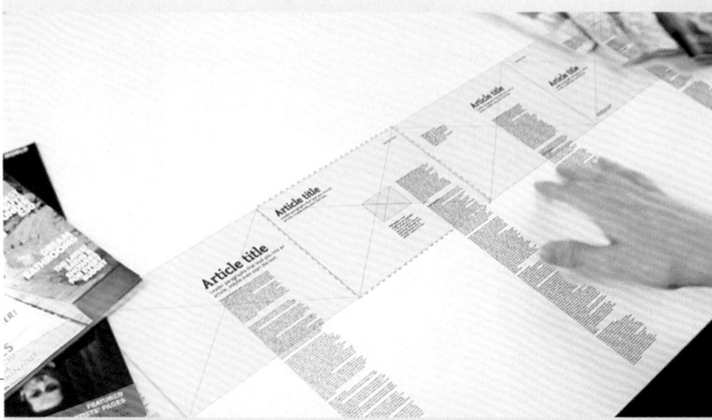

右边中间的图是影片截图，杰克·舒兹（Jack　Schulze）解释数字刊物版式设计的理念。

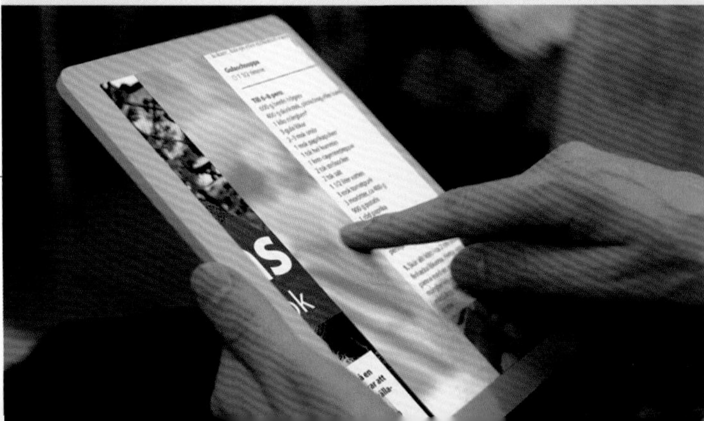

当读者用手指将版面左右滑动时，可以切换到不同文章，如果想阅读某篇文章，可以上下滑动阅读全文内容。这样的数字杂志设计如今很常见，而且，它们往往仍以网格线系统和样式表单作为版面设计的基础。

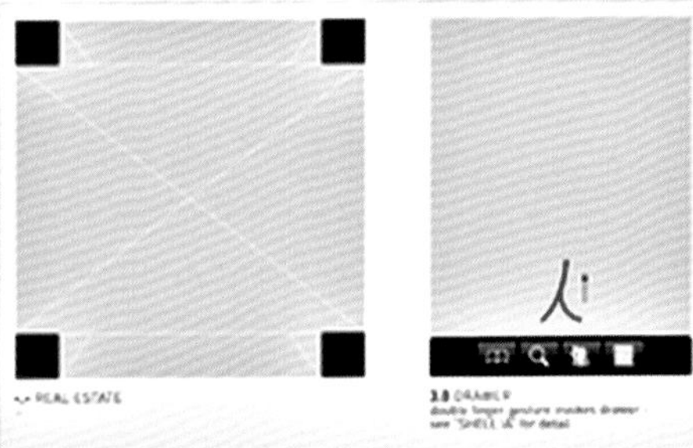

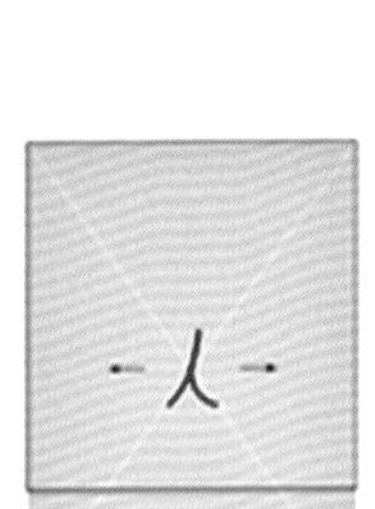

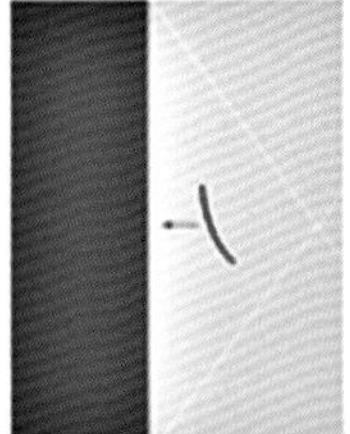

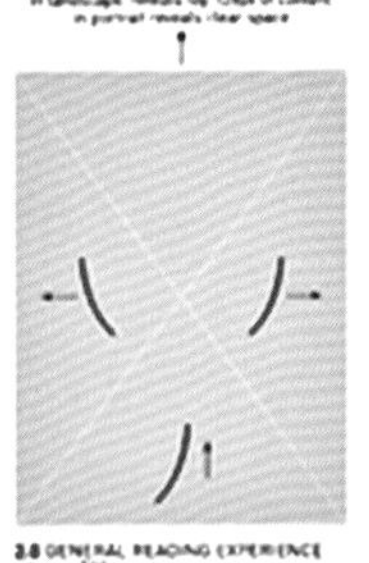

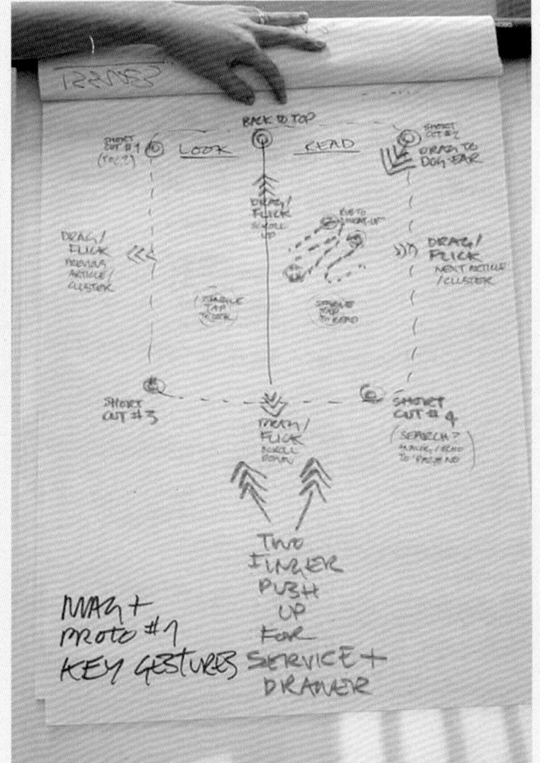

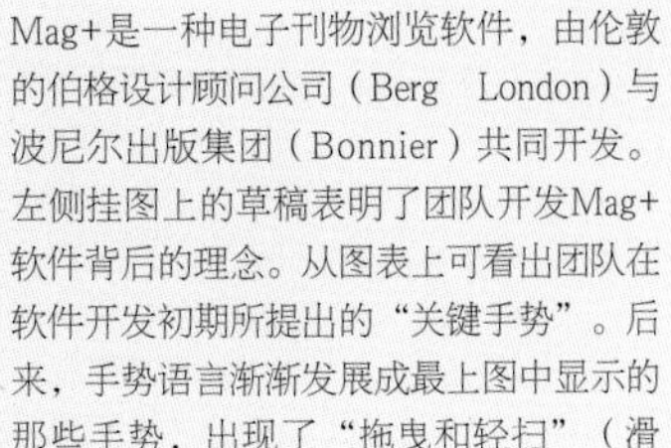

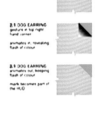

Mag+是一种电子刊物浏览软件，由伦敦的伯格设计顾问公司（Berg London）与波尼尔出版集团（Bonnier）共同开发。左侧挂图上的草稿表明了团队开发Mag+软件背后的理念。从图表上可看出团队在软件开发初期所提出的“关键手势”。后来，手势语言渐渐发展成最上图中显示的那些手势，出现了“拖曳和轻扫”（滑动），以及“拖曳角落”（页面顶端有往下折的小角，吸引读者继续往下一页前进）等触控动作。上图墙上的草图显示页面的水平以及垂直版面的规划逻辑，设计师必须按照这些新的阅读方式来设计版面，抛舍开纸质版本的设计思维。

更多信息。所有文案，包括功能按钮都必须容易理解、能吸引注意，才能引起读者的兴趣。

格状浏览系统能够起到双重导航的作用。首先它能够传达出内容的重要程度，重要文章会比次要文章占用更多格子。其次，还可以作为按钮，通过点击进入正文内容。正如波特解释所说：“格状浏览系统搭配主图，能够带领读者进入任何想要阅读的报道，浏览正文内容。在纸质版刊物中，重要的文章会使用大标题，在数字版本中，重要的报道会占用二至三个方格，读者可以上下滑动观看。”

网站

杂志网站可采用使用者已很熟悉的菜单栏和下拉选项卡菜单。之后，设计师可用文字层次和单元名称引导读者进入各个单元与页面。刊物网站与纸质版刊物的最大差异在于，网站并不是用来吸引读者长时间阅读的，此外，网页版的设计也与触控式的平板电脑版本不同。网站让设计师有机会展示品牌，以吸引读者去购买刊物。

平板电脑

平板电脑有触控屏幕，这让数字刊物的浏览体验更上一层楼。但设计师一定要记得，读者仍然希望能尽快找到内容，不想面对太多障碍。毕竟平板电脑有时反应不是很快，难免导致部分读者失去耐心。波特曾说：“报纸网站受欢迎是因为读者能很快找到信息，例如赛事结果，并深入阅读。读者一进来，就能看到熟悉的新闻标题栏表，相当简洁明了。然而，令人满意的数字杂志不多，读者在阅读数字杂志时期待的互动方式和纸质版是不同的。”

将活泼丰富的物品排列在版面上，并使用隐藏式的说明文字，让照片保持完整。当读者需要了解图片说明时，只要轻轻滑动画面中的对象，说明文字就会显示出来。上方的专题报道主题为玛丽莲·梦露的鞋子，只要点选不同的鞋子，就会出现不同的故事。

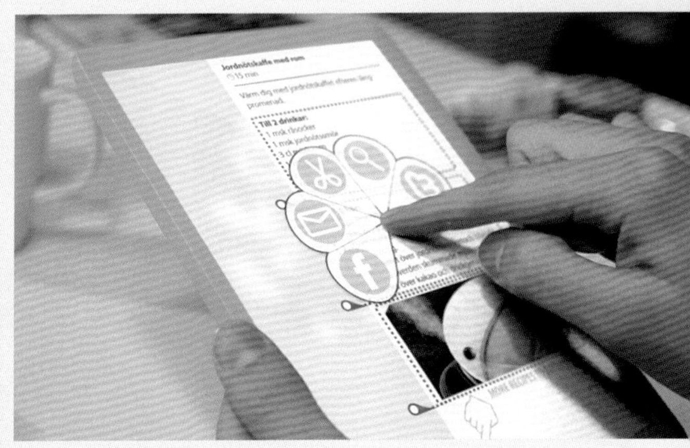

在平板电脑上，利用不同的触控选项来阅读更多报道或使用其他功能。例如手指轻轻一按就会弹出一个子菜单，就是图中这个放射状的小图示，能让用户用不同的平台来分享内容，用户可把杂志图片或其他内容直接发到社交网站上。以这个例子来说，如果按住屏幕，就可把食谱通过电邮发送给朋友，也可以在推特或Facebook上进行分享展示 。

App

报纸的App内容需要持续进行实时更新，因此相比于杂志，报纸的App更需要模板，有点类似于网站制作。如今市面上也有许多杂志App，比较简单的是直接把纸本版面转成PDF页面（或许还可嵌入外部网页与社交网站链接），也有非常多功能的聘请开发者量身定制的App。

除了直接把页面转成PDF，或斥重资开发全新版本的App，还可以利用很多现有制作App的插件来制作数字版本的刊物，在平板电脑上使用。理论上，要把杂志数字化并不困难，但在实际转换的过程中会出现很多复杂的状况。不同的平板电脑有各自的尺寸和默认文件格式。另外，平板电脑的屏幕多比纸质版杂志小，所以每个页面也需要重新调整。即使现在平板电脑的屏幕分辨率越来越高，但文字通常也仍要比纸质版更大，才能清楚易读。

由于成本与资源有限，小型出版商只能从转换PDF这种方式着手。但不少出版社另辟蹊径，开发出更适合读者阅读的版本。*Wallpaper**杂志的莎拉・道格拉斯（Sarah Douglas）谈到如何为杂志制作App时强调：

> “重点在于换个方式思考。数字版的运作有何不同?读者会以何种不同方式阅读?你也得从这些不同出发，转换思考方式。设计师需要带着读者的眼睛来思考，想想读者会怎么使用平板电脑，在平板电脑上页面可以很轻松地放大缩小，但智能手机屏幕不够宽，就受会到限制。网站版本理论上空间更大，但如果仔细研究用户阅读屏幕的方式，就会发现其实网站在屏幕上的显示也没有想象中那么大。多数的网站对浏览导航是通过菜单完成的，这在看列表或阅读一本书时不会造成太大阻碍，但对期刊阅读来说体验感就没那么好了。”

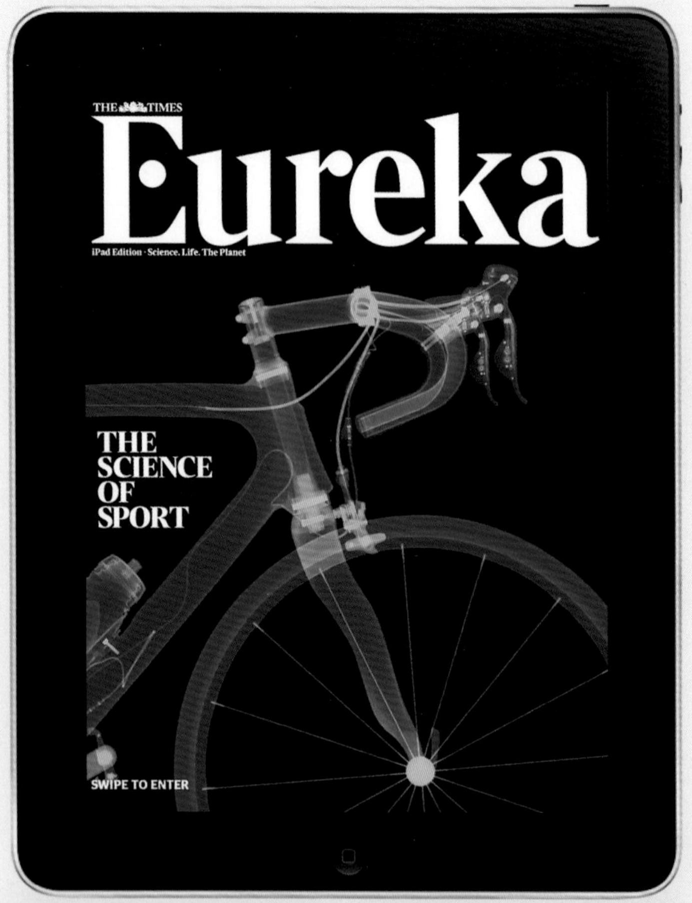

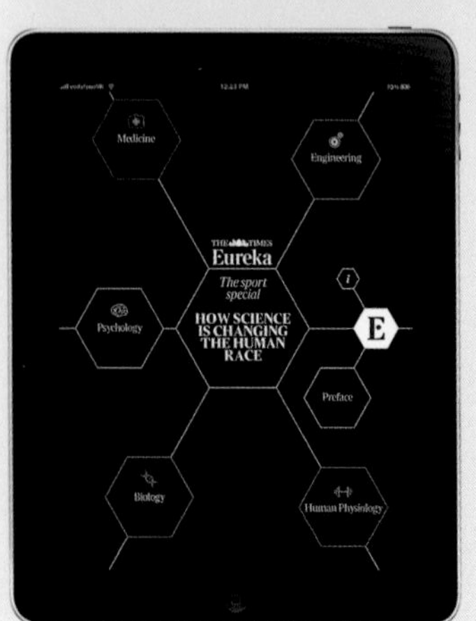

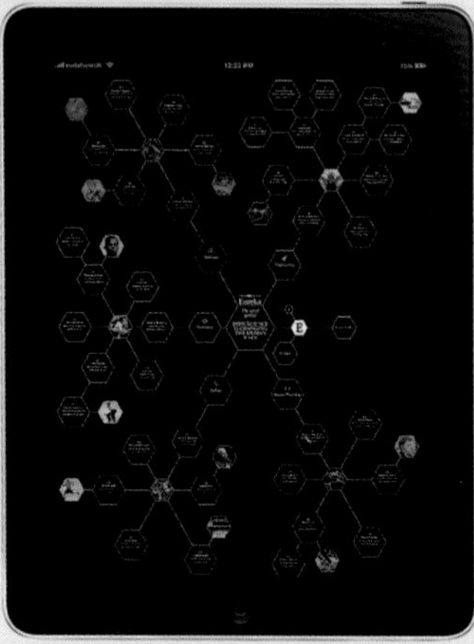

上图是《泰晤士报》出版的*Eureka*科学杂志平板电脑版。乔恩·希尔用报头来增加视觉冲击力，封面设计保持简洁。读者翻开封面之后可按下六边形进入内容。六边形的设计体现出杂志的科学质感。

杰瑞米·莱斯利（Jeremy Leslie）的数字刊物设计技巧

1. 最基本的原则是理解。清楚你要说什么、对谁说、怎么说。

2. 依靠分享观念与共同兴趣来汇聚读者。人人都想要有归属感，这一点永远不变，数字技术的发展让这一点更容易实现了。

3. 渠道和技术只是为故事服务的，不要为了满足形式需要而凑足形式元素，只采用能够增强故事的技术和方式，不要画蛇添足。

4. 互动元素虽然很炫，但一定要确保这些元素有用及有意义，而不是为做而做。

5. 从编辑角度来看，制作纸质版或数字版本的工作并没有极大区别，除非你刻意让两者天差地别。

6. 别忘了，杂志会创造出自己的世界，而读者在那个世界里能够忘我地阅读。

落版单

落版单是出版物制作过程中不可或缺的重要工具。刊物的落版单和电影的分镜有异曲同工之妙，参与制作的人只要看一眼落版单，就能明白页数、页面顺序、内容、印刷信息、正文与广告的比例与分页等。通常落版单是交由制作编辑或编辑室制作经理负责，并且根据制作过程中的变化不断更新。例如专题报道必须延长、缩短或删除，或是特定广告必须放到某篇文章的对页等。每当有变动出现，整个单元就得重新安排，才能维持刊物整体的平衡和节奏。因此版面一旦变动或调整，就要把新的落版单印出来给大家，让合作伙伴了解最新情况。编辑设计师可从落版单上看见哪篇专题能有几页的空间及目前的页面顺序。有些公司会使用数字化的落版单，每次更新后会将数字化落版单实时共享给所有参与者。

A	1	2	3	4	5	6	7	8	9	10	11	12	13	14	15	16
B	17	18	19	20	21	22	23	24	25	26	27	28	29	30	31	32
A	33	34	35	36	37	38	39	40	41	42	43	44	45	46	47	48
C	49	50	51	52	53	54	55	56	57	58	59	60	61	62	63	64
C	65	66	67	68	69	70	71	72	73	74	75	76	77	78	79	80
B	81	82	83	84	85	86	87	88	89	90	91	92	93	94	95	96

上图是一份96页刊物的落版单，可以看出页面顺序。整份刊物共使用A、B、C三张纸双面印刷，每面印16页，双面则为32页。如果你的刊物不是全彩的，就需要在落版单上清楚标示出彩色单元与黑白单元：彩色页面可用色底显示，或在周围用粗框标出来。上面的例子中，黄色页面代表全彩印刷，红色页面代表双色印刷，灰色页面代表单色印刷。彩色单元可能出现在刊物中的任何位置，但要符合每张纸上所分配的色彩区块。虽然也有20或24的倍数的印刷机，但多数印刷机打印与装订的页数都是16的倍数。通常印刷厂会要求处在同一色块表上的页面先送印，例如所有在A面的页面必须要比B面和其他纸张提前一两天下印。

落版单会显示刊物的各部分内容会出现在哪里。第一页通常是封面，而用来放广告的页面常会打叉，也可用清晰的“AD”标志来表示，因为有时候打叉可以用来标示已完成设计、校对或其他步骤的页面。最好有固定的命名或编号惯例进行落版单更新，在明显处标示出日期、时间、版本编号。右图是*M-real*杂志的目录页，很巧妙地展示出刊物在送印前落版单的模样。

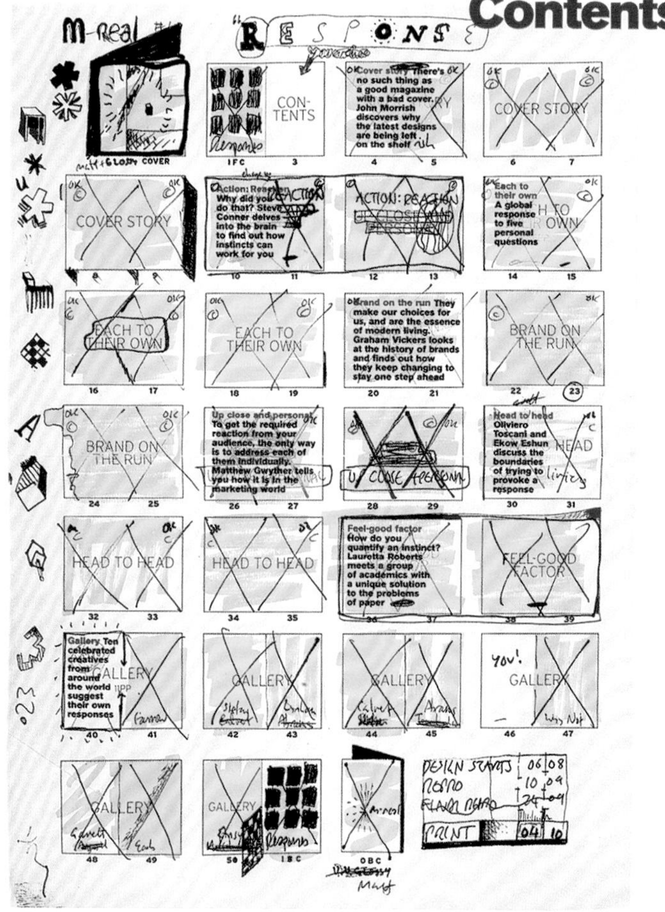

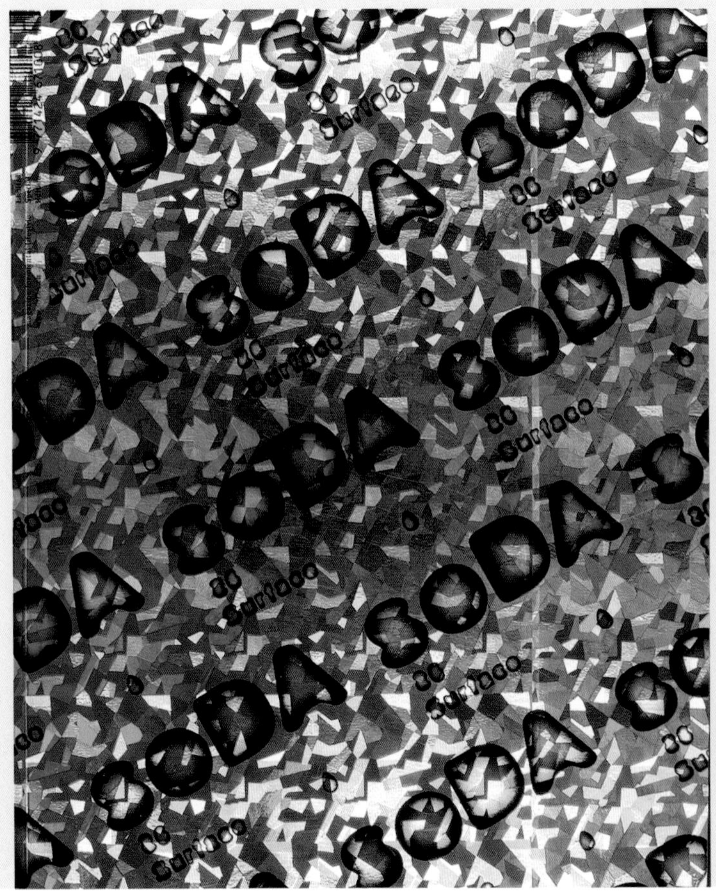

用纸选择

纸质版刊物如今面临着数字版刊物的竞争，必须尽量展现其优于数字版本的特色，纸张质感变得尤为重要。纸张的选择对刊物的感觉、调性、风格与外观很重要，会影响刊物内容的表现和传播。一般而言，选纸有两种方式。一种是请印刷厂与纸商协助，编辑设计师最好能不厌其烦地向两者取经。印刷厂能够提供良好的初步引导，依据设计师的制作需求来寻找最适合的纸张。例如，如果想要用薄的、有涂布、不会透、亮度高的纸张，印刷厂通常可建议几种具有这些特性的纸款。纸商也乐意提供印制好的书籍样本、空白假书与纸样给设计师，甚至会按照设计师的要求制作成一样的开本，让设计师更清楚刊物的重量与质感。如果要用到特殊的印刷工艺，例如使用烫金、压模等，也可以向印刷厂请教，印刷师傅的专业知识能够有所帮助。另一种是参考现成的出版物都用了哪些纸，找到符合自己需求的，根据出版物上列出的印厂信息，顺藤摸瓜，为己所用。

纸张考量

好的选纸要能兼顾每个环节的需求。如果主要考量的是如实呈现图片原色，最好使用明亮、高白、厚涂布，摸起来特别滑顺的纸张。这种纸张最能反射光线，不需要额外加重图片色彩。当然，还有其他因素要考量，例如厚涂布纸不利于文字阅读，重量也较重。如果想知道该如何权衡，下面的指南也许可以提供帮助。

涂布还是非涂布?涂布纸的反光性较好，吸墨性低，因此能细腻呈现图片：涂布越厚，图片越清晰。无涂布纸印刷显色通常较柔和，很适合美术作品或插画的呈现，也能让文字更易读。

*soDA*是一本小型独立杂志，依靠广告、订阅以及印刷厂或纸厂支持。纸厂或印刷厂很乐于借此机会推广特殊的印刷工艺和纸张。本期内容是讨论纸张的表面质感，封面是用激光全像卡制作，内页则使用金属油墨以及有涂布与无涂布的多种彩色纸张印制（对页左图）。*Flaunt*的封面常运用浮雕和压印工艺——这份2001年5月号的封面（对页右图）是模仿本，上面还有凸起的胶带及REM乐团和歌手蜜西·艾丽特（Missy Elliott）的Logo，非常立体。创意总监吉姆·特纳（Jim Turner）表示，封面采用这种风格的原因在于："如果杂志称为*Flaunt*，就多少得爱现、爱炫耀（注：faunt的意为"炫耀"）。我们得炫耀特殊油墨、印刷工艺，就是要那么花哨，才能名副其实。浮雕的地方摸起来很有趣，这是多数杂志没有的。大家都喜欢摸这封面。"

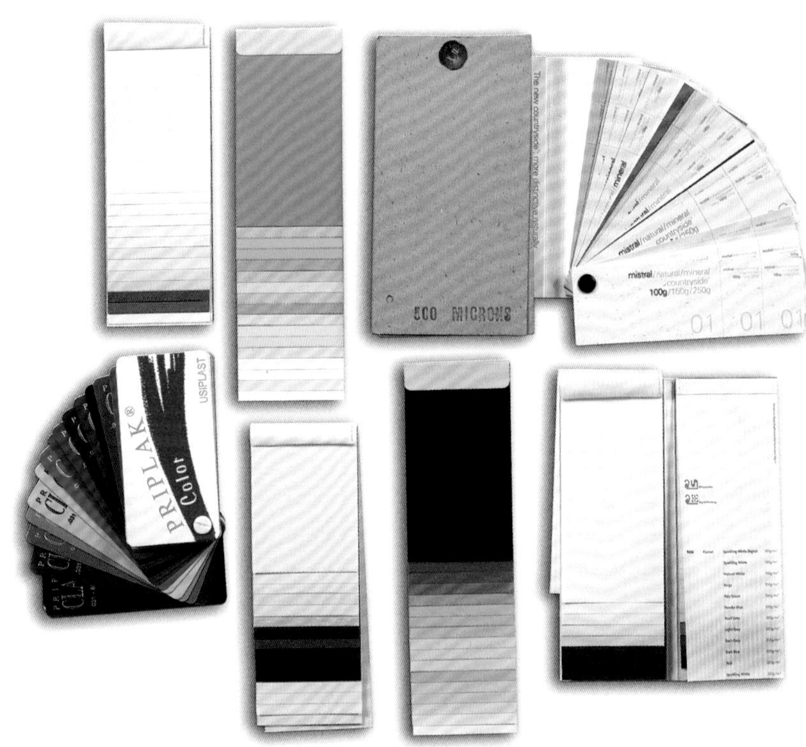

亮面还是雾面？亮面纸常用于印刷有大量图片的高质量刊物，但许多雾面纸的效果也很好，可以让刊物在竞争中脱颖而出。

厚还是薄？厚的纸张往往令人联想到艺术与"高雅"的书籍，但是薄纸张仍可用密度、亮度与涂布等方式做出雍容华贵的感觉。

密度还是不透明度？纸张的不透明度会影响其"透墨性（show-through）"，因此在挑选纸张时，可把纸张放在黑白条纹上测试透墨性。页面上的元素也会影响纸张的透墨性。如果影响很大，就要小心处理版面布局，通常不对称的网格线会比对称网格线更容易使纸张产生透墨现象。

重型还是轻型？重的纸张常让人联想到奢侈品，不过奢华感是有代价的——不光是纸张本身的价格高，还要考虑邮寄方便与便携性。如果刊物像电话簿那样

纸厂和供货商会不遗余力地说服设计师用他们的纸张，常常制作许多纸样册子与豪华的样本，展现某种纸张的重量与色彩。要注意的是，印制好的成品可能看起来很不一样，因此如果对某种纸张有兴趣，最好要求看一份这种纸张印制成的刊物，并让对方按照你的刊物尺寸与页数做一份样书。

报纸销量持续下降，因此报社纷纷绞尽脑汁吸引更多读者。许多报纸从不方便携带的大报版式改版成较受欢迎的柏林版式、紧凑（左图）或小报版式，例如A4版。缩小版式的理想做法绝不是只让内容“缩水”，而是用新的尺寸来重新设计合适的版面，重新研究栏位（数量、宽度与长度）、负空间、文字编排与其他设计元素（例如线条与页码）。马里奥·加西亚（Mario Garcia）用这样一个比喻来解释，就像从大房子搬到小公寓：“得重新评估你需要什么、想留下什么、愿意舍弃什么。”

重，可能会让读者望而却步。从读者的角度出发，就需要考虑刊物是否便携。

亮度高还是亮度低?纸张越明亮，反射蓝光越多，呈现图像的效果越好。不过也可能造成眩光，干扰读者阅读。此外，为了达到高亮度，就需要更多漂白剂，这会影响纸张的耐用度与印刷难易度。

再生纸还是原生纸浆?读者在阅读环保议题的杂志时，会期待刊物使用环保纸张，例如使用百分之百再生、无漂白的纸张，或是使用部分再生纸张。多数纸厂都有这种纸张供应。设计师也可以从环保机构查询各种关于环保纸张的定义，作为参考。

开本

开本就是刊物的尺寸，即页面的形状及大小。最常见的开本是A4。开本是由纸卷的宽度与轮转印刷机的滚筒大小决定的。由于美国和欧洲的滚筒规格不同，因此印刷品尺寸也有细微差异，美国的标准开本就比欧洲短。消费类杂志最好遵守常规，这样才能符合商场书刊陈列架以及一般信箱开口的规格，以便能够用便宜的邮费投递。虽然小批量的刊物可以定制，但仍应从读者的角度出发，并且大开本或形状特殊的杂志不利于归档与日后查阅。

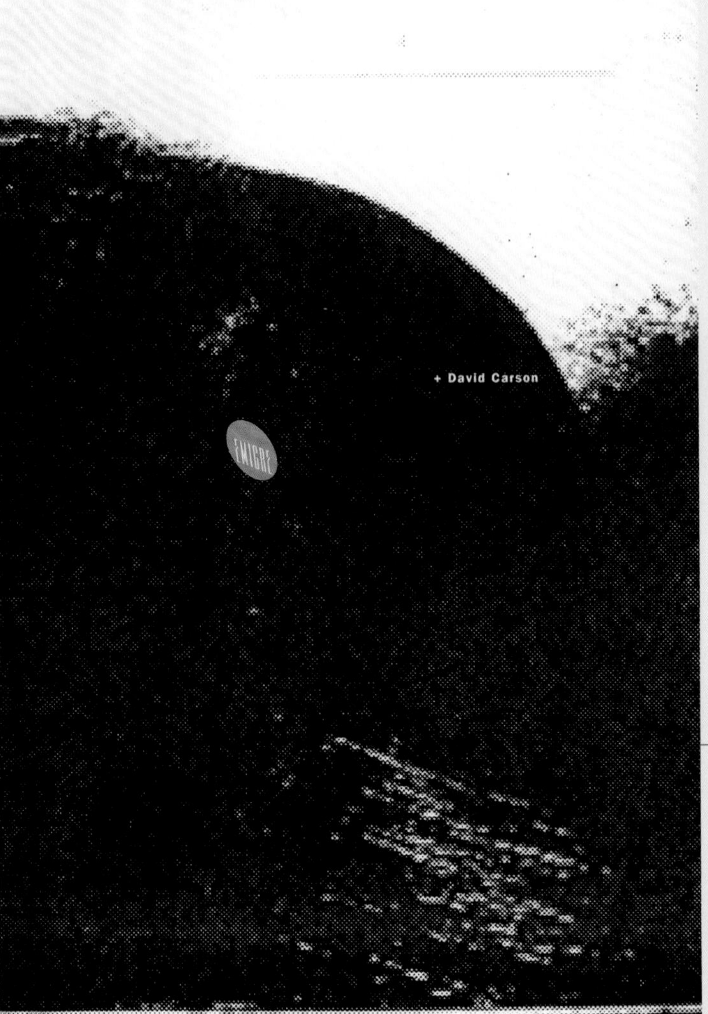

字体的选择与应用

任何刊物都应该为读者提供愉悦、易懂、舒适的阅读体验，这很大程度上是由排版决定的。读者虽然在读小说时能够习惯密密麻麻的页面，但是在阅读杂志时，仍然更喜欢富于变化、有装饰性元素与留白空间、经过精心设计的版面。字太小、太密、一成不变、沉闷的版面会令读者却步。编辑设计师须利用各种技巧，维持读者的阅读兴致。

此外，设计师还须考虑刊物的实用性。某些类型的刊物，尤其是日报和周刊，设计师需要精确地调整文章的长度和标题。最重要的一点是，字体要能反映刊物的识别与吸引力，这点要结合美学、情感与上下文背景思考。字体比其他设计元素更能引起读者联想，因此选择字体时就要仔细斟酌，到底需要什么功能的字体，还有字体呈现出来的风格是否与刊物相符。在设计*Flaunt*杂志时，李·科尔宾（Lee Corbin）会考虑到整期刊物的编排设计，他表示：“我会思考所有文章使用的照片、服装、内容、插图等，判断每个单元要用哪些字体，适合用抽象还是具体的手法来处理。所有的故事都在这期杂志上。我决定哪些版式将在哪些部分中使用，以及不同程度的抽象化。”

L'AMOUR ABSOLU

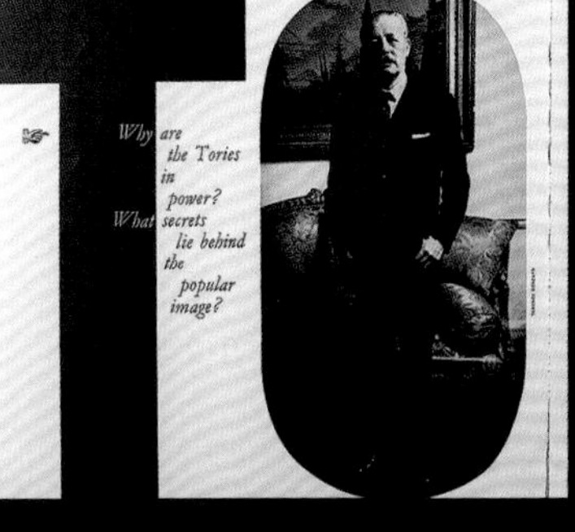

刊物格式应该考虑到读者对象以及目的，才更能明确该如何做出变化。左页右上图的*Tank*是种“微杂志”，每一期都会改变开本。左页下图的*Emigre*则会实验性地每期改变版式。宽大的版式很符合这一期的主题——大卫·卡森的设计作品。

这两个例子运用两种截然不同的标题字体，但是都很有效。最上图为*Vogue Paris*在跨页上使用鲜艳的红色衬线字来写出“L' AMOUR ABSOLU”字样，看起来热情又大胆，却不会显得粗俗。而*Above Town*正好相反，很有威风凛凛的男子气概，很好地反映了这篇文章的主题。

Intervju: Per Morberg

"Fingertoppskänslan är min största begåvning"

DAILY Mirror

FIRED!

Why kissogram Gerri had to go

EXCLUSIVE: BY SIR ALAN SUGAR

G'day China!

KYLIE HITS SHANGHAI

WE'RE IN!

Three Brit teams in Euro semis and Man United play AC Milan

BANK BANS THE POOR

You need £50,000 to get proper service

内文与标题的字号究竟该多大，并没有明确规定。大原则是展示性的文字信息要够醒目，吸引读者优先阅读。例如标题与导语的字号就应该比正文和图片说明的字号大，在页面占据明显的位置，而正文的字号也不能过小，让读者能轻松阅读。多尝试不同字号与行间距，然后把不同的版本打印出来进行对比。没有一种通用字号可以适用所有情况，重要的是判断哪一种看起来最适合刊物的读者群。别忘了，多数字体是为特殊需求设计的，因此它们通常会有各自适合的字号。但有时设计师会用各种方式使用这些字体，做出有创意或突兀的版面。相对地，新闻标题的设计不宜花哨，要简洁明了。这并不意味着标题不能用衬线字体——许多优质报纸会在标题上使用斜体与衬线字体（注：例如中文宋体），传达出无衬线字体（注：例如中文黑体）不足以传达的严肃感。

英国小报（常称为“红头报”）很少使用衬线字体，标题常全是大写字母，例如上图的《每日镜报》。虽然使用大写的无衬线字没什么特别之处，却能清楚地传达这份报纸的类型。小报很自豪能运用紧密的文字编排，把整个页面塞满故事，只留下小小的留白空间，看起来物超所值，即使在读者忠诚度越来越难以维持的今天，它也能在激烈的竞争中有一席之地。

相比之下，瑞典的《每日新闻》（最上图）引语与放大的首字都使用衬线字体，有一种文艺气息。水平的留白空间营造出更轻松的气氛与知性节奏。这些设计元素结合起来，传达出内容是理性、经过深思熟虑的。版面并不是把密密麻麻的信息一股脑堆在版面上，而是用设计暗示读者需要自己深入阅读来了解故事。

用版面布局来塑造图文的影响力

这个版面的主角是大标题——页面正中一个大标题作为页面重点，吸引读者目光。和小报一样，这种版面给人的感觉是大标题比正文还要重要。字体也没什么变化，留白空间小，文章长度短，让报纸呈现一种简单直接、看完就可以丢掉的感觉。

Dummy settings which use other languages to approximate text have the disadvantage of distracting attention to themselves.

THE HARD-HITTING NEWS HEADLINE

SECOND STORY HEADLINE FOR RELATED NEW ARTICLE

A simulation of actual copy, using ordinary words with normal letter frequencies.

第二个版面使用同样的图文，感觉却截然不同。标题比较柔和，留白空间较大，还有引语和小标题，提高了正文的优先级。这一版面设计让读者能更容易地进入正文阅读，但仍然用很大的标题与图片来吸引读者。这是高级杂志比较常见的处理手法，并且利用重复的元素为文章营造出知性感。

The softer-hitting news headline

Dummy settings which use other languages to approximate text have the inherent disadvantage of distracting attention to themselves.

Use even gibberish to approximate text

第三个版面还是一个以图片为主的版面。运用虚线隔开文字，让内文读起来像短篇故事，并用抢眼的导语营造出杂志精美的感觉。这样的刊物很容易浏览与翻阅。照片够大，能在读者的快速翻阅中引起注意。

The medium-hitting news headline

Dummy settings which use other languages or even gibberish to approximate text have the inherent disadvantage of distracting.

Dummy settings which use other languages or even gibberish to approximate text have the inherent disadvantage that they distract lots of attention to themselves.

This is dummy text. It is intended to be read but have no meaning. A simulation of actual copy with.

EVERY DAY *is a* HOLIDAY *story* STELLA WU

PART 1 *of* 4

WHEN BERNADETTE'S CAR COASTED TO A SPUTTERING STOP on the side of the 210 freeway, she was glad she had decided to wear panties under her Girl Scout uniform. It was a small but significant comfort in the face of her new, overwhelming inconvenience: hoisting herself into the unreasonably high cab of a tow truck would, now, be that much less mortifying. Bernadette drummed her fingers against the steering wheel and scowled at the dashboard as traffic screamed past. She called Maxwell, but he didn't answer; leaving him a vaguely explanatory message, and wondering where in the hell he could be, she called the Auto Club instead, frowning out the window as the operator spoke to her in annoyed, exasperated tones. "That's really not a safe place to be," the operator said peevishly upon learning that Bernadette and her car were cowering on the shoulder. "We don't ever recommend pulling over on the freeway. It's a lot better for us if you get off the freeway."

I'm gonna duct tape this hunting knife to the wall, right up at the level of my heart, and just run at it, full speed.

*Fishwrap*的版式与纸张每期都不相同。但平面设计师丽莎·瓦格纳·霍利（Lisa Wagner Holley）表示，通过文字编排可以将各期整合起来。文字编排是经过精心设计的，读者的体验是重要指标。其实，各期也看得出有相同的设计手法，例如使用引语来整合图片（用像对话一样的方式将内容传达给读者，更亲近地交流）。在上图的一期中，字体（Minion、Trade Gothic、Knockout与Young Baroque）也经过仔细挑选，才能彼此搭配。丽莎·瓦格纳·霍利解释道："文字非常重要，我们尽量让刊物友善、易读、美观。"

选择内文字体时，更要考虑到易读性与实用性，因为内文字体承担着传达刊物主要信息的使命。读者通常更习惯阅读衬线字体。因此一般而言，文字量大的时候，就会使用衬线字体，而穿插其中的短文（新闻、评论、资讯之类的）则用无衬线字体做些变化。衬线字体感觉较为正式，无衬线字体则比较轻松、有现代感。若字体造型流动、有曲线，类似手写字，则会传达出柔性氛围。边缘较刚硬、日耳曼风的歌德体，则传递完全不同的信息（但两者都不好读，文字量大的正文不宜使用）。编辑设计师应把字体看成一种图像，有时字体本身甚至可以当作视觉元素。字体是最有弹性的设计元素，也是营造刊物风格的重要角色。

报纸的字体的使用

虽然文字编排是所有出版物设计的基础，但报纸与杂志的文字编排逻辑大不相同。马克·波特解释说：“报纸最重视的永远是版面的易读性及内容的可读性。先满足这两项条件，再去思考如何用字体为报纸建立特色，做出美丽、有戏剧性的版面设计。”至于报纸的主要字体该如何选择，波特表示：“基于文字的易读性，以展示性文字来说，颜色与粗细的变化越来越重要。”波特只用专门为《卫报》设计的埃及字体（Egyptian）作为标题与正文字体，是很少见的。卫报埃及字体有超过两百种粗细变化，因此能灵活扮演不同角色，发挥不同功能，是非常特别的字体。

刊物在引进新字体时， 无论是委托设计的还是原本就有的字体， 创意总监必须确保字体与品牌、内容以及其他设计元素的关系尽可能契合。这需要依靠创意总监的直觉和对刊物的理解，波特说：

> “《卫报》使用的埃及字体是专门委托设计的。我们想要既有经典衬线字体的特性，同时又要有现代感与独特性。这种字体必须易读、灵活，有强烈个性，而埃及字体正好符合所有条件。埃及字体的粗细变化很丰富，因此我们不必用和多数报纸一样的字体，用衬线字体搭配无衬线字体。我们在多数版块只使用埃及字体，让《卫报》的文字编排与众不同。”

字体的表现力

在无法使用图片或图片沉闷的版面中，可以用更有创意的文字编排来弥补，就像中世纪彩绘手抄本以及近些年纯文本的广告海报。数据、图像或文稿越无趣，就越考验设计师的想象力与创意技巧。有自信的编辑设计可以大玩文字，例如将不同字体的文字放在一起、改变字体的结构，进行排列组合，创

THE SEATTLE PUBLIC LIBRARY by

OCTOBER 2004
FEATURES

eborah Jacobs, Alex Harris, Greg Maffei,
Larson, Gil Anderson, Gordon McHenry, Re
haas, Joshua Ramus, Natasha Sandmeier, Bj
els, Meghan Corwin, Carol Patterson, Mark v
of-Zogrotski, Keely Colcleugh, Rachel Doherty
arah Gibson, Laura Gilmore, Anna
Morrough, Kate Orff, Beat Schenk,
a Sutor, Chris van Duijn. Victoria
d with Florence Clausel, Thomas D
, Achim Gergen, Eveline Jürgens,
mes Peer, João Ribeiro, Kristina Sk
Waeltli, Leonard Weil, John Neshol
mer, Sam Miller, Tim Pfeiffer, Ste
agnesh Pariikh, Vern Cooley, Mary
hris Baxter, Jim Brown, Wayne Flo
ard, Mette Greenshields, Cassandra
n, Ed Kranick, Ken Loddeke, Howa
tthews, Damien McBride, Howard
Rice, Kathy Stallings, Page Swanb
almond, Atila Zekioglu, Anders Ca
rroll, Alistair Guthrie, Bruce McKin
olly, John Gautrey, Armin Wolski,
than Phillips, Jon Magnusson, Jay

METROPOLIS

THE MAKING
OF A LIBRARY

Most writing on architecture treats the profession as a realm of sole authorship, where One Big Name is heaped with praise, quoted extensively, and ultimately held responsible for whether a building succeeds or fails. The truth, of course, is so much more complex that even a book-length work wouldn't do the process justice. It isn't just that there are engineers and project managers and contractors who figure out how to make a design physically possible—or that underlings at big architecture firms often end up doing the actual renderings. The complexities arise from the very way ideas form—where they come from, what shapes them along the way, and how they are finally born. To call the Seattle Public Library's new central branch a "Rem Koolhaas building" is to ignore the three-month research project that preceded the design—the input from librarians, publishers, technology experts, business gurus, even the Department of Justice—not to mention the brainstorms and what-ifs of the architects and designers at Koolhaas's firm, the Office for Metropolitan Architecture, and its partner on the project, Seattle's LMN Architects. And, no, we're not forgetting the graphic delights of Bruce Mau Design, the engineering feats of Arup and Magnusson Klemencic Associates, the interior design and landscaping by Inside/Outside, the lighting by Kugler Tillotson Associates, or the furnishings and materials from more firms than we can list. We haven't the space to tell the whole story, but in this issue (itself the product of our own group effort) we gesture at it—and by scratching the surface hope to reveal both the enormity of this undertaking and the importance that collaboration plays in projects large and small.

拉滨担任设计杂志《大都会》的创意总监时，只要字体能和文章搭配得当，不怕尝试罕见的标题字体。每个单元的正文与图片说明都是用Bodoni Book与Trade Gothic字体，不过大标题的字体通常由文章内容决定，变幻出风格。这是《大都会》重要的品牌识别，也和多数刊物标准的“刊物标准用字”做法不同。上图就是一个好例子。这是一篇专题报道的开头，而这篇文章用超过20页的篇幅，谈论建筑师雷姆·库哈斯（Rem Koolhaus）新作品——西雅图公共图书馆（Seattle PublicLibrary）。版面的设计灵感来自这栋建筑物的概念，“这栋建筑物可说是众人合作而成的项目，不像一般知名建筑物，建筑师占尽所有风头，”拉滨解释，“我们在第一页把参与设计方案的所有参与者名单都列了出来，并且刻意裁掉了一部分，意味着还有更多参与者。因为这是谈设计的专题，所以操作自由度才能那么高。我认为，这种做法没多少杂志能接受。有时候甚至每一个专题的标题都使用不同字体，来配合文章调性，效果会很好。”

The New York Times Magazine DECEMBER

Blueprint: The Shock of the Familiar

The rules are breaking down. In a frenzy to move pro
design is exploding, mutating, multiplying.
Design anarchy! More morph! By Herbert Muscham

n Greek mythology, Morpheus is the god of dreams. But I also like to think of him as the deity who presides over design. Morph means form. Morphology is the study of shapes. Metamorphosis is the transformation of appearance. And these are dreamy times for design.

There has never been more stuff, and stuff must have a shape, an appearance, a boundary in two or three dimensions to distinguish it from other stuff. Even water, and perhaps eventually air, must arrive in distinctive bottles. The shapes are no longer content simply to arise. They explode, mutate and multiply. Design is now subject to the whimsical laws and seasonal revisions of fashion, the quick impulses of journalism.

There is no dominant style, no prevailing trend. There's just more and more stuff that has been styled, molded, carved, folded, patterned, cut-and-pasted, prototyped, mocked up, punched up, laid out, recycled and shrink-wrapped. Modernity has rendered the material world into some kind of plasma that is perpetually prodded and massaged into an endless variety of contours. Look around. Our designed world has the polymorphousness of clouds, the rapid, shifting, irrational play of dreams. Let's call it morphomania.

Is it a good dream or a nightmare? Design today is certainly a challenge for a critic who wants to make sense of, much less evaluate, the things designers shape. In the 18th century, it was believed that design could be evaluated according to universal laws. The purpose of cultivating taste was to educate the senses to their essence.

Today, the most powerful laws governing design are dictated by the marketplace. Catch the eye. Stimulate desire. Move the merchandise. Yet even in the market-driven world of contemporary design, there adheres a mythological dimension.

Design, that is, reaches into the psyche as well as the pocketbook. Like dreams, forms can hold momentous meaning, never more so than today, when images have gone far toward displacing words as a medium of communication. This issue of The Times Magazine tries to interpret the shapes that are bombarding our psyches.

We are all morphomaniacs now. Exhibit A: My medicine cabinet. Here's a partial inventory: •Colognes: Chanel Pour Monsieur — classic square bottle designed by Coco herself. CK One — frosted

Fontography *Each article in this issue begins with a capital letter from a different typeface, or font. We asked the typographer Tobias Frere-Jones to select 12 fonts that illustrate the evolution of type design, from the days of Gutenberg to the present. The 12 typefaces he selected (shown here by the letter A) are described on the title page of each article.*

A A A A A A A A A A A A

左图《纽约时报》杂志的版面是用各种字体组合而成的，包括由Jonathan Hoefler重新设计的Cheltenham、海史密斯与卡特（Mathew Carter）重新设计的Stymie，还有Garamond和Helveica字体。编辑设计师对于文字编排与字体应用的热爱，在这期谈设计的刊物中展露无遗。前艺术总监佛罗里克说：“这一页可说是整份刊物的导语，告诉读者接下来会看到更多关于设计与文化的想法。因此，它的任务就是作为刊物开场白，把文字编排的历史像电报一样简要地呈现。我们请字体设计师托拜厄斯·弗雷尔-琼斯（Tobias Frere-Jones）选出十二种具有历史代表性的字体，从古英文到摇滚风一应俱全。杂志中每篇文章的开头会从那些字体中挑选一个，做成首字大写，而所有字体的“A”排成一列作为索引。这种设计的优点在于信息层次很清晰，比例很经典，字母形状美丽又能传递信息。”

——珍妮特·佛罗里克（Janet Froelich），前艺术总监

太多出版社、编辑与设计师担心大段大段的文字会影响读者的阅读兴致。但排满的文本框其实看起来简洁优雅，可被当作形状来运用。左下的例子是大卫·希尔曼（David Hillman）在*Nova*的设计，大方在跨页上放满密集的文字字段，搭配引语或首字大写，而右页是大卫·卡森在*Émigré*杂志的做法，他只用文字对开页，就传达出一场非常知性的对话。

造出好看的对比。不妨参考卡洛斯·德拉蒙德·德安德雷德（Carlos Drummond de Andreade）、斯蒂芬·马拉姆（Stephane Mallarmé）、乔治·赫伯特（George Herbert）与伊恩·汉密尔顿·芬利（Ian Hamilton Finlay）的图像诗作品，也可以研究俄国构成派、包豪斯、达达主义派的作品。此外，*MCCall*杂志的奥托·斯多奇（ottoStorch）、*Harper's Bazaar*的布罗多维奇、*Queen*与*Town*的汤姆·沃尔西（Tom Wolsey）、*Nova*的哈利·佩齐诺提（Harri Peccinotti）、*The Face*的布洛狄、*Vogue*的贝隆、*Beach Culture*与*Raygun*的卡尔森，*Speak*的威尼茨基，*Zembla*的佛洛斯特，都曾在作品中展现精湛的字体应用技巧。

图像化的文字应用

字体最基本的功能是传达文字内容，但它还有其他作用。编辑设计可用字体来诠释与表达内容、传达意义、增加版面变化，还可搭配图像与其他设计元素来传递情感，做出有象征意义的设计组合。要实现这些效果，需要懂得一些处理字体的方法，才能展现不同趣味。运用不同粗细、行距、字号与排列方式，单纯的文字编排也能更具表现力，产生抽象或与字面不同的解读。某些常用字体（例如歌德体或打字机字体）能直接使读者产生文化联想，为内容所传达出的信息加分。

SPACE

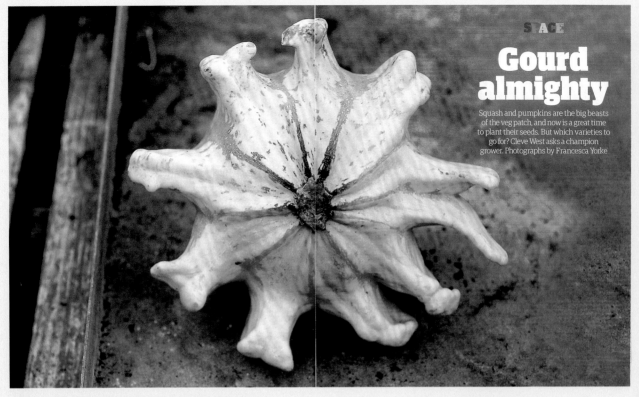

Gourd almighty

Squash and pumpkins are the big beasts of the veg patch, and now is a great time to plant their seeds. But which varieties to go for? Cleve West asks a champion grower. Photographs by Francesca Yorke

National

Counter-terrorism

Raided, arrested, released: the price of wrong intelligence

Only 23 of 895 arrests have resulted in convictions

Officials say terror risk means they must act on tip-offs

'Muslims are being picked on the basis of single-informant information'

'They said we were importing suicide jackets, grenades, rockets, machine guns'

Law

Fast-track summary powers 'to cross line on civil liberties'

Plan to keep minor offences out of court

Contested motoring cases may be dealt with by post

Writ large

In plain English, these people create problem

Marcel Berlins

几乎所有的报纸都采用衬线字体（serif）搭配无衬线字体（sans），因此当马克·波特邀请保罗·巴恩斯（Paul　Barnes）与克里斯蒂安·施瓦茨（Christian Schwartz）为柏林版式的《卫报》设计新版式时，原本也打算以重新设计的Helvetica字体与新版式结合。波特说：“但我们只看中埃及字体，虽然它刚开始只有三种基本的粗细：细字、标准与粗体。然而后来我们发现，只要把埃及字体作出充分的粗细变化，就足够灵活运用了。”如今《卫报》使用的埃及字体有超过200种粗细变化，字号从8号开始加大。

FICTION

FICTION

BY DAVID ROSE

You have been in this country some months now. You are settling in? You are finding work. Your work is improving? You are treated well.

In home I was stand-up comedian. Now, from accident to leg, sit-down comedian. But, I get no work, get no gigs (this is correct?). But, persons here are kind. They wish to know. They ask to me, you come on Eurostar? I say no, I come under Eurostar. Travel cheap, how you say, strapped for cash. How I lost leg. Straps too tight.

Was hard. Not able to move, just watch rail, stay still, count sleepers. Twenty, thirty. Not me. Too noisy for sleep, I travel third class. Under toilets.

You left your home, family, friends? It is a very different life now.

In home, my wife wear burka. They say to me, you Muslim? I say no, she most ugly woman. I leave her behind there. I leave rest of her there. Travel by own. Still people mistake. They say, you Afghan? If wife here, yes. She look like dog (this is correct?). I say no. Still they beat me.

I wish I am Mussulman. Hit them back (this Turkish joke). POW! (This prisoner of war joke.)

Even one person to travel, much money required. This person, that person, agent, lorry driver, all ask money. You say, Arm and a leg. I lucky. Get cheap. Leg only. How I lost leg. Not lost – stolen.

I stop persons in street – only talk – by they tell to me, hop it (this is right?). But no Persons here is kind.

You now enjoy the comforts of the West. You like shopping? You are free now. You like the freedom?

I like to shopping if money. But. Vouchers only. I always ask to shop, you will have vouchers? They look to me. Luncheon vouchers? (What are these luncheon vouchers?) They look. Say, you have card? Visa? Master? I say, goodbye.

Some shops only. But they say, exchange, exchange, going down. Make charge.

Sometime, three times, four times, show me things, very cheap. Say to me, back of lorry. I say no, not lorry, Eurostar. They laugh at me.

If enough in hat, I sometimes shop. If not, look, to windows only look.

All things here shiny, shiny. All wrapped, shiny. Cars shiny. Shoes, clothes. Musics. Blue musics. Stone musics. Say, get down dirty, but discs shiny, clean. But good, is good, is nice to buy. All things to be bought. Books. Many many books. Anything to be wanted. Special books, wrong books. I find book, *Kama Sutra*. But is all dots. How you say it? Braille. I say in shop, is no good to me, is no pictures.

Nonetheless, you bought it? You keep it under your locker. You take it out at night, when the room is finally quiet, dark, you run your fingers across the page, searching the texture, seeking with your fingers the recesses of memory, of your past, the contours of her body, breast and bone, skin over sternum the hollows and clefts, rests for your soul? Burrowing back?

If plenty in hat, I go time to places I was told, look for woman, but talk, only talk. Money to talk, but is alright, is good. They tell me to come, up the stairs, come please.

Doors up stairs is always chips, paint is peels, is curtains, is quiet, is smell. Is good, I feel home in such place.

You like, then, the flesh-pots of London?

Is times dark, is windows, look down, persons walk in street. is dark but lights, shops, is quiet here, is safe. Is alright. But times, is men here, is knives, is fights, in stairs, in back, in dark. Times is threats to me. Tell to me, Packy, cut off your dick. They mistake. How I lost leg. Same size (this is right?).

You have a regular job now. You pay your taxes? You work in London? You work in the carpet trade.

I have job now. Is good, all right. Rugmaker. People ask to me, you make wigs? Tell to them, yes, weave, not scalp. Others to scalp. Also die. Ground. Ground colours, earth colours, stir, mix, big pots, big stove, stir, mix, stir. Is hard. But, colours, is nice, sky colours, tree colours.

You work then, in the dyeing room? You blend the colours, heat the dye, keeping a constant temperature, continually checked, dip the wool.

You also cut and trim the canvas, then work on the weaving, work with the others, men and women, side by side, sitting, legs crossed – leg, rather – on floor cushions of worn leather. You push the hook through the canvas, hook in the wool, pull it back, making a loop of half an inch exactly, knot and cut it. Repeat.

This hurts your fingers, hour by hour? Your fingers are hardening now, calloused? Yet you soothe your soul through your fingertips, running them through the pile, tracing the pattern as it emerges, continuing it in your head, abstractly, the geometrics of your past, the tautological figures, cul-de-sacs of amazement, end stops, locked keys; you follow the grid.

At times, the design is ethnic, folk-art motifs: scarfed peasants, panniered donkeys, hens, birds, but stylized, almost ideographs. Hieroglyphs of happiness, mythical joy. Woven fictions of contentment. Is this correct?

[96] zembla magazine *spring/summer_two thousand and four*

spring/summer_two thousand and four zembla magazine [97]

by Greil Marcus

"It's Only Rock and Roll: Rock and Roll Currents in Contemporary Art" is a show that's been trundling around the country for the last few years; I caught it at the Austin Museum of Art this March. That it's a bad idea is an argument the Avengers were making more than twenty years ago.

The Avengers were San Francisco's best punk band—in their moments, they were, you knew, better than any other band playing that night anywhere in the world. They gave no quarter; their songs ranged from protest numbers that came off like street-preacher rants on the end of the world to explosions of rage they dared you to find contrived. Led by a teenage screamer named Penelope Houston, they broke up twenty years ago; an album made up mostly of live recordings, called *The Avengers Died for Your Sins*, was released just this February.

这两个跨页虽然呈现手法不同，但都是把字体当成图像元素来使用的创意典范。右图是*Speak*杂志，马丁·威尼茨基（Martin Venezky）将文字编排成蜉蝣一样的形态，透露出虚无缥缈的感觉，当作配图传达这期的主题——探索摇滚与当代艺术的关联。这种处理同样适合*Zembla*的设计（上图），这是一篇关于民族主义的文章，通过可视化文字设计，达到呼应主题的效果。

下图选自*Inside*杂志，杰弗瑞·多切蒂（Jeffrey Docherty）用字体变化对各个部分进行区分，为每个单元赋予特色，“这样才能彰显出彼此的不同，还能看出品牌元素。”他尽量避免使用过多字体，正文和标题各用一种，而第三种字体可以和前面两种形成对比。多切蒂说：“第三种字体可以改变氛围和读者的阅读情绪。这个专题谈的是德国科特布斯（Cottbus）图书馆，建筑师在建筑外部设计了字母纹样，我决定在标题上参考这项设计。我让字母彼此堆叠，效仿这栋建筑物的外部设计。”

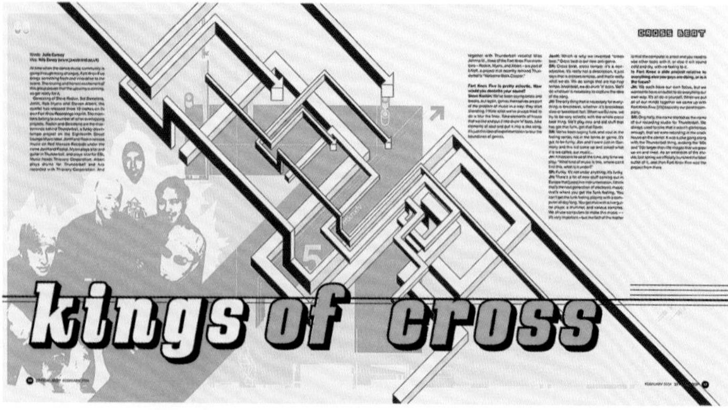

寻找活字

有很多种方式找到活字的现实参考，无论饼干切模、冰箱磁铁、意大利面的形状或发饰，都是字体实例。马丁·威尼茨基（Martin Venezky）到跳蚤市场和古董商店寻找；文斯·弗罗斯特大概拜访过英国每一间凸版印刷厂；艾伦·基钦（Alan Kitching）早期就是以绘制字体为业的。许多编辑设计师都喜欢花大把时间搜寻奇异的活字，凸版活字与木活字现已很少使用，就算被使用，也多用在展示文字而非内文上。现代的设计师最常通过印刷厂、在线字体目录与字体公司来获取字体，只要上网购买即可马上使用。如今设计师已经很少使用实体活字，但实体活字仍能为设计师带来不少灵感。

右图跨页以截然不同的方式把文字当作图片来用。中间是*Inner Loop*独立舞蹈杂志，超大的标题与处理方式刚好契合杂志狂热、无秩序的调性。杂志前艺术总监伊万·科特雷尔（Ivan Cottrell）说：“我使用截然不同的两种标题字体，一种是一板一眼的军事风格，另一种是手写风格，这两种字体搭配得很好，塑造出*Inner Loop*的独有风格，也能传递受访者各自不同的声音。这些字体很有力量，强烈的图像元素横放在矢量图上，显得很抢眼。”左下图则是费尔南多在*Vanidad*杂志时的字体运用，图文并置得很巧妙，标题成为美丽的装饰。“Belle DeJour（白日美人）”的字母被拆开来放，与人物因网络日记而出名的应召女郎身份相呼应。画面右半边的脸部特写和最左边的标题完美平衡，营造出一种温柔和谐却又强有力的布局。

定制字体

和其他创意产业一样，字体的设计和使用也会跟风，以至于最后各种刊物看起来都差不多。想要脱颖而出，最简单的办法就是请人为自己的刊物定制专属字体，这样不仅能够有独一无二的品牌识别性，也能更好地传达出品牌特性。*Flaunt*、*Another Magazine*与《卫报》都定制了自己的专属字体。*Flaunt*杂志的李·科尔宾（Lee Corbin）表示，在杂志改版的时候，自己决定在新版中加入一种定制字体，认为Berthold Akzidenz Grotesk和Century Schoolbook字体搭配使用，可以产生更好的效果。

“在去年的《纽约时报》特刊中，我们用Gotham与Hoefler Text作为标准字体。这两种字体都是出自弗里尔·琼斯与霍夫乐设计公司（Frere-Jones & Hoefler），搭配效果很棒。Gotham字体的辨识度很高，但不会太刺眼，加粗后又能够成视觉焦点。我们把Gotham与Hoeler Text搭配起来，因为Hoefler Text的外观看起来很经典，又是个庞大的字体集，变化丰富、选择颇多。”

使用了新Logo，就需要采用新的字体，科尔宾设计了两种不同的字体，有不同的粗细以及更多变化。

“首先是杂志的Logo，在构思设计时，我准备用根据几何图形设计的超粗字体。我尝试利用一些符号，例如十字、X、三角形和圆圈作为字母。我在Logo上做的尝试可让字体更有特点，也强调了新的Logo，但这种做法需要格外谨慎，以保证Logo和字体都能够经受住时间的检验。”

最后，在使用字体时还有一些注意事项：印刷技术的发展迅速取代了一些专业分工（例如排字工），而这些工作就需要由设计师和计算机来完成。但是QuarkXPress、Adobe、InDesign这些软件的默认设置不一定符合刊物的需求和调性，所以设计师务必要清楚如何调整字距、连字符、行间距、字母间距、补漏白与平均字距等。同时，页面构图与拼版

（原本是要在现实空间里实际挪动各种元素，例如展示文字、排版毛样和图像）等工作都可以使用电脑操作完成。这样是好是坏，人们各有争论。在二维的数字环境中制作，最后需要在三维世界里呈现的纸质刊物，无疑在成品与情感效果上可能都会打折扣。所以如果想要弥补这样的缺憾，可以先多尝试纸张、颜色、油墨、照片及各种元素。尽量把版面印出来看，因为纸质刊物和计算机中呈现的版面差异很大。一定要在纸质页面上校对，而不是在电脑上校对，也坚决不能只靠电脑程序的拼写检查功能。所有的标题、展示文字与图片说明都要仔细阅读，毕竟审稿编辑重点关注的是正文内容，没人想到再去检查展示文字，可能会导致展示文字错漏频出，也要确保所有的连字符与字间距合适。

FICTION

FICTION

by Graham Buchan

He returned after lunch to the bright, buzzing open office and there, concentratedly present, unmissable, in the darting cross-fields of indoor activity, like a refugee on the doorstep, in the middle of his mouse mat sat a ripe plum: large, polished, obscenely full, purple.

He glanced round. No one acknowledged his question. He held it weightily in his upturned palm. It had the promise of a breast. It seemed to want to rupture and squirt at him. It said: "Bite me. Suck me. I'll make a mess of your mouth."

He put it aside with a glance. Later, he lowered it into a drawer. It stewed its concentrate in the darkness. In an idle moment he heard it speak: "I came from a tree. I was brought here. Another hand held me. I was licked by a tongue."

At day's gold end he pondered. He looked down at it preening itself in its burrow. He considered. He lifted it like a scientist, enveloped it in tissue. He made space in his briefcase.

The custom was to drink and talk for an hour. The conviviality of colleagues. A little crude banter. He thought above the alcohol: Who? (You can't take a plum out in a pub.)

The crowded return commute. The door, the stair. He was eager to unwrap it and look, as if it were smuggled pornography. But it was squashed, a mess. The frail paper had soaked into the split flesh. He lifted it out in lumps. He licked his fingers before rinsing. In the sink the stringy remains said: "You were curious, but you need courage for love."

[104] zembla magazine autumn_two thousand and four

autumn_two thousand and four zembla magazine [105]

这些放大的木版印刷风格的字体，在跨页上能清楚看出很好的组织纹理，选用如此

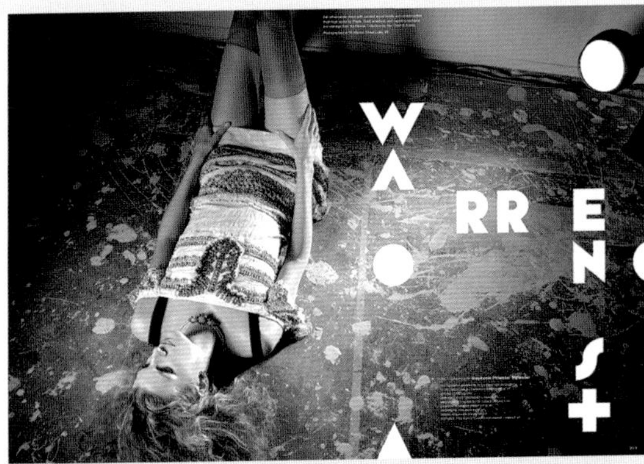

2006年，*Flaunt*杂志设计了自己的专属字体（最上图），艺术总监李·科尔宾表示："在当时现有的字体中，没有一种能够充分传达我想要在新一期刊物中呈现的感觉，所以设计了杂志专属的新字体。"他也制作了许多替代字母，应用到新版的*Flaunt*杂志，这样就"能为刊物设计出更独特的组合。这些字体源自我改版时设计的Logo字体，在读者翻开每一期的封面开始阅读杂志时，这个字体能强化他们对新版设计的印象"。在这两种粗细不同的字体中，粗体用得较少，每个字母的周围空间也较多。"*Flaunt*的新Logo就是由这些字母组成的，而运用这些字母，也强化了新的标识。细字用得更多，也更有试验性。因为它看起来不那么重，可以放得比较大。"

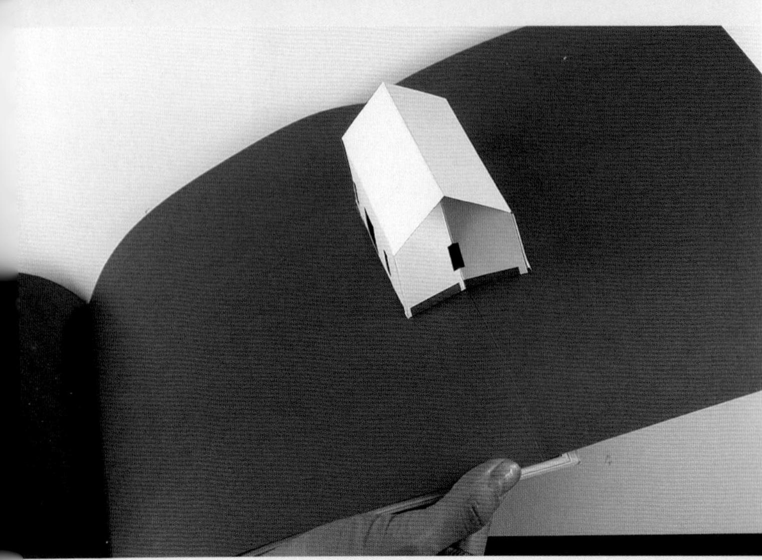

最上图是独立杂志*Amelia's Magazine*，封面由插画艺术家罗伯·瑞恩（Rob Ryan）的剪纸作品包起来，给读者带来了一个限量版的艺术作品封面。下面的*Esopus*杂志沿用这一理念，进一步运用拉页，设计出能够“弹出（pop-up）”的立体书。

设计技能与后期问题

编辑设计师需要具备各种技能，包括但不限于版面设置、平面设计与图像处理软件的操作（例如知道如何使用Adobe Creative Suite的InDesign、Photoshop和Illustrator等软件，或是QuarkXpress）。是否熟练使用软件，对设计师能否充分实现自己的视觉创意影响很大。此外，设计师要能够随机应变，处理突发状况，才能应对刊物在付印前排山倒海而来的修改与变动。设计师也要懂得更多相关专业的技术，例如屏幕校准、色彩管理、印前与印制流程，以及如何委托插画与摄影等的事宜。

软件

早期的排版软件主要是配合传统生产流程，随着软件的发展，现在很多传统生产流程中的工作都由计算机和编辑设计师承担了。原本需要由专业人士、排版工人、印前与校对人员等处理的各项工作，现在都落到编辑团队身上。这些程序不仅复杂，需要的能力也很多，编辑设计师要多学习、多试验，才能胜任。

屏幕校准

电脑屏幕通过光来呈现色彩，也就是“加色”的原理，但印刷品是油墨印刷，也就是“减色”原理。这两种呈现色彩的方式很不一样，为了让屏幕呈现的颜色尽量接近印刷成果，要使用屏幕校准软件来修正屏幕显示出来的色彩。苹果电脑用户可以用BERG Design设计的共享程序SuperCal进行校准，效果很明显。也可以使用电脑自带的色彩平衡功能，Photoshop软件的帮助菜单中也有很清楚的步骤，引导如何使用控制面板校准功能。

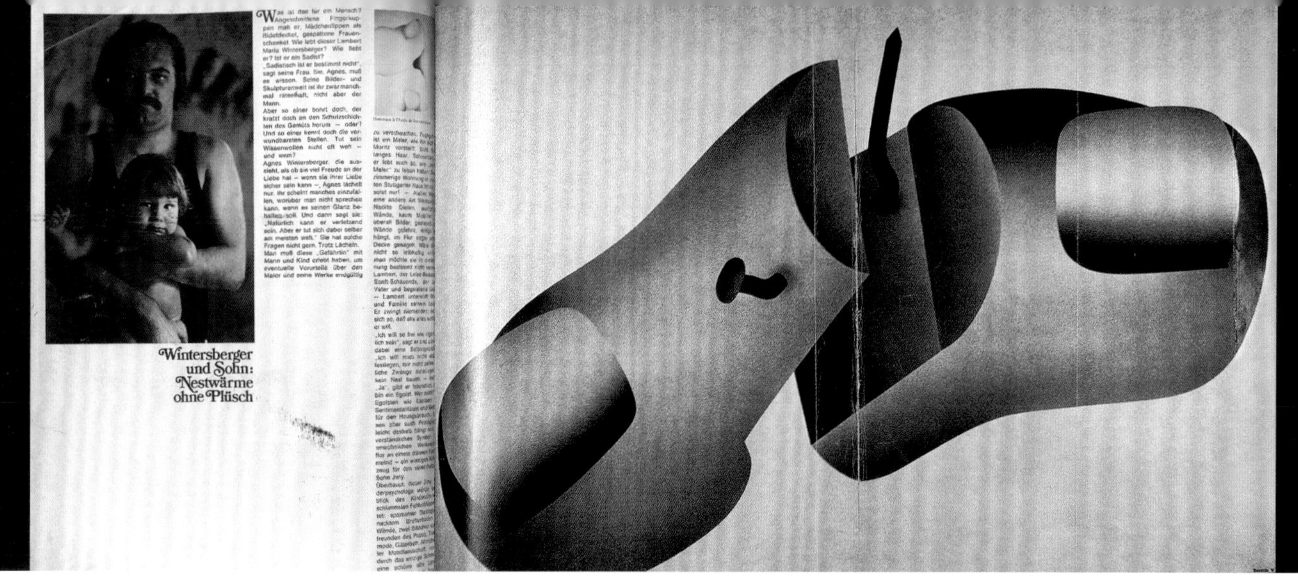

Wintersberger und Sohn: Nestwärme ohne Plüsch

上图的*Twen*常使用折页，呈现出效果很好的视觉元素（包括游戏、艺术复制品与一流的图片故事）。

印刷

最好的印刷厂是那些经常印制和你的刊物类似的印刷品的印刷厂。所以可以在和自己刊物类似的印刷品中查找、查看印刷厂名称，或联系刊物的制作编辑进行询问。此外还有一些因素也需要考虑到，例如这间印刷厂能不能接受你的印量?能否处理你用的纸张和开本?能不能从印刷厂取得ICC配置文件，让你在桌面系统应用?印刷厂能否配合你的出刊时间?费用能接受吗?在决定之前，最好先比较三四家印刷厂的报价。不过最重要的还是沟通是否顺畅，和印刷厂建立良好的长期合作关系绝对会有巨大的回报。印刷厂有你永远无法掌握的知识、经验与技能。好好培养双方的关系，才能获得更多帮助。

色彩管理

控制印刷物的色彩是件复杂的工作，眼睛看到的、屏幕显示的、印刷喷墨头产生的颜色都不一样。印刷业拟定出一套颜色管理系统，可帮图像提供配置文件（称为国际色彩协会，Ilnternational Colour Consortium或ICC配置文件），这样图像从原作到屏幕检查、分色、印前、校样、制版和印刷等流程中，所有工具都可以进行校准和调整，确保色彩精准一致。如果你的刊物没有ICC配置文件，最好在配色时采用“安全”色（可以参考软件内置的色域警告，颜色从RGB转换成CMYK时，如果出现颜色差异，就会出现警告信息），不要过于依赖屏幕上显示的颜色。在这种情况下，潘通色卡图册（Pantone Swatch Book）来补色是比较安全的做法，该图册应每年更新一次，以免褪色失准。要注意的是，并非所有的潘通色卡图册颜色都能用CMYK复制：如果你想要用的潘通色卡图册颜色在排版软件中出现色域警告时，可能需要改用五色印刷。这时最好询问印刷厂该如何处理。

打样

印刷打样通常被称作“确认校样”，目的是确认印刷品图文色彩质量的实际状态。打样种类很多，美国印刷厂只免费提供PDF屏幕校样，其他种类的校样都要另外付费。不过，花钱制作打样是值得的，尤其是当封面与内页有大量全彩图时。

最常见也最经济的打样是数字样，也就是用大型数字喷墨打印机（例如爱普生）打印的印在刊物实际用纸上的样张。至于传统样则是用实际印刷时的油墨来印制。当封面使用了比较复杂的效果，如特别色、烫金、压花、UV等时，就有可能会要先打传统样来看，但即便是传统样，也未必和实际印刷的成果完全一样。

一般复印机或打印机印出的打印稿，其色彩一定不准，无法当作印刷的颜色参考，但可以用来检查字体与位置等，对喜欢以纸质样代替屏幕校对的人来说特别有用。最后还有PDF校样（也就是“软打样”），但这种校样也没办法看出实际输出色彩，多半只用来校对颜色之外的元素，除非是用校准过的Eizo屏幕来看样。

图片的获取、评估和使用

在处理照片时，要先发掘这张照片最与众不同的特点，是取景，或是色彩（完美的蓝天，或是艳红的洋装），还是构图、光线、色调的细腻程度 。无论是什么，好的图片必定有它脱颖而出的原因，在处理这张图片时就要时刻记住它的优势，这会影响版面形状、比例或结构，通常也是版面上最重要的元素。必要的话，在屏幕校样阶段就可以开始和印刷厂讨论，如何让图片特质尽量被彰显出来。印刷厂比最厉害的设计师还懂色彩，知道要怎样控制印刷过程以达到理想的色彩层次。不过编辑设计师可以先考虑以下要素。

原稿的完善程度：图片的色域越广越好，也就是亮部与暗部应该横跨整个色域，而中间色调也要尽量丰富明确。最好在印前阶段就确定亮部、暗部及中间色调，可达到最佳的输出效果。

在CMYK模式下，亮部应设定为：

C：5%；

M：4%；

Y：4%；

K：0%。

图像能反映出年代或事件，文字未必能做到这一点。寻找或创作这种图像的能力，是编辑设计师的关键技能。珍妮特·佛罗里克回忆起“9·11事件”后的余波时，曾讨论到这一点。她说，“9·11事件”发生在星期二早上，《纽约时报》杂志则在每周五就已经完成，这就意味着“9·11事件”要刊登在九天后的杂志上。我们有三天时间撤掉9月23日原本的内容，转而报道“9·11事件”。那时大家还深陷在事件刚爆发的噩梦之中，但我们必须要向前看，其中一个想法是请艺术家与建筑师构思纪念碑。艺术家保罗·迈达（Paul Myoda）和朱利安·拉弗迪埃（Julian LaVerdiere）的工作室曾经就在双子塔，他们提出了一个命名为“光之塔（Towers of Light）”的构想，想象双子塔的原址发出两道很强的光束，指向天空。我将弗雷德·康拉德（Fred Conrad）拍的照片进行后期处理，照片拍摄时间是灾难发生的当晚，曼哈顿下城漫天尘雾和碎片，设计师用Photoshop做出那两道想象中的光束。我们用这张图当封面，一年后，曼哈顿下城开发公司（Lower Manhattan Development Corporation）实现了这个封面的构想，使它成为“9·11事件”最动人的纪念碑，从八十公里外都能看见。看见杂志封面化为真实的纪念碑，缅怀着恐怖事件，那种骄傲又敬畏的复杂感受非文字可以表达。

The New York Times Magazine

SEPTEMBER 23, 2001 / SECTION 6

Remains of the Day By Richard Ford Colson Whitehead Richard Powers Robert Stone James Traub Stephen King Jennifer Egan Roger Lowenstein Judith Shulevitz Randy Cohen William Safire Andrew Sullivan Jonathan Lethem Michael Lewis Margaret Talbot Charles McGrath Walter Kirn Deborah Sontag Allan Gurganus Michael Ignatieff Kurt Andersen Jim Dwyer Michael Tolkin Matthew Klam Sandeep Jauhar Lauren Slater Richard Rhodes Caleb Carr Fred R. Conrad Joju Yasuhide Angel Franco Joel Sternfeld Katie Murray Steve McCurry Carolina Salguero Lisa Kereszi Jeff Mermelstein William Wendt Andres Serrano Richard Burbridge Paul Myoda Julian LaVerdiere Taryn Simon Kristine Larsen

中间色或伽马系统可调整与改善图片整体的明亮度或暗度，但调整时不能影响亮度。设计师可用Photoshop的曲线功能调整影像，把曲线调高或调低50%，直到明亮度达到理想程度。

在屏幕上的色彩显示如何？假设屏幕已正确校准，那么屏幕所呈现的效果应该非常接近印制成品。如果你对屏幕显示的画面不满意，就可以在送印之前调整。Photoshop有很多修图功能，最基本的是非锐化屏蔽（Unsharp Masking），也是多数专业人士的修图方法。一般非锐化屏蔽的设定值是总量160%，强度为2像素，阈值9。只要调整好这些设定，几乎所有的照片都可以更漂亮。

寻找图片

优秀的刊物设计师会不断通过代理商、毕业展、其他刊物或媒体、获奖书籍或光盘来找新的摄影师与插画家。目前多数图片制作者都有在线作品集，但仍建议亲自约见交流。安排充足的时间好好看摄影师或插画家的作品集，再询问对方关于作品的问题及合作方式。你提供的设计纲要会影响他们的创作成果，因此要讲清楚你想要什么，并好好沟通。无论你的设计纲要多么清楚，都要和你委托的人亲自交谈，确保他们明白你的需求、截稿时间、费用和一些行政要求（发票、费用、付款、税务事项）。最后，如果是委托拍摄，要确保流程能安排得井然有序。

使用网络图片

许多网站提供免费内容，有些网站则要付费使用。不过，图库的图片几乎人人都能用，因此很有可能会撞图。大型图库的网站上会有关于版权清理的法律信息供参考。

社交媒体催生了一系列在线照片分享网站。有些提供免费的照片，有些则是基于订阅的用图。Flikr、Pixable和Snapfish都是著名的图片分享网站，它们为设计师提供了大量的图片来源。

千万要确认图片的分辨率能印制、版权清晰，图片来源标明正确，避免侵权影响声誉。

信息图表

随着互联网与数据采集技术的发展，现代人对信息图表的兴趣也随之大增。信息图表让设计师能变换内容的呈现方式，让复杂的信息容易被吸收理解。曲线图、图解、图片与数据结合运用，能更清晰地讲述故事。

信息图是21世纪视觉信息文化的完美体现。事实上，信息图表大约在75年前就出现了，当时设计师兼印刷工托马斯·梅特兰·克莱兰（Thomas Maitland Cleland）为商业杂志《财富》设计了这种呈现形式，以一种全新的方式整合文字与视觉。从那时起，像《广播时报》《连线》和《彭博商业周刊》这样的杂志，将信息图表改进得更为精美，应用也更广泛。

Radio Times是电视节目信息列表杂志，戴维·德来弗（David Driver）利用许多技巧与风格，传达各式各样的信息，包括阿波罗宇宙飞船（Apollo）与联合号太空舱（Soyuz）宇宙飞船如何对接（左页）、管乐团如何运作（最左图），以及影集《警网铁金刚》的纽约警方辖区如何划分。理查德·德雷波（Richard Draper）的信息图表也深受欢迎，他以图像的方式解释不好理解的信息，例如下图“地下运动”。德雷波用蒙太奇的手法，将跨页上的图框文字和插画整合起来，增强视觉效果，这样一来，即使文字量很大也能吸引读者。

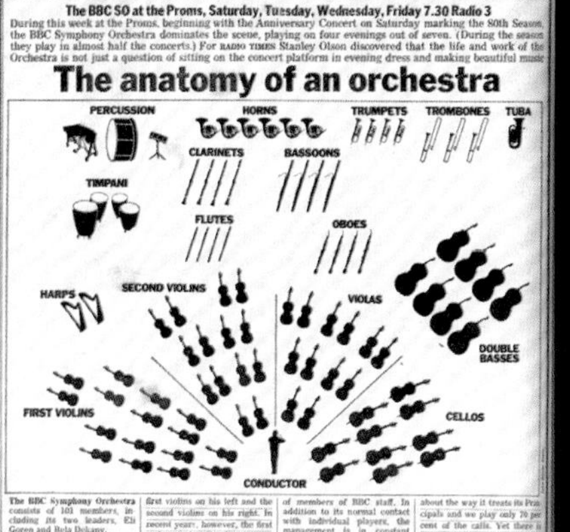
Page 10

The BBC SO at the Proms, Saturday, Tuesday, Wednesday, Friday 7.30 Radio 3

During this week at the Proms, beginning with the Anniversary Concert on Saturday marking the 80th Season, the BBC Symphony Orchestra dominates the scene, playing on four evenings out of seven. (During the season they play in almost half the concerts.) For RADIO TIMES Stanley Olson discovered that the life and work of the Orchestra is not just a question of sitting on the concert platform in evening dress and making beautiful music

The anatomy of an orchestra

The BBC Symphony Orchestra consists of 101 members, including its two leaders, Eli Goren and Bela Dekany.

It was founded in 1930 and was first conducted by Sir Adrian Boult who continued as its conductor for 20 years. He was succeeded by many notable conductors: Sir Malcolm Sargent, Rudolf Schwarz, Antal Dorati and Colin Davis, now the Chief Guest Conductor. Pierre Boulez took over in 1971 and will be succeeded in September 1975 by Kempe.

‘The layout of the orchestra (as shown above) is partly traditional and partly a matter of the conductor's preference. When Sir Adrian Boult conducts he asks to have the first violins on his left and the second violins on his right. In recent years, however, the first and second violins have sat together, with the violas in the middle and the cellos and basses on the right. There are valid arguments on both sides.’

PAUL HUBAND
GENERAL MANAGER OF THE BBC SYMPHONY ORCHESTRA

The members of the BBC SO are on contract and salaried. Except during the Proms – the most onerous part of their schedule – they work five days a week.

Unlike some of the other London orchestras, the BBC SO has a management consisting of members of BBC staff. In addition to its normal contact with individual players, the management is in constant touch with the Orchestra's Committee of seven elected members.

‘This gives an opportunity,’ says Huband, ‘for a frequent exchange of views between the two bodies on all matters – material and musical.’

In Friday's Prom the soloist in Nielsen's Clarinet Concerto is Colin Bradbury, one of the Section principals.

‘In 1956 I left the Royal College of Music and joined the Sadler's Wells Orchestra. In 1960 I joined the BBC SO as principal clarinet. The Orchestra is quite reasonable about the way it treats its Principals and we play only 70 per cent of the calls. Yet there is little relationship between the money and job satisfaction. I spend innumerable hours in rehearsal and I earn as much as one might earn in a day or two in the recording studio. The amount of preparation is enormous and one cannot miss rehearsals. But during the interval at a Prom last year someone came up to me and asked me what I did during the day!’

COLIN BRADBURY
PRINCIPAL CLARINET, BBC SO

➜ 13

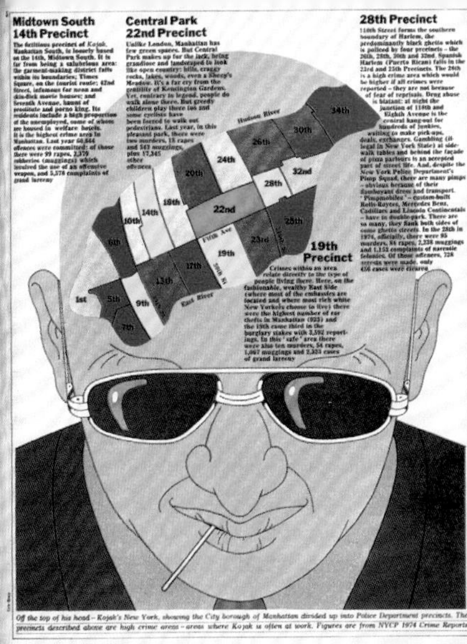
Midtown South 14th Precinct

The fictitious precinct of *Kojak*, Manhattan South, is loosely based on the 14th, Midtown South. It is far from being a salubrious area: the garment-making district falls within its boundaries; Times Square, on the tourist route; 42nd Street, infamous for neon and skin-flick movie houses; and Seventh Avenue, haunt of prostitute and porno king. Its residents include a high proportion of the unemployed, some of whom are housed in welfare hotels. It is the highest crime area in Manhattan. Last year 60,844 offences were committed: of these there were 99 rapes, 2,379 robberies (muggings) which involved the use of an offensive weapon, and 5,578 complaints of grand larceny

Central Park 22nd Precinct

Unlike London, Manhattan has few green spaces. But Central Park makes up for the lack, being grandiose and landscaped to look like open country: hills, craggy rocks, lakes, woods, even a Sheep's Meadow. It's a far cry from the gentility of Kensington Gardens. Yet, contrary to legend, people do walk alone there. But children play there too and some cyclists have been forced to walk as pedestrians. Last year, in this pleasant park, there were two murders, 13 rapes and 543 muggings, plus 17,345 other offences

28th Precinct

110th Street forms the southern boundary of Harlem, the predominantly black ghetto which is policed by four precincts – the 26th, 28th, 30th and 32nd. Spanish Harlem (Puerto Rican) falls in the 23rd and 25th Precincts. The 28th is a high crime area which would be higher if all crimes were reported – they are not because of fear of reprisals. Drug abuse is blatant: at night the junction of 116th and Eighth Avenue is the central hang-out for hundreds of junkies, waiting to make pick-ups, deals, exchanges. Gambling (illegal in New York State) at sidewalk tables and behind the façade of pizza parlours is an accepted part of street life. And, despite the New York Police Department's Pimp Squad, there are many pimps – obvious because of their flamboyant dress and transport. ‘Pimpmobiles’ – custom-built Rolls-Royces, Mercedes Benz, Cadillacs and Lincoln Continentals – have to double-park. There are so many, they flank both sides of some ghetto streets. In the 28th in 1974, officially, there were 95 murders, 84 rapes, 2,238 muggings and 1,152 complaints of narcotic felonies. Of these offences, 728 arrests were made, only 436 cases were cleared

19th Precinct

Crimes within an area relate directly to the type of people living there. Here, on the fashionable, wealthy East Side (where most of the embassies are located and where most rich white New Yorkers choose to live) there were the highest number of car thefts in Manhattan (923) and the 19th came third in the burglary stakes with 3,592 reportings. In this ‘safe’ area there were also ten murders, 54 rapes, 1,067 muggings and 2,323 cases of grand larceny

Off the top of his head – Kojak's New York, showing the City borough of Manhattan divided up into Police Department precincts. The precincts described above are high crime areas – areas where Kojak is often at work. Figures are from NYCP 1974 Crime Reports

The World About Us, Sunday 7.25 BBC2

You'll be familiar with the insect life that scuttles away when you lift up a stone, of course, but were you aware of how much is going on beneath the ground's surface – from moles and earthworms to springtails and false scorpions? Richard Draper's drawing shows you a sample of them and here Richard Mabey writes about the special problems of making Sunday's programme about what's alive and well and living under our feet

Underground movements

FOR MOST of us, the notion of ‘the ground’ is as some kind of foundation to things. It is a baseline, a place to begin at – or end. It's the surface on which we're advised to keep our feet firmly planted, and also the one above which most of the life we regard as important begins.

But, really, life grows out of the soil more than outside it. That hidden cycle of life and death that creates soil and keeps it fertile, is something we cannot afford to ignore.

It is this underground movement that is the subject of Suzanne Gibbs's **World About Us** film. It is, she believes, one of the most challenging subjects she has worked on. It is not just arrogance or claustrophobia that has caused us to pay so little attention to life beneath the soil: it is, quite simply, the most difficult environment of all to explore – black, heavy, clinging and impermeable. Underwater filming is like floating on your back by comparison.

It has needed the combination of some improbable technological devices to piece the picture together. To film moles, for instance, inside their pitch dark tunnels, Suzanne Gibbs went for advice to the Endoscopy Unit at the Bristol Royal Infirmary, who use flexible, tubular fibre-optic lenses for viewing and filming inside the body. But this was only the beginning of a solution.

‘Moles are not only incredibly difficult to find when you want them, but are interested in nothing but eating,’ says Suzanne. ‘When we were driving our star mole between locations, we had to stop at every lay-by and dig up a fresh supply of worms. As for our fibre-optic lens wriggling down its tunnel it was too great a temptation. It assumed it was a particularly long worm, and tried to bite the end off.’

But moles are really at the edge of the underground movement. It is the myriads of smaller creatures, the mites and springtails and worms that feed off living and dead plant material and thus create soil, that are at the heart of it. And myriads is no exaggeration. A recent count of the creatures in just one square metre of beechwood leaf litter and topsoil revealed 320,000 mites, 51,000 springtails, 170 millipedes, 140 woodlice, 75 centipedes, 25 spiders, and eight false scorpions. As for worms, they are so abundant and active that when Darwin weighed the casts thrown up over one acre during the course of a year, they amounted to over ten tons.

The problems of investigating the lives and interrelations of these creatures is not just that they are microscopically small, but that they live in the very fibre of the soil. Says Suzanne Gibbs: ‘Many of the small creatures that live underground are related to aquatic animals, and actually live in the thin films of water between solid particles. Often extracting them from the soil means you end up with dead animals. We filmed a technique developed at the Agricultural Research Station at East Malling in which the animals are separated by a series of washing and filtering processes.’

But Suzanne Gibbs was not out to produce some kind of freak show of underground monsters: ‘We were most concerned to show the complex relations between the different soil creatures. We always tried to film creatures actually playing out their roles in the underground community – moving, breeding, eating, or being eaten. The false scorpion, for instance, which has pincers like a lobster, is at the top of the food chain in the soil. He eats springtails, which in turn feed off mites. And the mites in their turn prey on still smaller animals like the nematodes that feed on tree roots.’

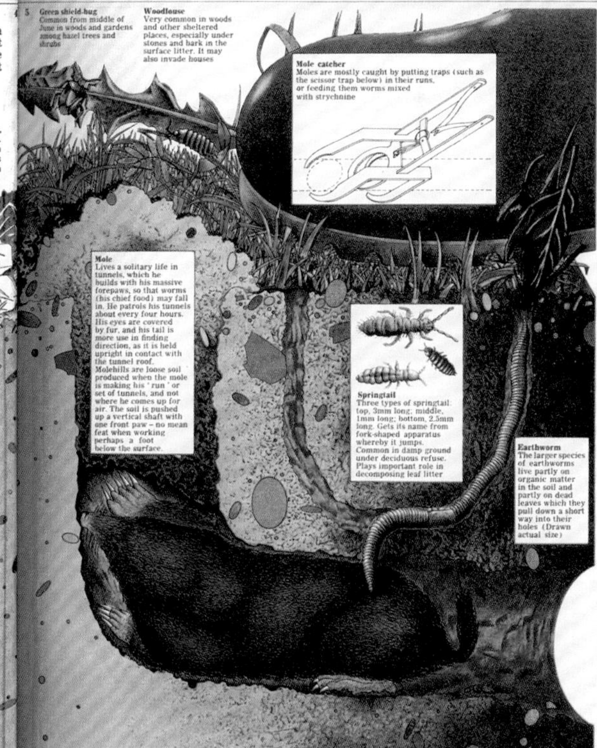

右图巴西《圣保罗日报》中，保罗用信息图表形式来说明飞机在空中相撞的事件，更形象化地解释这起悲剧。图表列出飞机的航线、位置及诸多无法以照片说明的信息。这些图表的作用在于提供额外信息，而不是取代照片，它们能让读者对事件有更全面的了解。

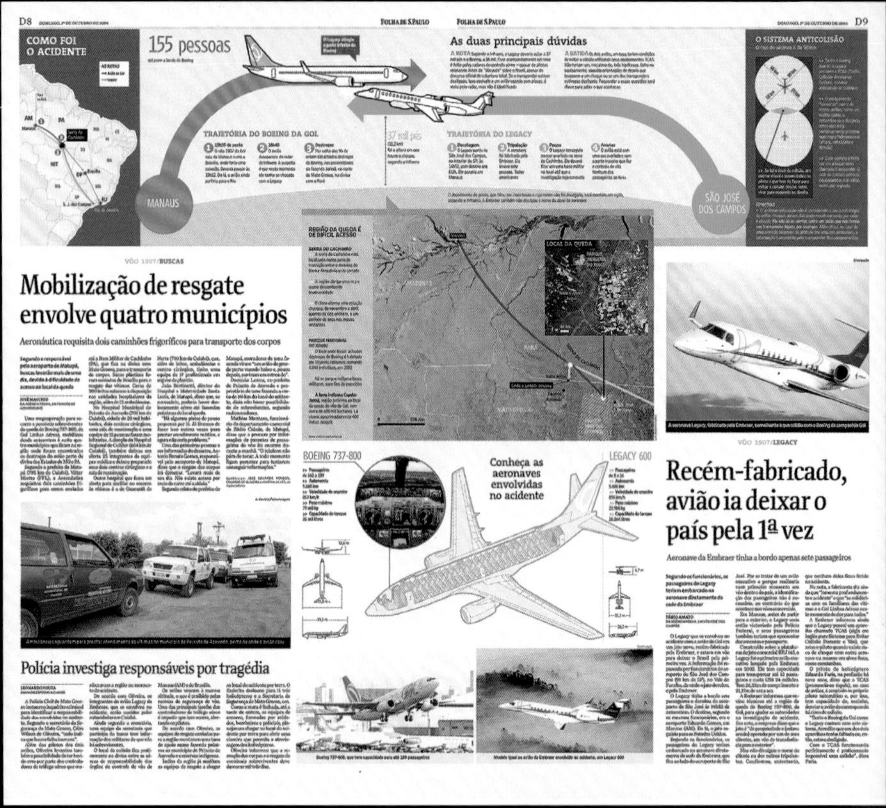

在右图的这张跨页中，《卫报》以信息图表形式解说军火交易，将武器直接画出，读者可以马上理解。

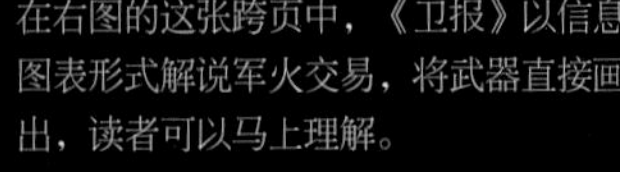

右页上图：2012年伦敦奥运会期间，《泰晤士报》设计了密密麻麻的信息图表。这张图表设计得像跑道。金牌、银牌与铜牌的列表搭配了深入分析的详细资料。这是《泰晤士报》的特长，因为他们的资料库中有丰富的数据与图片可利用。

右页下图：《泰晤士报》的赛前分析图上有决赛的日期信息。小图标、色彩、文字与图片的运用，让信息展现丰富的层次。信息图表有助于强化内容，帮助读者对文章留下更深刻的记忆。

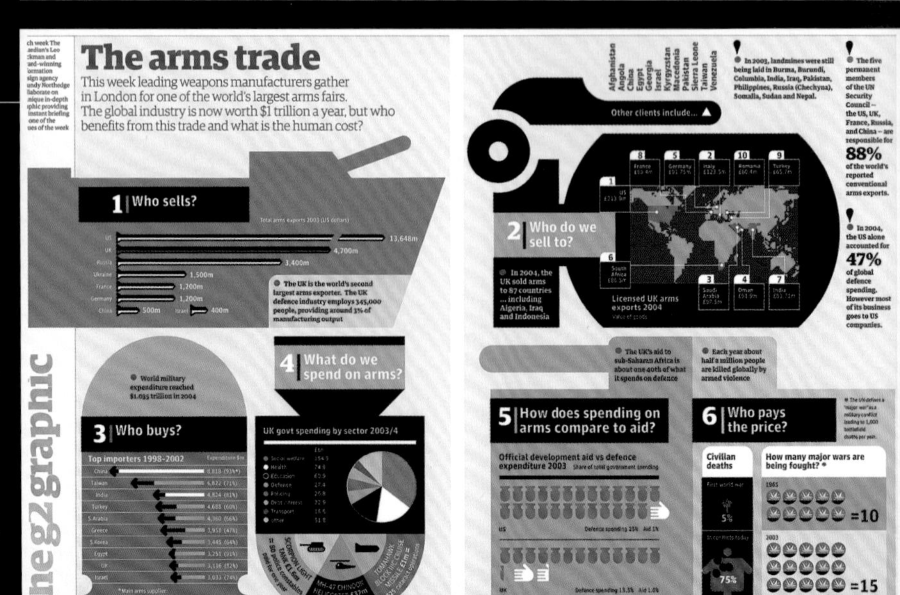

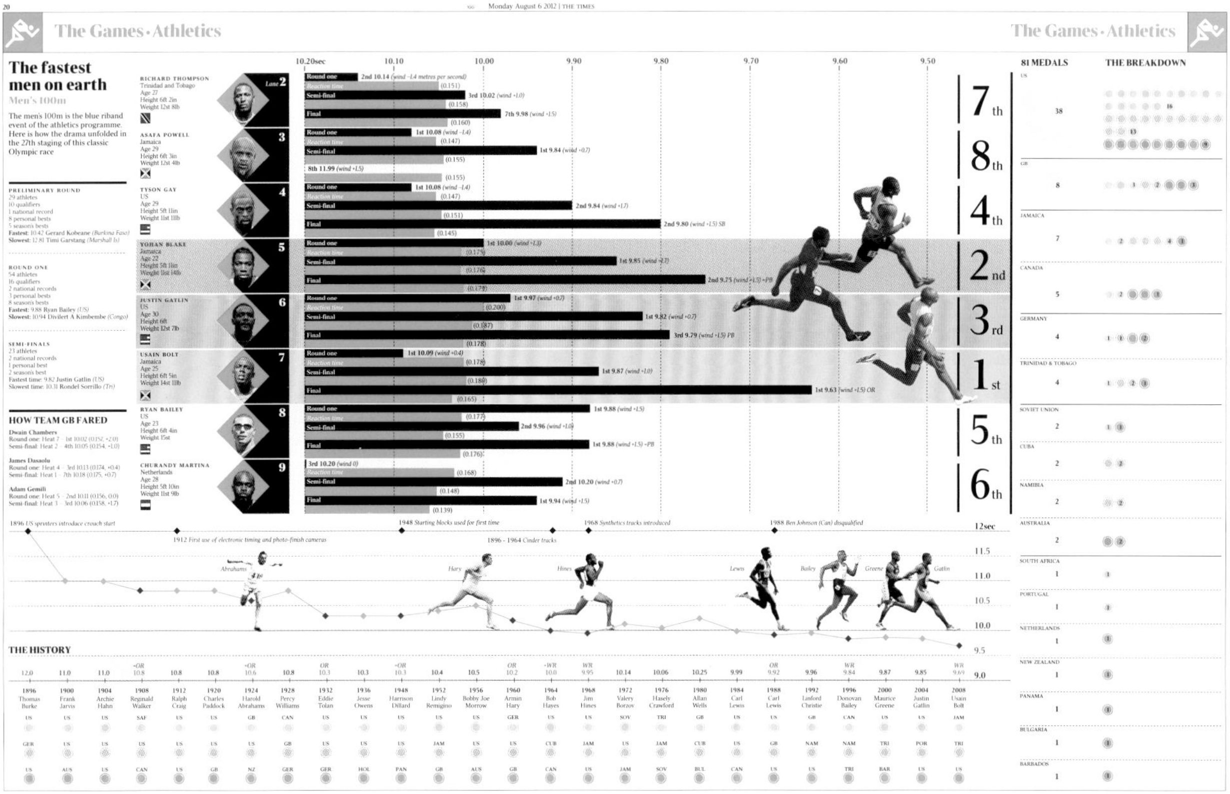

20   Monday August 6 2012 | THE TIMES

The Games · Athletics

The fastest men on earth

Men's 100m

The men's 100m is the blue riband event of the athletics programme. Here is how the drama unfolded in the 27th staging of this classic Olympic race

PRELIMINARY ROUND
29 athletes
10 qualifiers
1 national record
8 personal bests
5 season's bests
Fastest: 10.42 Gerard Kobeane *(Burkina Faso)*
Slowest: 12.81 Timi Garstang *(Marshall Is)*

ROUND ONE
54 athletes
16 qualifiers
2 national records
3 personal bests
8 season's bests
Fastest: 9.88 Ryan Bailey *(US)*
Slowest: 10.94 Divilert A Kimbembe *(Congo)*

SEMI-FINALS
23 athletes
2 national records
1 personal best
2 season's best
Fastest time: 9.82 Justin Gatlin *(US)*
Slowest time: 10.31 Rondel Sorrillo *(Tri)*

HOW TEAM GB FARED

Dwain Chambers
Round one: Heat 7 – 1st 10.02 (0.157, +2.0)
Semi-final: Heat 2 – 4th 10.05 (0.154, +1.0)

James Dasaolu
Round one: Heat 4 – 3rd 10.13 (0.174, +0.4)
Semi-final: Heat 1 – 7th 10.18 (0.175, +0.7)

Adam Gemili
Round one: Heat 5 – 2nd 10.11 (0.156, 0.0)
Semi-final: Heat 3 – 3rd 10.06 (0.158, +1.7)

Scale: 10.20sec · 10.10 · 10.00 · 9.90 · 9.80 · 9.70 · 9.60 · 9.50

Lane	Athlete	Round one (reaction time)	Semi-final (reaction time)	Final (reaction time)	Place
Lane 2	**RICHARD THOMPSON** Trinidad and Tobago, Age 27, Height 6ft 2in, Weight 12st 8lb	2nd 10.14 *(wind -1.4 metres per second)* (0.151)	3rd 10.02 *(wind +1.0)* (0.158)	7th 9.98 *(wind +1.5)* (0.160)	7th
3	**ASAFA POWELL** Jamaica, Age 29, Height 6ft 3in, Weight 12st 4lb	1st 10.08 *(wind -1.4)* (0.147)	1st 9.84 *(wind +0.7)* (0.155)	8th 11.99 *(wind +1.5)* (0.155)	8th
4	**TYSON GAY** US, Age 29, Height 5ft 11in, Weight 11st 11lb	1st 10.08 *(wind -1.4)* (0.147)	2nd 9.84 *(wind +1.7)* (0.151)	2nd 9.80 *(wind +1.5) SB* (0.145)	4th
5	**YOHAN BLAKE** Jamaica, Age 22, Height 5ft 11in, Weight 11st 14lb	1st 10.00 *(wind +1.3)* (0.175)	1st 9.85 *(wind -1.7)* (0.176)	2nd 9.75 *(wind +1.5) =PB* (0.179)	2nd
6	**JUSTIN GATLIN** US, Age 30, Height 6ft, Weight 12st 7lb	1st 9.97 *(wind +0.7)* (0.200)	1st 9.82 *(wind +0.7)* (0.187)	3rd 9.79 *(wind +1.5) PB* (0.178)	3rd
7	**USAIN BOLT** Jamaica, Age 25, Height 6ft 5in, Weight 14st 11lb	1st 10.09 *(wind +0.4)* (0.178)	1st 9.87 *(wind +1.0)* (0.180)	1st 9.63 *(wind +1.5) OR* (0.165)	1st
8	**RYAN BAILEY** US, Age 23, Height 6ft 4in, Weight 15st	1st 9.88 *(wind +1.5)* (0.177)	2nd 9.96 *(wind +1.0)* (0.155)	1st 9.88 *(wind +1.5) =PB* (0.176)	5th
9	**CHURANDY MARTINA** Netherlands, Age 28, Height 5ft 10in, Weight 11st 9lb	3rd 10.20 *(wind 0)* (0.168)	2nd 10.20 *(wind +0.7)* (0.148)	1st 9.94 *(wind +1.5)* (0.139)	6th

1896 *US sprinters introduce crouch start* · 1912 *First use of electronic timing and photo-finish cameras* · 1948 *Starting blocks used for first time* · 1896 - 1964 *Cinder tracks* · 1968 *Synthetics tracks introduced* · 1988 *Ben Johnson (Can) disqualified*

12sec · 11.5 · 11.0 · 10.5 · 10.0 · 9.5 · 9.0

Abrahams · Hary · Hines · Lewis · Bailey · Greene · Gatlin

THE HISTORY

	1896	1900	1904	1908	1912	1920	1924	1928	1932	1936	1948	1952	1956
Time	12.0	11.0	11.0	=OR 10.8	10.8	10.8	=OR 10.6	10.8	OR 10.3	10.3	=OR 10.3	10.4	10.5
Winner	Thomas Burke	Frank Jarvis	Archie Hahn	Reginald Walker	Ralph Craig	Charles Paddock	Harold Abrahams	Percy Williams	Eddie Tolan	Jesse Owens	Harrison Dillard	Lindy Remigino	Bobby Joe Morrow
	US	US	US	SAF	US	US	GB	CAN	US	US	US	US	US
	GER	US	US	US	US	US	US	GB	US	US	US	JAM	US
	US	AUS	US	CAN	US	GB	NZ	GER	GER	HOL	PAN	GB	AUS

	1960	1964	1968	1972	1976	1980	1984	1988	1992	1996	2000	2004	2008
Time	OR 10.2	=WR 10.0	WR 9.95	10.14	10.06	10.25	9.99	OR 9.92	9.96	WR 9.84	9.87	9.85	WR 9.69
Winner	Armin Hary	Bob Hayes	Jim Hines	Valery Borzov	Hasely Crawford	Allan Wells	Carl Lewis	Carl Lewis	Linford Christie	Donovan Bailey	Maurice Greene	Justin Gatlin	Usain Bolt
	GER	US	US	SOV	TRI	GB	US	US	GB	CAN	US	US	JAM
	US	CUB	JAM	US	JAM	CUB	US	GB	NAM	NAM	TRI	POR	TRI
	GB	CAN	US	JAM	SOV	BUL	CAN	US	US	TRI	BAR	US	US

The Games · Athletics

81 MEDALS		THE BREAKDOWN
US	38	16 · 13 · 9
GB	8	1 · 2 · 3
JAMAICA	7	2 · 4 · 1
CANADA	5	2 · 1
GERMANY	4	1 · 1 · 2
TRINIDAD & TOBAGO	4	1 · 2 · 1
SOVIET UNION	2	1 · 1
CUBA	2	2
NAMIBIA	2	2
AUSTRALIA	2	2
SOUTH AFRICA	1	1
PORTUGAL	1	1
NETHERLANDS	1	1
NEW ZEALAND	1	1
PANAMA	1	1
BULGARIA	1	1
BARBADOS	1	1

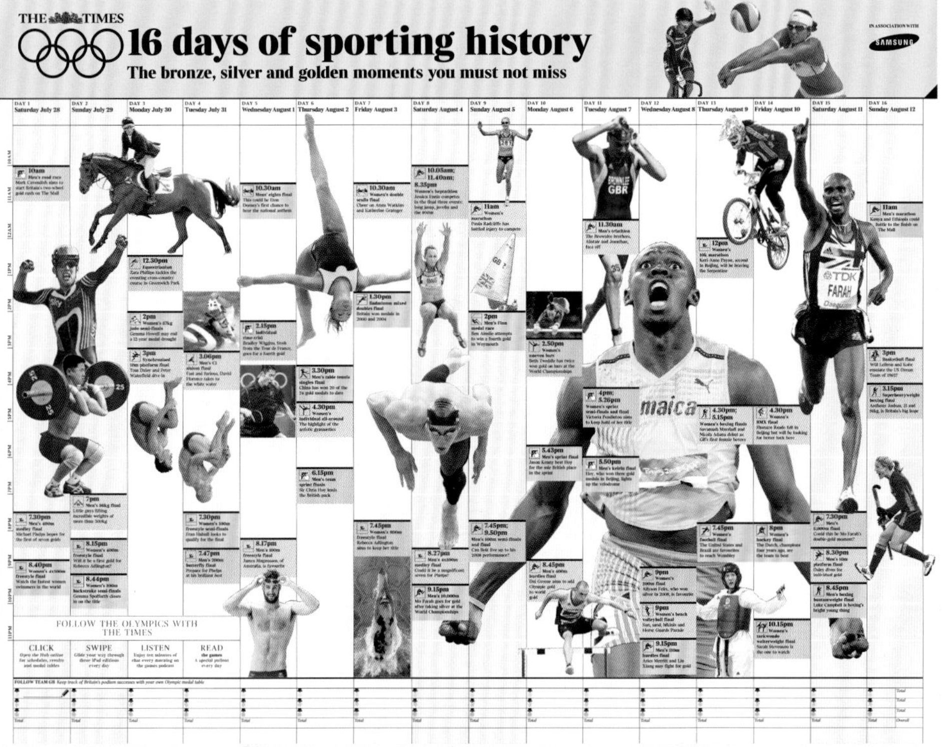

风格一致，不等于单调

编辑设计师最具挑战性与最享受的任务，就是创造出独特、有个性的刊物，在清楚说明它属于某知名品牌之余，又要避免各期外观或质感都一样，以免缺乏变化与新意。如何做到这一点?运用灵活的网格线系统，分页时确保类似跨页之间有其他页面穿插并发挥创意，灵活运用设计元素，这些都是必备技能。

日报或新闻周刊的截稿期限仓促，生产周期短，设计要以功能性和易读性为导向。因此，设计师要拟定解决方案，并建立能快速制作版面的网格线和设计系统，加快排版与送印过程，设计的过程也要井然有序。不过，只要比较几种类似的刊物，就可以发现在有秩序的架构之下，依然有很多机会可以探索新奇的结构，展现不同设计，呈现出全然不同的成果，右图的欧美新闻周刊就是一个很好的例子。

年刊、季刊或月刊的架构可以更为松散，图片的预算也更多，在处理网格线、字体与图片方面也更灵活，因此也可以尝试更多版式。不过，这种做法也有缺点：自由度越大，维持品牌调性就越难。因此，设计师需要学会权衡，在那些需要保持不变的元素（品牌和识别）与那些会随着每期刊物变化的元素之间找到平衡。

德国新闻周刊*Stern*的封面、专题报道和图片跨页让新闻更生动，符合大众期待。周刊的裁剪手法十分大胆，上图中，杂志的边缘将面部对半切开，简洁明了的三栏网格线与一直流行的航拍图，都能吸引广大读者。

右图是英国财经杂志《经济学人》的封面，诙谐又很抢眼，和专题报道区的平静有序呈现出鲜明对比。这也说明了，杂志的每一期、每一页都要想办法保持创意，抓住读者的心。

The Economist

Why Britain has soured on immigration

Ending Iran's spin cycle

Has America's housing bubble burst?

The limits of air power

A step forward for stem cells

AUGUST 26TH–SEPTEMBER 1ST 2006 www.economist.com

WHO
KILLED THE
NEWSPAPER?

£3.25

34

9 770013 061169

大学毕业后，恰好碰上独立杂志兴起的浪潮。当时桌面计算机与出版软件出现，推动桌面出版革命，在数百个专业领域中催生成千上万的杂志。罗瑞里决定也投身其中。

罗瑞里的第一份刊物是关于音乐影片的行业刊物，用他自己的话来说，那份杂志“惨不忍睹”。他的第二份刊物则是大众文化杂志*Speak*，这份杂志被公认为这类型中的佼佼者。这份杂志的成功，很大程度上要归功于罗瑞里与出版物的艺术总监马丁·威尼茨基（Martin　Venezky），这两人可以说是不打不相识。从一开始， 罗瑞里就对杂志的设计有强烈的想法：

> “在当时，报摊的杂志不计其数，我希望*Speak*在视觉上脱颖而出。艺术总监需要愿意挑战既有形式，也需要让作品保持战，因为得到我的认可并不容易。”

罗瑞里和马丁·威尼茨基（Martin　Venezky）吵得不可开交，还曾对簿公堂，闹得满城风雨，但罗瑞里其实很了解设计师对杂志的影响：

> “艺术总监必须要有求知欲、对杂志内容好奇。他不能只想循规蹈矩，或不顾杂志内容，把杂志当成自己的作品来炫技。”

罗瑞里深知威尼茨基具有这些优点：

> “他会阅读、思考，失败大过成功吧，但我知道和马丁共事之后，恐怕无法再和其他设计师合作了。”

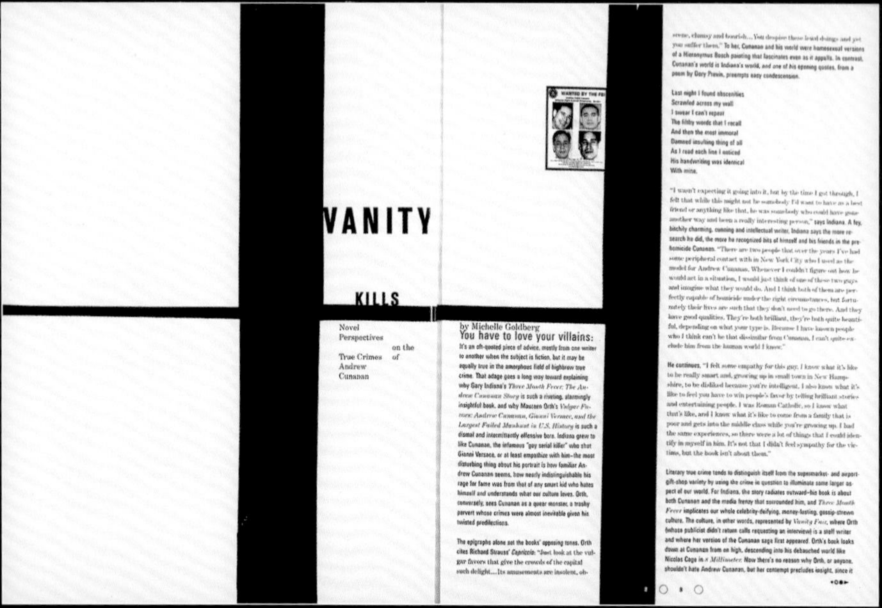

“编辑与艺术总监要主动积极地密切合作、彼此尊重，还要能争论，有时也要愿意认输。”

——马丁·威尼茨基，*Speak*艺术总监

刊物的标准风格与摄影

杂志是连续出刊的，所以每一期都要保持熟悉的外观，易于读者辨认。想要制作出独特但具有连贯性的外观，可从几个层面着手。首先，开本、用纸与识别设计要一致，当然这些元素也不会每期都变化。再下来，就是一些零碎的视觉元素——例如照片的风格走向，或一些规则，如某些类型的标题总是超过两行等。

维持风格的最重要工作，就是好好设定样式表单，让页面上的每个元素都有详细的排版说明。部分章节模板、样式表是可以预先设计好的，包括字体、大小、颜色、各项参数以及应用于页面上的每个元素（标题、引文、正文、图片出处、脚注等）。这么一来，只要套用设计好的样式，就能自动呈现一致化的细腻设计。

IN PERSPECTIVE

FOXY SPACES

Leisure in Copenhagen? The city has plenty to offer for the design-conscious crowd. Think Arne Jacobsen and the Little Mermaid. But until recently the city hasn't contained a proper designer hotel. Now there is plenty of room for leisure activities after the opening of Hotel Fox in April. Twenty-one artist groups from around the globe have designed the rooms individually, creating unique spaces and experiences in each room. The result is impressive.

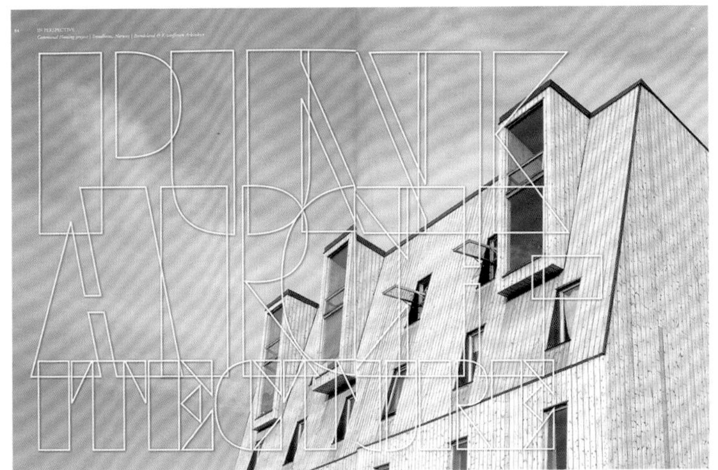

上图是*Inside*杂志“侧写（in Profile）”单元的两个跨页。这两份跨页的版面虽然看起来采用了不同的设计，但是都大胆利用字体与设计元素。许多潜在因素会影响版面设计的流程，例如每篇文章的页数、字数、图片裁剪，甚至是刊物中的广告。在安排杂志架构时，也必须考虑到色彩与文字流动的方式。

改版的时机与原因

设计师可以调整版面元素，让刊物呈现新风貌，以保证出版物与时俱进，避免与时代脱节（例如会适应当下读者的喜好，把粗的字体以同一个字体集的一般字体或细字体代替）。不过，即使是设计得再好、历史悠久的刊物，也会有过时、生机不再的那天。在这样的情况下，就需要考虑大幅变动或改版。许多刊物会在销售下滑或经济衰退时改版。在经济衰退时期，广告收入下降，他们需要提高销量，然而这些策略也很冒险，毕竟改版在吸引新读者的同时，也会失去一些现有读者。

> **编辑设计为报道提供被阅读与诠释的架构，它不仅要建构刊物的文章结构（以及隐含的选辑结构），也要对个别报道做特殊处理，因为有些报道内容可能会违反刊物的主张。**
>
> ——马丁·威尼茨基，*Speak*艺术总监

改版的最佳理由，就是为了符合并且能反映读者需求，例如锁定十六岁读者的杂志，过了五年之后，由于大环境的时尚、品位与风格已出现很大变化，不得不改版才能跟上潮流。重要的是，不能只顾着讨好读者，而是要参考文化趋势与变化，让刊物能融入大环境。别忘了，改版不能在真空状态下进行。如果视觉趋势会随着时间转变，刊物的其他构成元素也会改变，因此要自行检查内容与调性，并谨慎处理，以确保没有任何元素脱节，而整体刊物也能呈现连贯风貌，明智地向前发展。*Flaunt*杂志在2006年就是采取这种态度彻底改版，还换了全新Logo与专属字体。李·科尔宾解释：

> “整份杂志都做了重新设计，具有更多变化与新的识别性。必须要舍弃很多旧有的视觉特性，才能为新想法腾出空间。杂志会不断变化，何况我们是月刊，因此要能顺应环境变化，让杂志发出新声音。”

虽然杂志经常改版，但是报纸却很少大刀阔斧地彻底改版。对每日出刊的刊物来说，大幅改版会劳师动众，弄得人仰马翻。波特解释道：

> “报纸往往是受到市场刺激才会改版。一般而言，编辑都会觉得眼前的报纸已经足够好了，只有销售开始下滑之后，他们才明白不改版不行。但有意思的是，现在全世界新闻读者群的阅读量都在下滑，而来自电视与网络的压力，迫使记者与设计师质疑自己是否不懂现代读者的需求。以前从来没有过这么多报纸同时在改版。”

网络媒体、信息传播方式和人口变化，都会对编辑设计产生影响：开本变小、页面更趋一致、浏览方式也越来越简单。正如马里奥·加西亚所言：“互联网创造出拥有满手信息却没耐性的读者，他们想要纸质刊物有清楚的文字层次、便捷的浏览方式，能让视线在纸质页面上快速移动。”

过去五年，印前技术也出现变化。为了应对最新变化，应该上网看在线教程，自学一些新知识，也可以向印刷厂请教。如果想更系统地掌握相关知识，了解正确的印刷条件，可参考安柏洛斯（Ambrose）与哈里斯（Harris）的著作《视觉设计大全：印前制作、字体编排、平面设计、插画》（*The visual design box set: pre-press and production, typography, graphic design, illustration*），由Fairchild出版社在2012年出版。

In Review TRANSPORTATION: Stokke Stroller (*reviewed by Ayse Birsel and Bibi Seck*) ..... PRODUCT: Push Braces (*reviewed by Michael Wiklund*) ..... SOFTWARE: Wild Divine Computer Game (*reviewed by David Womack*) ..... BOOK: *Disruptive Pattern Material* (*reviewed by Tom Vanderbilt*)

Seg 12 Fev Edição Lisboa

Segunda-feira 12 Fevereiro 2007
Ano XVII, nº 6163
Portugal: 0,90€ (IVA incluído)
Espanha: 2,00€ (IVA incluído)
Director: José Manuel Fernandes
Directores adjuntos: Nuno Pacheco e Manuel Carvalho

Público

Cinema
Eastwood mostrou uma obra-prima

Pele do corpo
Mais de 200 bactérias, um verdadeiro zoo

YouTube
Tem lixo e tem arte, como encontrar os melhores vídeos?

Barack Obama
Quem é o homem que quer mudar a América

Sim 59% Não 41% Abs 56%

Agora sim

"Sim" obteve mais um milhão de votos do que no referendo de 1998

Abstenção acima de 50 por cento não torna o voto vinculativo

PS promete alterar depressa a lei e combater o aborto clandestino

● Desta vez o "sim" à despenalização do aborto venceu. Numa noite que foi curta no apuramento de resultados, confirmaram-se as expectativas formadas pelas sondagens. O "sim" ficou muito próximo dos 60 por cento dos votantes, num referendo que voltou a não atrair a metade de eleitores que a lei exige para que se torne vinculativo. De qualquer forma, houve imediato consenso entre vencedores e vencidos em considerar que a abstenção não é um obstáculo à passagem a lei do resultado da consulta popular.

O Partido Socialista compromete-se a tratar o tema com carácter de urgência em sede parlamentar. Nos vencedores da noite surge, de forma incontornável e destacada, José Sócrates, que imprimiu ao PS uma posição sem equívocos a favor do "sim". Do lado perdedor, surgem na linha da frente a Igreja católica e Marcelo Rebelo de Sousa, na sua qualidade de "pai" político de um referendo que, pela segunda vez, não conseguiu atrair mais de metade dos eleitores.

A análise do mapa de resultados confirma um país dividido ao meio, com o Sul a favor do "sim" e grande parte do Norte a votar "não".

Deste referendo fica ainda a novidade da participação dos movimentos cívicos, cuja mobilização, de ambos os lados, animou uma campanha onde os partidos não foram os principais protagonistas. Agora a palavra é-lhes devolvida na Assembleia.

Tudo sobre a noite do referendo → páginas 2 a 17

Balanço 'independente'

MAI fez ajuste directo ilegal de estudo dos fogos

● O Ministério da Administração Interna contratou através de um ajuste directo no valor de 375,7 mil euros a empresa que fez "pela primeira vez" o balanço "independente" das operações de combate aos incêndios florestais do ano passado. Três juristas que analisaram o despacho do secretário de Estado consideram que o mesmo é ilegal. → Nacional, 20/21

Este espaço é seu.
Envie a sua mensagem de confiança a quem mais gosta para vidaconfiante_publico@axa.pt, indicando o seu nome, idade e localidade e o nome do destinatário. Quem sabe, será publicada aqui.
AXA
PROTECÇÃO FINANCEIRA

Eleições em França

Ségolène Royal propõe pacto de esquerda

● Ségolène Royal, a candidata socialista que aspira a ser a primeira mulher Presidente de França, anunciou ontem um plano em 100 pontos que assinala uma clara mudança de política à esquerda. Prometeu maior intervenção do Estado na economia e um modelo de redistribuição ainda mais generoso, tentando demarcar-se do liberalismo de Nicolas Sarkozy. → Mundo, 24

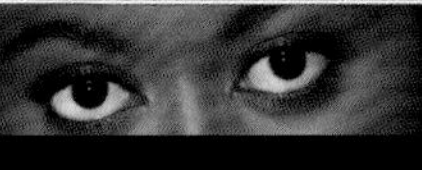

FIXE A PRESTAÇÃO MAIS BAIXA
Transfira já o seu crédito e reduza a prestação até 50%

Opposite: Rockwell's Design for the Kodak Theater in Hollywood – the Academy Awards' new home. Left, David Rockwell.

A LIFE IN THE THEATER

Theater, with a lower-case 't,' has fueled David Rockwell's drive for the past 20 years. But it's collaboration, the new and the scary, and the moment of not having the answer that keep the creative engines revved for this unpretentious architect who wears his passions as comfortably as his trademark flannel shirt.

THE ROCKWELL GROUP'S LOFT on New York's Union Square West has that off-Broadway feeling. Its tiny, industrial-grunge lobby is set like a stage with Rockwell icons: a giant, molded wing chair from Grand Central Station's Food Court, a Victorian red-velvet chaise from Bar Bat and, for high-tech contrast, a wall clock with a laser-light doggy that ticks off the seconds. Just inside, there's a birch-twig bow from Mohegan Sun Casino—or is it from Nobu? A richly layered collage of fabrics, images, grass, beans, flowers, pottery, stones, branches and ephemera—stuff of projects past and present—fills every cranny of this humming atelier, where theater begets Theater.

David Rockwell choreographs it all from the disarmingly simple office he shares with an assistant and a few defining objects—a photo of himself at age 12 with his mother and four brothers, costumed for a family musical; a snapshot of his grinning 2-year-old son Sam wielding crayons; and a collection of fine old kaleidoscopes, whose glass shards swirl into endless, mirrored patterns. "Kaleidoscopes let you see the familiar in an unfamiliar way," he tells a visitor. "That's pretty much what we do, too."

It was Rockwell's signature hospitality and entertainment-design projects—Planet Hollywood, Monkey Bar, The All Star Café, among many others—that first catapulted him to success in the mid-'90s. And his firm still turns out such dazzling entertainment venues as the 3,300-seat Kodak Theater in Hollywood, completed in time for this year's Academy Awards, or the 115,000-square-foot "Casino of the Sky" for Mohegan Sun's already vast complex in Connecticut.

But these days, The Rockwell Group is working its magic in unexpected places, such as the Carl Sagan Discovery Center at the Children's Hospital at Montefiore in Bronx, N.Y., or McCann-Erickson's new 18-story corporate headquarters in midtown Manhattan.

2003年，*I.D.*艺术总监科比·贝内兹里（Kobi Benezri）与他的前任艺术总监尼科·施韦泽（Nico Schweizer）合作共同操刀该杂志改版（左图）。“当我们动手设计时，就知道*I.D*必须更新外观。上一次改版是1992年由布鲁斯.莫（Bruce Mau）进行的。新设计和新刊物内容相辅相成，例如报道更广泛的设计领域、以不同角度讨论先前谈过的主题、增加新单元，观点更具批判性。我们要确保设计能具有同样的态度，呈现出丰富客观的信息。同时，我们不打算以哗众取宠的设计元素，强行给人留下印象。版面的设计目标很清楚，我们设法保持优雅，必要时加上我们的特色。”新的特色包括新字体：正文改用Gerard Unger设计的现代感字体Conrantc取代Scala字体。而Meta字体也在日后随着发展被替换成各种各样的字体。例如*Scape*的图文跨页及新的后段评论，“Crit”版块等都是新增的单元（左上）。波特2007年重新设计葡萄牙日报《公共报》时，也参考了2005年《卫报》的改版风格。

马里奥·加西亚曾为世界各地的很多报纸做了改版设计，推出全新版式。2005年，他重新设计了《观察家报》，从大报版式改成了柏林版式。他认为改变版式不是问题，不会有负面影响。由于空间变小，所以要更清楚地强调文字层次。刊物在考虑版面保留什么，去掉什么时，要根据重要性有一个排序。要好好研究现今的读者，探索视觉与刊物过去的发展，评估什么必须留下，什么可以删除。总之要搞清楚什么是真经典，什么不值得保留。报纸经常把旧材料看成宝贝古董，但其实只是破铜烂铁，不值得收藏。在重新设计《观察家报》时，加西亚运用文字编排，保留大报的优雅感，同时以配色营造年轻活泼的感觉。调查显示，读者更喜欢活泼感。他为不同单元选择不同的配色，确保配色可和页面元素相配。在改版时，他是从新闻的角度出发的，因为他坚持："读者看报纸是为了内容，而不是外观。设计的目的是强化内容。"

加西亚的十条改版法则

1．改版没有万能公式，每个刊物有自身的情况，要因刊制宜，因此要找出最适合的方式，重新思考产品定位。

2．要全面了解对该版刊物的期待、目标读者与改变幅度。我总是说，有些刊物改版只是小打小闹，但有些则非常深入彻底。

3．改版时，要重新思考刊物的四大故事结构，有所规划：文字编排、页面、架构与色彩。

4．第一步就应该思考故事结构，编辑如何在刊物中讲故事?应该创造多少说故事的技巧与样式?如何强调内容的重要性与层级?

5．就文字编排来说，至少要测试三种衬线字体与无衬线字体的组合，再从中选择最好用、最匹配的。

6．就页面架构来说，至少设计两种格线系统，并搭配不同的栏宽，或许最终也会同时采用两种模板。

7．在开始配色时，至少使用二十多种色彩组合，包括最暗到最明亮，还有介于中间的色彩，之后再从中设计出简单的配色（不要超出十种色调）。

8.一定要确保刊物的易读性，现在的读者耐心有限，没办法耐着性子一直阅读。所以要尽可能设计简单易读的浏览策略，这是改版流程中最重要的事情。

9．检查整份刊物的分页结构，也就是内容出现的顺序。什么时候要采用或不用某些元素?是否需要改变顺序?

10．和编辑、记者密切合作，他们会在刊物视觉变化的过程中，从新闻视角提供相关的建议。

内容来源：《材料、工艺、印刷：平面设计的创意解决方案》，作者丹尼尔·梅森（Daniel Mason）

NEW YORK

Die Narbe

Fast fünf Jahre nach dem 11. September sind auf dem Platz, auf dem das **World Trade Center** stand, lediglich die Untergeschosse fertig. Eine Rampe führt direkt in die Mitte des so genannten Fußabdrucks des Nordturmes. Rechts, etwas oberhalb, ist noch der halbe footprint des Südturms zu sehen. Im Winkel zwischen Rampe und Straße ist die Gedenkstätte im Bau. In der äußersten Nordwestecke von Ground Zero (auf dem Bild links unten) soll bis 2009 der neue Freedom Tower entstehen und mit seiner Höhe von 1776 Fuß (541 Meter) an das Jahr der amerikanischen Unabhängigkeitserklärung erinnern. Auch den Bezugstermin 2010 glauben die Bauherren einhalten zu können. Das benachbarte World Financial Center am Ufer des Hudson mit seinen türkisgrünen Dächern strahlt schon seit fast vier Jahren in neuem Glanz.

FOTO: VINCENT LAFORET/POLARIS/STUDIO X

medizin

„Warten Sie mal, ich schalte meinen Gehirnschrittmacher ein“

Helmut Dubiel ist Hochschullehrer. Er leidet an **PARKINSON.** Tief in seinem Kopf sitzen zwei Sonden. Er steuert sie mit einer Fernbedienung. Ein kleiner Stromstoß – und er kann sprechen. Ein größerer Stromstoß – und er kann gehen. Zumindest eine Zeit lang

Von **ARNO LUIK** und **VOLKER HINZ** (Fotos)

STERN 35/2006 135

避免版面单调的秘诀与技巧

尝试改变构建版面结构的方式。

试试看以色块或去掉背景的图片来建构页面，这样能在文字、图片与其他元素周围搭配留白空间与几何元素。

多参考别人的设计，把每一种版面当作单独的作品。

*Stern*杂志在专题报道页面使用简洁的三栏网格线（左图）以及广受欢迎的航拍图（上图），这吸引了广大的读者。

工作坊5

完成与展示作品

目标

继续完善在工作坊4制作的跨页，提升品质。

操作步骤

详细检查版面上的所有元素，再用高质量的纸张打印出来。

1. 仔细检查所有细节，例如基线是否对齐、是否出现寡行（若有排列得不美观的部分，则需要重新分行）。如果有照片就要注明出处来源，并检查整篇文章是否有错别字，这对版面上的任何文字来说都很重要。作品集不出现错别字，才能展现出重视细节的态度。

2. 一旦完成校对，先用便宜的纸张把版面打印出来，再检查文字是否太大（在屏幕上常容易把字号设得太大）。之后，如果希望作品集里的作品颜色很正，就用优质的彩色打印机与高质量纸张打印出来。这份跨页作品也存一份PDF文件，加入在线作品集。不过，文件不要存得太大，以免下载花费太多时间。

3. 这些设计法则可以反复练习，扩展到不同的设计作品与平台。为你的杂志网站设计首页时，可进一步规划浏览方式，增加互动工具，甚至实际操作看看。想想看，你的杂志是不是在任何平台上都可以很好地呈现并供人阅读?别光把PDF放到网站上，就觉得这是在线杂志了。如果字号太小，难以阅读，就无法达成阅读目的。别忘了，编辑设计师是要服务内容，包括文与图。不过要对自己有信心，别担心与众不同。即使作品不如你喜欢的设计师设计得那么美，但只要有自己的价值，就值得放在作品集中。

这份名为《逃脱》（*Escape*）的旅游杂志是由学生创作的，制作者是本科生桑德拉·奥图凯特（Sandra Autukaite）。一开始她决定用网络上的图片作为素材开展视觉设计。她为照片加上简单醒目的图框，显示出摄影是这份杂志的主要元素。到了最后阶段时，奥图凯特决定自己拍照，如此一来，她就掌握了所有素材的版权。虽然学生的练习作品不太可能出版，不过还是需要注意版权问题，尊重别人的创作成果。

ESCAPE

AMPHIBIAN ISSUE

JULY

27 TUTORS
500 STUDENTS
THOUSANDS OF ALUMNI
ONE ORGANISED MESS

BRINGING ORDER
TO THE CHAOS

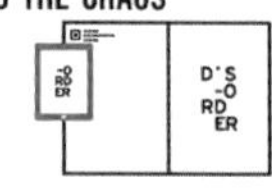

"STRUCTURE
HELPS ME THINK"

这份作品是团队合作的成果，刊头采用了手写字体，并和在学校拍的照片拼贴起来，做出“乱中有序”的刊物。设计者本·西尔弗敦（Ben Silvertown）还把这份刊物送到印刷厂印出（右图），也模仿了平板电脑的版面，并在作品集中加入影片。其他团队成员也有功劳，每个人都把这项设计放在自己的作品集，并注明所有参与者的姓名。这是团队设计的惯例。

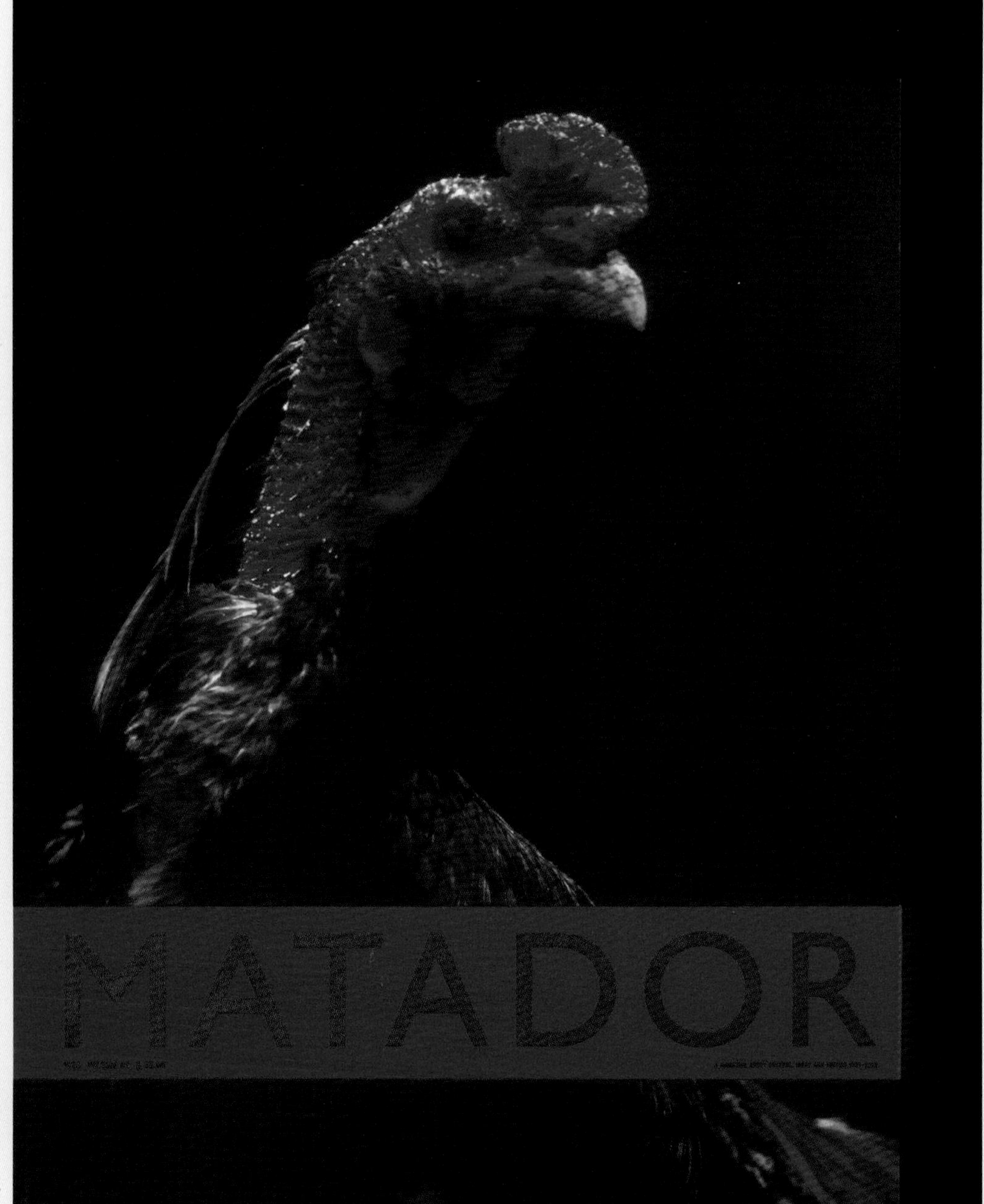
MATADOR

第七章　回顾过去，展望未来

所有设计工作者都要能洞察潮流趋势、文化变迁与当下的时代精神，这些特质对编辑设计师来说更重要。编辑设计师必紧跟潮流，走在视觉时尚和文化的前沿，以更广的视野来探讨文化、设计趋势与新的媒体技术。此外，编辑设计师也经常能够通过回顾历史上影响力深远、具突破性的作品，或从设计大师身上得到启发，获得编辑设计方面的灵感和洞察。过去八十年来，设计大师通过各种印刷技巧、制图风格、格线结构、文字编排与符号设计，创作出既如艺术品般赏心悦目，又能反映社会文化的刊物。历史上这样优秀的设计师非常多，本章主要介绍其中几位佼佼者。当代设计师在研究前辈大师作品时，应该注意观察和思考下列几点。

1.设计动机及隐含原则。
2.成功刊物与当下时空背景的关系，为何在特定的背景下做出特定的设计。
3.如何通过分析过去的作品，洞悉未来潮流趋势。

回顾过去
——动机和隐含原则

编辑设计师在研究作品时，经常只关注当代作品。研究当代作品当然有必要，能从中了解文化变迁，文字编排、配图、摄影、纸张的最新趋势。然而，研究过去的作品同样重要。设计师在回顾过去时，应该关注什么？当然是设计方法和设计逻辑，但是如果不了解作品背后的设计原则，是很难理解这些作品的。这些设计方法和逻辑，与当时的艺术、文化运动密不可分，又反过来在政治和社会背景中体现。举例来说，20世纪30年代的平面设计受到包豪斯学派的影响，重视机械化与功能性，这同时也呼应当时西方社会工业化及东欧社会主义兴起。20世纪70年代中期，纳威·布罗迪（Neville Brody）的文字编排与几何样式就源自俄国构成派的反压迫精神。现代编辑设计师通过对这些背景和原则的探索和理解，能够获得一套工具，用来处理自己对设计和当下文化发展关系的理解，更好地了解并满足自己的读者。向设计大师取经的方式，不是盲目模仿，而是理解他们的作品，从中学习其成功的原因。

理解作品所处的时空背景

想要了解一份刊物，就要全面了解它所处的大环境（也就是上文所说的动机和隐含原则），之后再重点关注个别作品的构成细节，了解为什么这份刊物要为读者设计这样的版面。可以试着拆解版面上所有的元素，分析它们各自的功能，再研究它们组合起来的效果。以*Face*杂志为例，纳威·布罗迪深知另类文化刊物可以借鉴非主流文化的影响与风格，为当代元素增添前卫的感觉，建立刊物的定位。他明智地选择了一种恰当的视觉风格，让政治品格尚未受到污染、文化立场清晰的年轻人，一看就明白这样设计的根源与意义。他通过版面上的文字编排、形状及几何元素（线条、方块与比例）来传达这些原则，产生了非常好的效果。作品新颖、大胆、惊喜不断，完全符合读者的偏好。

以史为鉴，继往开来

接下来会列出许多影响力深远的设计师与刊物。从这些例子可看出他们所处时代的知识、道德与文化氛围，会深深影响设计运动，也进一步左右编辑设计的风格。设计过程不是闭门造车，也不会因为时代精神转变后就失去意义。过去的作品或许不再流行，但如果能分析并了解其设计原则，总能从中受益。

与过去不同的是，现在与他人分享设计灵感变得更容易。人们可以轻易取得数字资料和影像信息。更重要的是，很多设计师有自己的社交媒体，例如推特、Facebook和博客，并常常更新、分享新作品。设计师有很多机会向优秀的业界设计师学习。许多编辑设计师与艺术总监乐于分享、不藏私，为设计界带来源源不绝的活力。接下来，我们将看到许多值得追随与学习的当代先锋及启发了一代代设计师的历史巨匠。他们都是推动编辑设计走过一代又一代的大功臣，每一位都可以给人以启发。

名人堂：设计大师与出版作品

M. F. Agha

Mehemed Fehmy Agha（多称为M.F.Agha）是“艺术总监”这一职位的开山鼻祖。他是出生于俄国的土耳其裔设计师，深受构成派影响，曾设计过德国版*Vogue*。当年康泰纳仕到欧洲寻找设计人才，想把现代欧洲风格引进自家刊物时，发掘了他。1929年，Agha担任康泰纳仕旗下的旗舰刊物——美国版*Vogue*的艺术总监，他果然不负众望，让刊物名声大噪。后来他又负责设计*Vanity Fair*与*House & Garden*。Agha的设计清新有活力，总能为刊物带来新气象。他率先使用无衬线字体，也大胆尝试印刷与摄影的新技术，例如蒙太奇、双色套印与全彩摄影，并尽量用照片取代时装插画，成功做出以照片为主的版面。甚至设计出血图片的大跨页，大气的做法让读者眼睛一亮。他邀请塞西尔·比顿（Cecil Beaton）和爱德华·韦斯顿（Edward Weston）在内的顶尖摄影师，也委任过马蒂斯（Matisse）和毕加索（Picasso）创作，这些都是美国杂志界的首创，领先其他杂志很多年。

Agha把欧洲现代主义的风格原则运用到*Vanity Fair*中，并引进美国市场。他的字体运用既简单又有系统，自由使用各种设计元素与页面留白，让版面鲜活、有变化。他善于利用设计元素的位置与大小，例如破天荒地把小小的标题放在页面底部的留白空间，使它们像漂浮在页面中一样。他将传统的装饰性元素全部去掉，版面变得豁然开朗。元素间的大小比例关系是他用来建构视觉风格的主要手段。

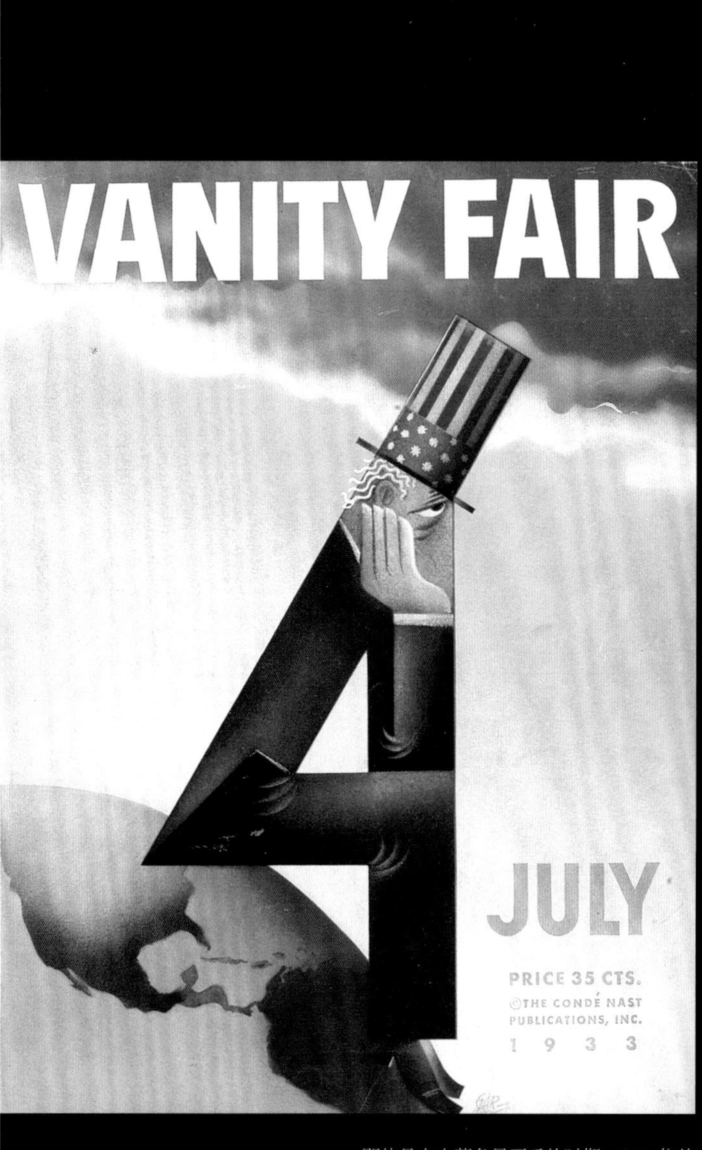

即使是在大萧条最严重的时期，Agha依然能够做出抢眼又优雅的设计，幽默而清晰地传达信息。

阿列克谢·布鲁多维奇（Alexey Brodovitch）

布鲁多维奇来自俄罗斯，1934年到1958年担任《时尚芭莎》的艺术总监，探索出许多日后被视为艺术总监必备能力的技术。他重新定义了艺术总监的角色，除了顾及版面设计，还要负责刊物架构，并委托合适的人来为刊物量身定制所需要的素材。布鲁多维奇受到达达主义与构成主义等现代主义运动与艺术设计风格的影响，为杂志引入了不对称版面，带来简洁有动感的风格。在此之前，美国刊物都是充满无意义装饰元素的静态页面。20世纪20年代在欧洲时，他就开拓单纯的“现代”图像风格，日后更致力于采纳当时革命性的新观念（例如早期的抽象表现主义），到了20世纪50年代，他的风格已然是优雅的代名词。布鲁多维奇善于处理留白空间与低调色彩、精准干净的文字编排（以Bodoni字体为主）及戏剧性的照片与精彩的跨页。

布鲁多维奇善于发掘与培养新摄影人才，例如欧文·佩恩（Irving Penn）和理查德·阿维顿（Richard Avedon）。他还为美国读者引入了欧洲前卫摄影师与艺术家，例如卡桑德尔（A.M.Cassandre）、达利（Salvador Dali）、亨利·卡蒂尔-布雷松（Henri Cartier-Bresson）与曼·雷（Man Ray）。他把摄影作品当作主视觉，使版面明亮宽敞、充满动感与表现性，而这些都是如今我们习以为常的杂志风格。他非常重视时尚摄影的故事性，让模特出外景，精心安排画面中场所、人物行为和服装之间的关联性。

布罗多维奇作品最核心的就是对创新和试验的渴望：这是一种本能，排斥理性与教条，不断变化，与时俱进。

放風箏*

We're not being flip. We mean it literally. When the April wind is up and you feel if you stay indoors another minute you'll burst—go fly a kite! We can just hear you thinking, "Girls don't fly kites!" And we say, "Why not?" Why should one of the most exhilarating pastimes ever invented be limited strictly to males? For pure fun, it can't be beaten. It's a wonderful excuse for a spring picnic (see the party story on page 109), an unfailing source of delight at the beach in summer. Store-bought kites cost as little as they weigh; making a kite of your own may be just what your creative instincts yearn for. So get yourself a kite. Tie your spirits to the tail and send them soaring into the upper air. You'll soon understand the enchantment kite-flying has exerted for something like twenty-five centuries.

The greatest "fliers" of ancient times were the Asiatic peoples—Chinese, Japanese, Koreans—and primitive tribes like the Maoris. The latter believed that kites protected them from evil spirits and often kept them flying from their roofs all night. Sometimes these kites were equipped with reed whistles which emitted a shrill note in the wind. The Korean child's counterpart for a letter to Santa Claus was a paper containing birthday requests, sent aloft via kite. It was assumed that if the kite got up high enough, the gods would read the paper and take note. The Chinese believed they could transfer their own troubles to the shoulders of the gods through the medium of a kite. An hour's flying would leave a hitherto harassed Chinese cheerful and relaxed! Kite-flying has always been so popular in China that shopkeepers run to the street with their kites in between customers! And every year the ninth day of the ninth month is known as Kites' Day, when every Chinese male takes a holiday to fly his most magnificent kite.

Everyone knows about Ben Franklin's kite experiment to prove that lightning contained electricity. Kites have been used for scientific purposes since the middle 1700's—for testing weather conditions, taking aerial photographs, etc. And they have been employed in many interesting ways during wartime. Centuries ago, a Korean general sent a kite, with line attached, to the opposite bank of a river. A cable followed the line, forming the nucleus from which a bridge was built. The Japanese developed a man-carrying kite, invaluable in scouting the enemy's position. Many armies used to employ kites for signaling purposes. Now some airplane lifeboats are equipped with kites carrying a radio antenna which automatically signals an sos.

As the illustration shows, kites can be made in an amazing variety of shapes and sizes. Probably the most familiar to you is the two-stick, diamond-shaped kite (see directions on how to make and fly it on page 120). A kite similar to the "airplane" shown on the next page was used by the Navy for target practice—because of a special rudder it could be made to do nose dives and loop-the-loops in the sky. The box-kite can be flown to great heights and, due to its greater sturdiness, is the type most often used for carrying scientific instruments. The decorative Chinese and Japanese kites will fly at the slightest puff of air. The centipede floats with grace, the eagle's tail flutters, the fly's eyes spin round on their tiny bamboo axes as it soars through the sky.

Almost any paper can be used to make a kite, but tissue or a light wrapping paper is best. The sticks should be of spruce or bamboo and light cotton twine will be strong enough to fly a medium-sized kite. Tails provide wind-resistance and, while not always essential, are picturesque. They may be made by tying bits of paper or rag onto a piece of twine or by braiding thin strips of paper or rags.

Ready to fly? We'll wager you and your gang will find it's more fun than you've had in a month of windy days, and it won't be long before all the teens in town will be following your example, migrating to the open fields, and claiming their own share of the kite-lines.

*Go fly a kite

BY BARBARA MOENCH

104

西佩·皮涅尔斯（Cipe Pineles）

现在人们对摄影师、插画家、艺术家和编辑设计师能够去解读故事文本，并以自己个性化的方式阐释与呈现设计。其实这是西佩·皮涅尔斯（Cipe Pineles）的首创。1946年，西佩·皮涅尔斯为《十七岁》（*Seventeen*）杂志中一篇小说搭配了专属配图，从此编辑设计师才敢纷纷尝试大胆地以个性化方式来诠释文字内容，以确保版面的特殊性与完整性。

初期，皮涅尔斯在康泰纳仕工作，是Agha的下属。1942年，在两人共事五年后，皮涅尔斯开始担任*Glamour*的艺术总监，成为首位有自主权的女性艺术总监。从*Glamour*开始，时尚摄影开始追求起艺术价值，并大气地使用跨页满版图与戏剧化的图文比例。刊物利用四色和双色页的区分，巧妙安排阅读节奏。皮涅尔斯结合现代主义的抽象结构，将图文呈现变得更有趣味，同时赋予刊物浓浓的个人色彩。但直到*Seventeen*杂志的配图创意（上图），她才真正有了属于自己的风格，那是当时的第一份少女杂志。

皮涅尔斯与*Seventeen*的创刊编辑海伦·瓦伦丁（Helen Valentine）认为，杂志的读者是严肃、有知识的青少年，他们需要的也是严肃、有知识的内容。于是皮涅尔斯给读者提供当时最能激发思考的艺术品，例如提倡激进政治主张的西摩·切瓦斯特（Seymour Chwast）与本·沙恩（Ben Shahn）的作品。她率先用不同字体来区分单元，并将美国的具象字体引入到编辑设计中，用实物形状代替字体来创造视觉上的双关语。同时，擅长利用手写字，并加以特殊效果如刮擦、撕破等，增添读者阅读时与文字的互动，为故事增加意义和表达方式。她的作品呼应当时美国艺术界的趋势，放弃具象的表达方式，转而探索概念主义与抽象艺术，探索多种媒介的表达。

1950年，皮涅尔斯在*Charm*杂志中扩大了对字体创意的实验与探索。这份深度刊物的标语是“给工作的女性”，也就是在变动不安的世界中扮演重要角色的女性们。皮涅尔斯将现代写实主义融入杂志，让杂志变得耳目一新。杂志内的时尚摄影常用城市与高速公路作为背景，呼应美国的工业发展，并使用富有地方色彩的字体将版面风格本地化。最重要的是，皮涅尔斯善于寻找艺术家与摄影师，并以尊重他们的专业为前提进行合作，她也因此成了最了不起的艺术总监之一。皮涅尔斯几乎得过所有的设计大奖，这也说明艺术总监与合作伙伴保持稳定关系与良好沟通有多重要。

汤姆・沃尔西（Tom Wolsey）

许多大名鼎鼎的现代主义设计大师都是欧洲人，在战后逃离故土，到美国开创新事业、开始新生活。不过，英国籍的汤姆・沃尔西却是个例外。他在20世纪60年代初期设计的裁缝杂志*Man About Town*（日后简称*About Town*，又改称*Town*），将瑞士现代主义的设计理念应用到原本华丽、花哨、古典的英国杂志中。他拒绝使用格线，大胆结合粗衬线字体（slab serif）搭配有现代感的无衬线字体（例如Haas Grotesk），给读者带来强烈的视觉感受，并选用同样视觉强烈的配图，在版面上横向甚至斜着放置。所以他设计的版面从来不会单调乏味，而是充满趣味和动感。沃尔西对摄影有着优秀的鉴赏能力，委托过许多20世纪60年代最好的摄影师，包括唐・麦卡林（Don McCullin）与特伦斯・多诺万（Terence Donovan）。也曾为了如实呈现这些摄影作品，找来了最佳印刷工作团队，为20世纪60年代的杂志设计建立起标准和风格。

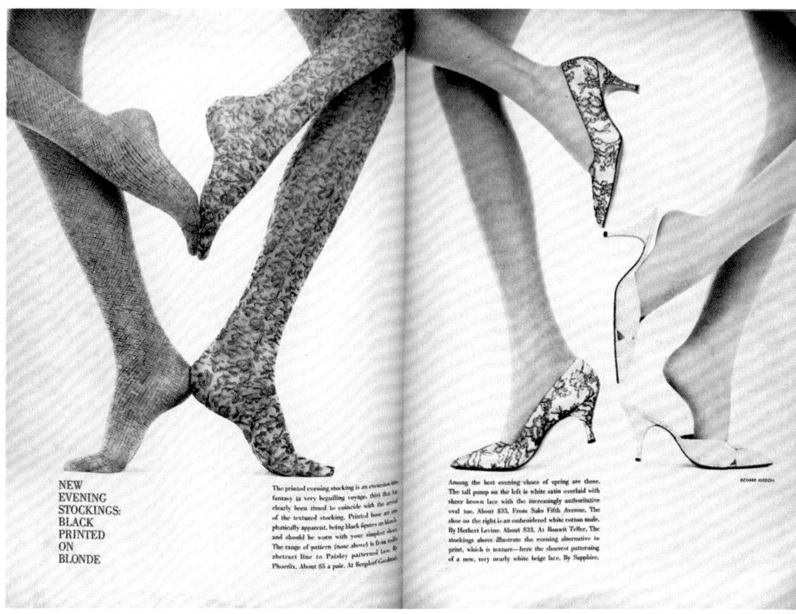
NEW EVENING STOCKINGS: BLACK PRINTED ON BLONDE

Among the best evening shoes of spring are these. The tall pump on the left is white satin overlaid with sheer brown lace with the increasingly authoritative oval toe. About $33. From Saks Fifth Avenue. The shoe on the right is an embroidered white cotton mule. By Herbert Levine. About $33. At Bonwit Teller. The stockings above illustrate the evening alternative to print, which is texture—here the sheerest patterning of a new, very nearly white beige lace. By Sapphire.

亨利·沃尔夫（Henry Wolf）

亨利·沃尔夫来自奥地利，曾在1952年到1958年担任《时尚先生》的艺术总监。凭借一种创新风格，彻底改变了这本杂志的设计。1958年，他接替阿列克谢·布罗多维奇（Alexey Brodovitch）的位置，成为《时尚芭莎》的艺术总监，工作三年后，在1961年创办了自己的杂志*Show*。

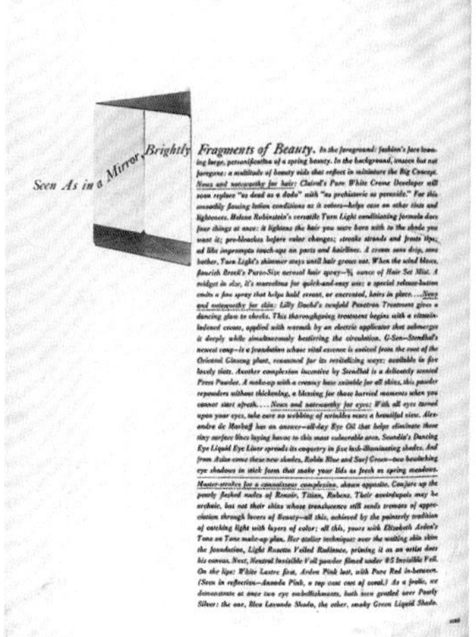
Seen As in a Mirror, Brightly

Fragments of Beauty.

沃尔夫擅长通过将严谨的排版与富有表现力、引人注目的布局结合起来，设计出极富表现力的版面，营造出色的视觉美感。沃尔夫在担任《时尚芭莎》艺术总监期间，在阿列克谢·布洛多维奇奠定的设计原则基础上，选用简单流畅的字体，构成宁静简洁的版面，让整份刊物呈现稳重节奏的同时，又不失流畅感。沃尔夫原本更擅长的是以图像为主的开放式跨页和封面设计，不擅长设计非常冗长的专题报道版面；不过面对后者，他也能成功地设计出文字井然有序、视觉生动活泼的漂亮版面。

沃尔夫的名言是“杂志不应该只是反映潮流，更要开创引领潮流”，他将自己擅长的广告设计运用在所设计的封面上，在*Show*杂志的设计上体现得尤为明显。这些精巧的封面展现出超现实的氛围，原创性极强。具有敏锐洞察力的沃尔夫，在20世纪50年代把欧洲现代主义介绍给美国人时发挥了重要作用。

威利・弗莱克豪斯（Willy Fleckhaus）

德国设计师威利・弗莱克豪斯（Willy Fleckhaus）只凭着两份刊物就声名鹊起：*Twen*（右图）与《法兰克福汇报》（*Frankfurter Allgemeine Magazin*，简称FAZ）。这两份杂志被认为是第二次世界大战后编辑设计领域最有影响力的刊物。弗莱克豪斯的天才之处在于，采用风行于第二次世界大战后平面设计的“国际风格”来展现20世纪60年代的生命力。在严谨的格线系统上，采用局部裁剪、具有视觉爆发力的大图，来展现爆炸性的能量。他把这种简洁、有影响力的风格带到了《法兰克福汇报》，在任职的十年，保持一贯的风格，制作简洁的跨页，并用实验性的手法处理插画和图片，而这种形式主义版面立刻在欧洲引起仿效风潮。弗莱克豪斯也为Suhkamp等出版社设计书籍封面，这些图书封面延续了他的典型风格，一样的沉着、低调奢华。

*Twen*杂志

*Twen*杂志于1959年创刊，是一份大胆前卫的刊物，经常为具有发人深省的内容搭配充满情欲色彩的图片。旨在吸引那些希望与父辈不同，寻找属于自己语言和风格的新读者群体。当时西方国家的青少年亚文化兴起，*Twen*杂志也顺势而为。弗莱克豪斯为青少年亚文化设计出新的图像风格，为以理性格线系统和简单文字编排为特点的瑞士版式搭配美国出版业大胆有趣的视觉风格。他设计了一套以12个单元组成的格线系统，供*Twen*的大开本（265mm×335mm）使用。这套格线可以有二、三、四、六栏等不同的灵活用法，甚至可以将横向单元跨越数栏，版面富于变化。通过灵活多样的坐标来固定这套版面系统，从众多杂志中脱颖而出。*Twen*出类拔萃的版面与视觉效果，可以说就是归功于这套格线系统。弗莱克豪斯通过激进的图像和超大特写，让版面构图产生特殊造型和超写实的风格。他追求设计的戏剧性和视觉上的颠覆感、与众不同，完全符合杂志需要的风格。

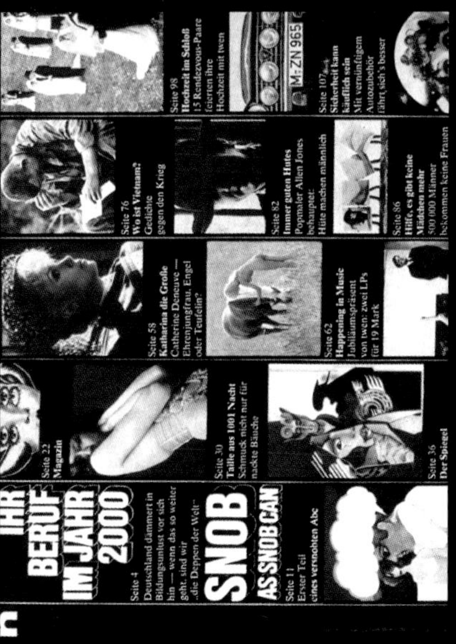

IHR BERUF IM JAHR 2000

Seite 4
Deutschland dämmert in Bildungsunlust vor sich hin — wenn das so weiter geht, sind wir die Deppen der Welt

SNOB AS SNOB CAN

Seite 11
Erster Teil eines vernobten Abe

Seite 22
Magazin

Seite 30
Taille aus 1001 Nacht
Schmuck nicht nur für nackte Bäuche

Seite 36
Der Spiegel

Seite 58
Katharina die Große
Catherine Deneuve — Ehrenjungfrau, Engel oder Teufelin?

Seite 62
Happening in Music
Jubiläumspräsent von twen: zwei LPs für 19 Mark

Seite 76
Wo ist Vietnam?
Gedichte gegen den Krieg

Seite 82
Immer guten Hutes
Popmaler Allen Jones behauptet: Hüte machen männlich

Seite 86
Hilfe, es gibt keine Mädchen mehr
500 000 Männer bekommen keine Frauen

Seite 98
Hochzeit im Schloß
15 Rendezvous-Paare feierten ihre Hochzeit mit twen

Seite 107
Sicherheit kann käuflich sein
Mit vernünftigem Autozubehör fährt sich's besser

PROVINZ PLAYBOYS

Julie der Zeitzünder

HAPPENING IN MUSIC

*Twen*杂志为了凸显这些版面上当时最吸引人的新闻摄影，常用全黑的页面搭配极简文字。弗莱克豪斯虽然是文字记者出身，但他认为图像在讲述故事方面更有力量。这是令人惊奇、有戏剧感的新刊物，也反映出刊物当时所处的社会与文化环境。

*Nova*杂志

1965年，激进女性月刊*Nova*在英国创刊，并把自己定位为男性杂志的女性版。这份杂志旨在为女性读者提供远超出时尚与化妆品范围的内容，既要有知识分量，又要概念新颖。艺术总监哈里·佩奇诺蒂（Harri Peccinotti）和编辑丹尼斯·哈克特（Dennis Hackett）决心通力合作，通过设计来反映这种前瞻立场，并效仿M.F.Agha和阿列克谢·布罗多维奇（Alexey Brodovitch）在20世纪50年代发展起来的美国表现主义风格（Expressionist Style）。尤其是*Nova*的封面，采用意想不到的图片、空间与文字组合，反映极具争议的主题（如种族歧视、虐待、性与政治等），视觉效果相当抢眼，表现力极强，连同文字编排也相当醒目，例如将普通的Times字体放大到占满一半页面来展示引语。1969年到1975年停刊前，担任*Nova*设计的大卫·希尔曼（David Hillman）深受佩奇诺蒂影响，也采用大胆、具有挑战性的视觉元素，尤其是摄影来讲述故事，很能反映出这段时期社会、性别与政治的动荡氛围。

希尔曼让编辑设计不仅能展现刊物的识别性，还能充分传达刊物的内容、调性与立场。希尔曼身兼副主编与艺术总监，勇于打破陈规、承担风险，利用设计来诠释杂志不妥协、有个性的特色。同时，也能关注故事，邀请很多摄影师通过作品来诠释故事，让他们提出不同观点，甚至相左的立场。正因如此，*Nova*杂志能够在不违背刊物理念与内容调性的情况下，不断开辟新的领域。

People who think they look like other people

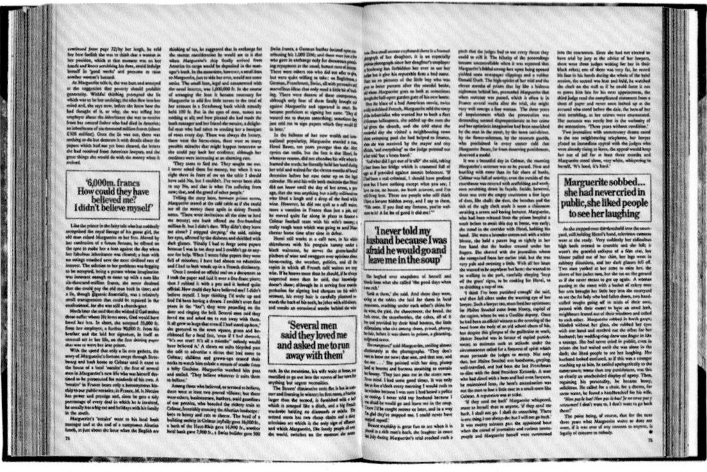
'6,000m. francs How could they have believed me? I didn't believe myself'

'I never told my husband because I was afraid he would go and leave me in the soup'

'Several men said they loved me and asked me to run away with them'

Marguerite sobbed... she had never cried in public, she liked people to see her laughing

*Oz*杂志

*Oz*杂志最初在澳洲创刊时，是一份具有讽刺色彩的前卫迷幻杂志，由主编理查德·内维尔（Richard Neville）和副主编理查德·沃尔什（Richard Walsh），及身兼画家、漫画家、歌曲作家与制片人的马丁·夏普（Martin Sharp）携手推出。正是夏普推动了该杂志的设计方向。1967年到1973年，*Oz*由夏普主导设计，同时也在伦敦发行，成为一份嬉皮杂志。在英国版的*Oz*获得艺术界好评之际，也招来不少争议，甚至在1970年遭到起诉，打了一场英国法律史上最漫长的淫秽案官司。夏普在*Oz*英国版中，善于采用先进的印刷工艺、纸张与油墨，做出相比当时其他杂志更有实验性与冒险性的封面。例如封面和封底连贯的设计、华丽的拉页海报，或是用金属油墨印刷，甚至使用锡箔一般的材质来当封面。夏普不断挑战并超越印刷技术的极限，为*Oz*提供丰富的视觉隐喻，探索当时对色情书刊、自由意志主义、淫秽与激进思考的既定概念边界，拓展刊物的可能性。

*Oz*杂志的开本经常变化，夏普也乐于探索刊本的可能性，但不管开本如何改变，出色反映亚文化转变的能力是始终如一的。然而，这份杂志的反独裁，充满服用药物后的混乱与实验色彩，随着时代对亚文化的接受程度逐渐稀释，最终被其所努力反抗的主流文化吸收。*Oz*不仅运用封面来展现这种转变，也通过乔恩·古德柴尔德（Jon Goodchild）的设计更进一步。古德柴尔德与夏普等其他撰稿人合作，把设计工作室当作实验室，用拼贴画与松散的文字编排来自由设计版面，让刊物的编辑设计摆脱了当时主流瑞士风格的限制，也在平面设计界产生了深远影响。

内维尔·布罗迪（Neville Brody）

1981年，在内维尔·布罗迪加入*The Face*之后，立即为这份杂志建立了一种根植于20世纪建构主义、达达主义和表现主义艺术运动作品的设计美学，但又不仅仅是模仿。杂志的文字编排带有象征主义色彩，以趣味的版面实验带领读者体会工业化后的印刷技术是如何影响平面设计的。版面上朴素、粗大的几何形式与符号，充满表现主义的味道。因为布罗迪是有明显政治倾向的设计师，政治反动色彩强烈的象征主义与表现主义对他来说格外有吸引力。*The Face*反权威、后朋克的政治色彩及视觉识别，和布罗迪充满实验性的个性化设计完美匹配，他们都具有反抗精神，定义了当代设计的外观与感受。布罗迪打破了传统字体的构建方法，将文字当作图像来使用，灵活多变、具有可塑性，同时又能够作为意义的载体。

*The Face*杂志

内维尔·布罗迪在20世纪80年代为反文化杂志*The Face*做设计时，彻底改变了字体在刊物中扮演的角色，对平面设计影响深远。布罗迪善于利用字体来传达意义，常在一段文字中使用不同字体，表现出内容的不同，编排方式也呼应俄国构建主义的前卫作风。他在页面上以图像与符号营造视觉连贯性，并用大胆的手法来处理图片，通过局部放大、裁剪甚至加框等方法，强化视觉效果，呈现时代错置、反秩序、极度个性化的刊物特色。随着杂志逐渐成熟，布罗迪对字体和图像的应用就更加出神入化，与读者保持同步，同时又能提供创新的设计解决方案。布罗迪也把相同的风格应用到*City Limits*与*PerLui*杂志，不过就没那么成功了。*The Face* 杂志依然是编辑设计中非常抢眼的代表作。

THE FACE No. 70
FEBRUARY 1986 90p US $2.95
THE FACE
DESTROY!

THE FACE JANUARY 1986

INTRO

BULLETIN

The **Absolute Beginners** industry starts in earnest mid-January with **David Bowie**'s single of the theme song (P7), the reissue of **Colin MacInnes**'s three novels, and the paperback of **Tony Gould**'s MacInnes biography. Your last chance to bone up. The film itself opens Easter '86 Open now: the western **Silverado** which has SDP jester **John Cleese** playing it straight as a less-than-honest lawman Bulletin also recommends William Hurt in **Kiss Of The Spiderwoman** (interview P20) and two superior splatter films: **Return Of The Living Dead** (not a George Romero vehicle but a homage to same) and the born-again comic schlocker **Re-Animator** (reviews P88) Films on TV over Xmas include Warren Beatty's **Reds**, Dustin Hoffman as **Tootsie** and ITV's ratings clincher **Minder On The Orient Express** Also note: a repeat of the **Max Headroom** pilot, explaining his origin, The Tube's **Live Aid** lookback, and Billy Connolly and Robbie Coltrane's alternative New Year's Eve show **At Last It's Hogmanay** At London's ICA until Jan 19, an exhibition devoted to the pioneering Viennese architect **Adolf Loos** Great, edgy, rock orchestral version of Detroit Emeralds "Feel The Need" on **Bryan Ferry**'s "Windswept" EP, with other tracks dating from his self-imposed exile in 1977 West Country stylists should check out **Fitz**, newly opened and sartorially smart menswear shop in Bristol's The Berkeley Centre At the Royal Court a powerful new play by **David Mamet** (Glengarry Glenross etc): **Edmond** has Colin Stinton and Miranda Richardson odysseying in low-life New York Seasonal recommendation: **The Jesus And Mary Chain** LP "Psychocandy". Not simply a post-punk thrash, there are agreeable pop songs below the surface rubble A literary side order to our Food feature (P27): **The Cook**, a novel by Harry Kressing, described as "a Kafkaesque cookbook" **Do-Do's** returns to Buzby's (Charing Cross Rd, London WC1) Dec 17 for a Christmas thrash. And if you miss that, there's **Raw**'s Xmas party (Dec 21 at the YMCA) And, finally, a panto: a new production of Mozart's **The Magic Flute** at the ICA in The Mall, London SW1, until Dec 29...

A double bill of STEVEN BERKOFF. Fresh from the set of the new Prince movie *Cherry Moon*, the actor-playwright is back on the London stage after a two year absence. The Donmar Warehouse Theatre has Berkoff in two one-man shows, reading Edgar Allen Poe's TELL-TALE HEART and in his own previously unperformed HARRY'S CHRISTMAS.

INTRO IS EDITED BY KIMBERLEY LESTON

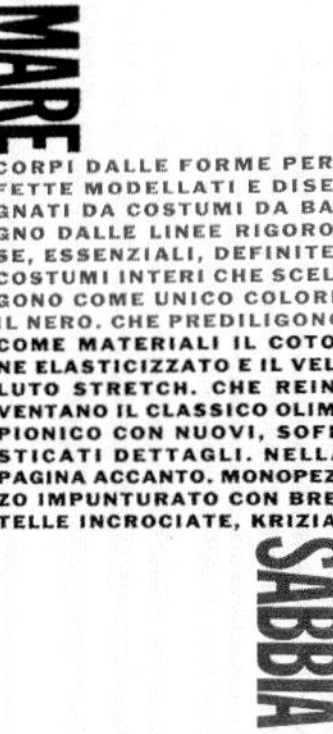
MARE

CORPI DALLE FORME PERFETTE MODELLATI E DISEGNATI DA COSTUMI DA BAGNO DALLE LINEE RIGOROSE, ESSENZIALI, DEFINITE. COSTUMI INTERI CHE SCELGONO COME UNICO COLORE IL NERO. CHE PREDILIGONO COME MATERIALI IL COTONE ELASTICIZZATO E IL VELLUTO STRETCH. CHE REINVENTANO IL CLASSICO OLIMPIONICO CON NUOVI, SOFISTICATI DETTAGLI. NELLA PAGINA ACCANTO. MONOPEZZO IMPUNTURATO CON BRETELLE INCROCIATE, KRIZIA.

SABBIA

166

法比恩·巴伦（Fabien Baron）

法比恩·巴伦从年轻就在家接受艺术总监的熏陶。法比恩在巴黎应用美术学院（Ecole des ArtsAppliqués）就读一年后，便随同父亲工作。他的父亲在多家法国报纸担任艺术总监，包括由让-保罗·萨特创办的激进左翼报纸《解放报》（*Libération*），后来，法比恩又到*Self*与美国的*GQ*杂志任职。他在1988年加入意大利版*Vogue*，之后在《时尚芭莎》与莉斯·提伯里斯（Liz Tiberis）合作时开始大放异彩。法比恩以强烈而独特的艺术总监风格闻名，一举打破所有规则，甚至包括商业规则。他在为某一期意大利版*Vogue*设计封面时，曾委托阿尔伯特·沃森（Albert Watson）等摄影师拍摄抢眼的抽象肖像，而不是时尚杂志喜爱的人物特写，让当期杂志在店面陈列时变得分外抢眼。

法比恩通过尽量减少设计元素，以原色搭配大量的黑色色块做出抢眼效果。同样，他也只和固定的几个插画家合作。在意大利版*Vogue*杂志上，他只使用过曾设计*Interview*杂志Logo的马茨·古斯塔夫森（Mats Gustafson）的插画。黑白摄影作品是他的最爱，通过精心裁剪，这些摄影作品呈现出令人难忘的效果。

更重要的是，法比恩将建筑元素融入编排方式中。例如将专题的首页用庞大的标题字占满，搭配极简的风格与大量留白空间，展现新颖的现代美学。他会以文字造型呼应图片，将版面上所有元素与文字排编相互组成特定的形状、线条或色彩，就像是建筑中的柱石一样承担起建构的功能。种种视觉彼此呼应杂志的精神理念，与之产生联系。在*Interview*杂志中，针对不同的受访内容，他也采用这种方法，精心挑选合适的字体来展现受访者的个性。法比恩为这两份杂志打造出连贯、活泼的形象。即使用了十多年，他仍在*Vogue Paris*杂志（上图）延续这种做法，传达出时尚的流动性与活力气息。

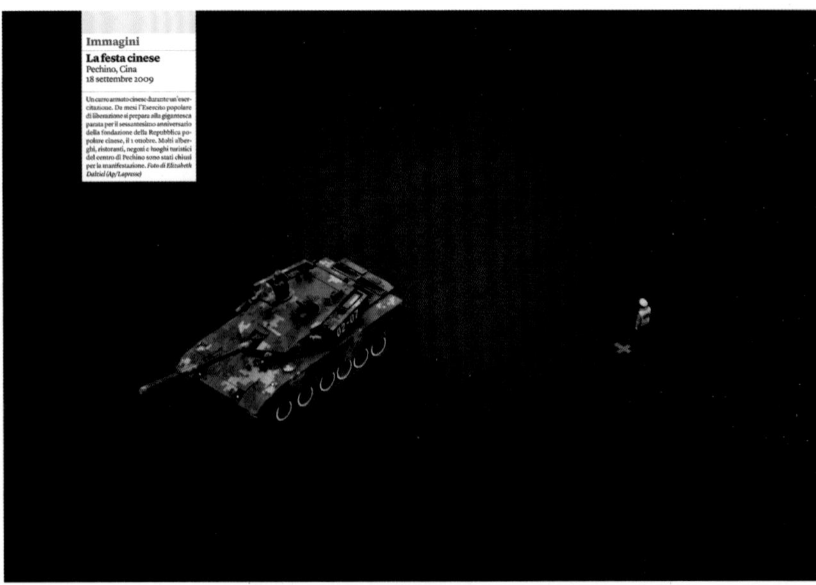

马克·波特（Mark Porter）

马克·波特曾担任*Colors*艺术总监，之后也负责《标准晚报ES》（*Evening Standard's ES*）及最初两期*Wired*杂志的设计。他在1996年到2010年担任《卫报》的创意总监，负责《卫报》的改版工作，深受好评。当时所有的报纸都在试图改版，而《卫报》的改版堪称典范，曾赢得D&AD的黑铅笔奖、出版设计师协会奖。波特在《卫报》任职期间也担任该报旗下周末杂志*Weekend*的艺术总监，并在理查德·特的协助下，连续开展了六次大刀阔斧的改版。

波特对细节充满热情，重视优质图片的使用，眼光敏锐，这让他的设计很受欢迎。他虽是语言学专业毕业，却成为非常优秀的视觉记者，认为自己更注重视觉。波特目前经营自己的设计事务所，近年作品包括《公共报》与《金融日报》（*Financieele Dagblad*）的改版。《公共报》是葡萄牙全彩日报，简洁明了的设计广受好评。整份报纸以图片为主，小版式让报纸更具现代风格，虽然是每日发行的报纸，却具有杂志的设计质感。

荷兰阿姆斯特丹的《金融日报》原本为纸质版刊物，但也有网站与数字版本。波特根据刊物在不同平台上的呈现需求，调整各个版本的设计逻辑。

盖儿·安德森（Gail Anderson）

盖儿·安德森是高产、行事低调的设计大师之一。她的作品启发了许多当代编辑设计师。安德森出生于布朗克斯区，父母是牙买加移民。她从纽约视觉艺术学院（School of Visual Arts）毕业后，进入兰登书屋工作，之后入职《波士顿环球报》。然而令她真正声名鹊起的，却是在1987年到2002年为《滚石杂志》（*Rolling Stone*）设计的作品。当时她与知名设计师弗雷德·伍德沃德（Fred Woodward）合作，以装饰性文字设计出美得像专辑封面一样的页面。安德森在2008年获得美国平面设计协会（AIGA）预发的奖项。知名艺术总监与作家史蒂夫·海勒（SteveHeller）曾采访她，说她的作品是“折中主义的文字编排（typographic eclecticism）”。在她手中诞生的《滚石杂志》风格相当细腻、重视手工质感，经常以委托设计的插画作为页面的主角。虽然安德森曾获得出版设计协会与美国平面设计协会的诸多奖项，但她仍谦虚地默默努力，把功劳归功于宝拉·舍尔（Paula Scher）和弗雷德·伍德沃德对她的影响。

离开《滚石杂志》之后，安德森加入了娱乐广告公司SpotCo，在那里为影视娱乐行业设计了令人惊叹的戏剧海报。她把有趣的图像与优秀的想法结合起来，并运用插图与文字编排技巧，创造出令人印象深刻的设计，为百老汇许多音乐剧打上了鲜明的个人设计印记，例如《Q大道的木偶》（*Avenue Qpuppet*）。她也为美国邮局设计邮票，担任邮票顾问委员。她作品中一贯的特色是强烈而优越的观念、娴熟的技巧以及耐心的表达，以当代手法搭配新旧文字造型，设计出引人注目的作品。她向史蒂夫·海勒（Steve Heller）（两人曾共同写过许多书籍）表示，许多想法是在SpotCo产生的：

> “我在做每一个项目时，都会寻找十几个好想法，不能只有一两个点子，”安德森解释他的作品如何赢得观众的注意力（与金钱）的。“做了七种设计之后，就会明白要探讨一个问题有许多方式。我很享受这个过程，即使知道最后只能留下一个想法。这是一个释放创意的过程。”

安德森目前在纽约视觉艺术学院教授设计课程，并经营着自己的设计公司。

费尔南多·古提耶雷兹（Fernando Gutiérrez）

谈到贝纳通（Benetton）公司的*Colors*杂志，一般人会想到创意十足的创刊者奥利维耶罗·托斯卡尼（Oliviero Toscani）与泰伯·卡尔曼（Tibor Kalman），但古提耶雷兹也功不可没。古提耶雷兹在2000年担任创意总监，沿用卡尔曼以图像来报道的原创概念，成功地运用图片来讲故事。古提耶雷兹多才多艺，作品种类繁多。自行创办公司，涉足范围包括图书出版、宣传推广与编辑设计，所有这些都是在他从伦敦印刷学院毕业后的短短七年内完成的。他在不知不觉中，彻底改变了西班牙和其他地区刊物的编辑设计视觉走向。古提耶雷兹先从一家政府部门的青年杂志入手，采用双重格线设计。而他最著名的设计，是西班牙时尚杂志*Vanidad*与*Matador*，两者分别是以文字和照片为主的杂志。此外他还主导了《国家报》的设计、以年轻读者群为主的增刊《诱惑》及《国家报》周末增刊*El País*，在全国拥有高达120万的读者。虽然这些刊物是由低档的印报机印制，但他善于发挥开本与纸张的优势，丝毫不受局限。虽然是报纸的增刊，但古提耶雷兹却积极地以独立杂志的规格来做设计，凭借活泼、富装饰性的视觉风格带动销量飙升。

从这些刊物中可以明显看出，古提耶雷兹能利用设计元素传达刊物的特性，为读者提供适当的体验。不仅如此，他的作品融合文化与民族色彩，让杂志脱颖而出。以*Matador*这份杂志来说，他计划到2022年时正好出版29期，每一期按西班牙文的字母顺序，向一种字体致敬。这份杂志的特色是高品质的内文纸、精美的印刷和大开本。这些元素与版面搭配得非常有戏剧性（右下图），独树一帜。他设计的刊物充满西班牙特色，然而当需要寻求不同文化认同时，也会特别注重版面的本土化。

文斯·弗罗斯特（Vince Frost）

在大卫·卡森之后，没有一位设计师能像文斯·弗罗斯特那样展现字体的表现性。两人的作品或许截然不同，但是都深刻理解编辑设计（尤其是文字编排）的主要任务，运用视觉手法忠实传达刊物的内容与识别。两位设计师的作品偶尔会被批评是“为设计而设计，并且造成阅读上的障碍”。不过，弗罗斯特能不断通过生动、刺激的设计，吸引读者注意。

在20世纪90年代中期，弗罗斯特担任《周六独立报》的艺术总监，也负责《金融时报》的周末杂志《金融时报商业周刊》的设计，两份刊物皆展现出知性气质。弗罗斯特喜欢简洁干净的设计，很懂得如何用凸版活字与木活字，做出与内容相关的装饰元素，摆脱字体的调性与风格限制。弗罗斯特不太使用花哨多变的设计手法，而是专注于让每个元素更简洁醒目。以另类时尚刊物*Big*（上图）为例，弗罗斯特与凸版印刷大师阿兰·基廷（Alan Kitching）合作，让字体呈现出像是摩天大楼、对话框、面具与各种物体的趣味效果，呼应一旁令人惊奇的照片，宛如纽约知名摄影师威廉·克莱恩（William Klein）的作品。而在英国文艺杂志*Zembla*的例子中，弗罗斯特在每一页以明亮、有活力、玩世不恭与无法预测的惊喜设计把文字当图来玩，让读者“享受文字的乐趣”。

不过，弗罗斯特不仅能在单独页面上展现不同特色，也能够掌握整体刊物的流动与节奏，使整本刊物成为令人期待、惊喜的佳作。*Zembla*的版面以令人大开眼界的照片（多为黑白）搭配凸版活字，产生出乎意料的美观效果，非专题报道的版块（信函、评论、新闻等）也十分用心设计。

珍妮特 · 佛罗里克 （Janet Froelich）

珍妮特 · 佛罗里克曾先后担任《纽约时报》与*Real Simple*的艺术总监，以其高超的设计能力和艺术品位闻名，她对摄影与设计有着无比的热情。在漫长的职业生涯中，佛罗里克和各种各样的创意人才合作，为自己的设计寻找最棒的图片和最棒的创意。实际上，她原本就是画家出身，曾在美国库伯联盟学院与耶鲁大学学习美术。

佛罗里克在1986到2004年担任《纽约时报》杂志的艺术总监，并在2004年升任创意总监，期间曾斩获出版设计师协会、艺术总监俱乐部、报纸设计师协会等机构颁发的六十多项大奖。她的团队帮《纽约时报》杂志赢得了2007年出版设计师协会的“年度最佳杂志”奖项。2004年，她曾任纽约时报时尚杂志*T: The New York Times Style Magazine*的创刊创意总监。这份杂志经常斩获各种设计大奖，她为杂志带来了赏心悦目的时尚摄影与内容编排。2006年，她还担任纽约时报运动杂志*Play: The New York Times Sports Magazine*与纽约时报房地产杂志*Key: The New York Times Real Estate Magazine*的创意总监，后来又担任*RealSimple*的创意总监，带领时代公司（TimeInc）的团队设计这份生活、饮食与居家杂志。在*Real Simple*杂志时，从纸质版到App版、数字版本以及网站的设计与产品包装，都由佛罗里克全面负责，这得益于她对设计有清晰的愿景和把控，追求卓越的摄影呈现以及文字编排。*Real Simple*的图片既美观，又非常有视觉冲击力，往往凭借一张优质的图片，就能很好地传达故事，即使在移动设备上缩小后，也很吸引人。

2006年，佛罗里克获得艺总监协会（Art Directors Club）颁发的杰出贡献奖。她在多年的设计职业生涯以及在纽约视觉艺术学院的任教生涯中，启发了许多年轻的设计师。她是出版设计师协会与艺术总监协会的董事，也是美国平面设计协会纽约分会的会长，通过这些组织，提升了大众对设计的重视。

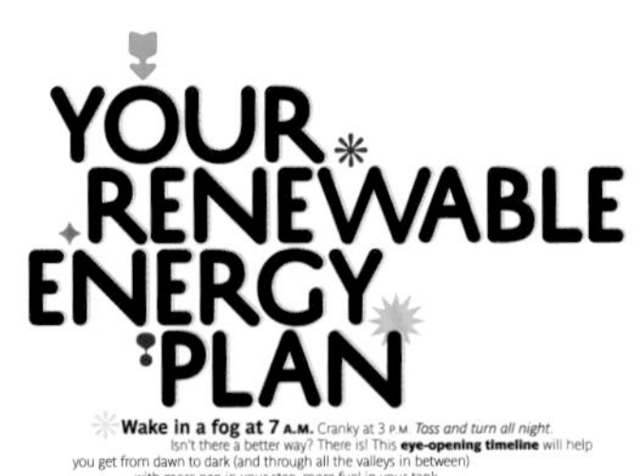

佛罗里克对美术的热情为她带来了很多灵感，并让她能够和很多艺术家和设计师顺畅合作，设计出令人难以忘怀的优秀版面。

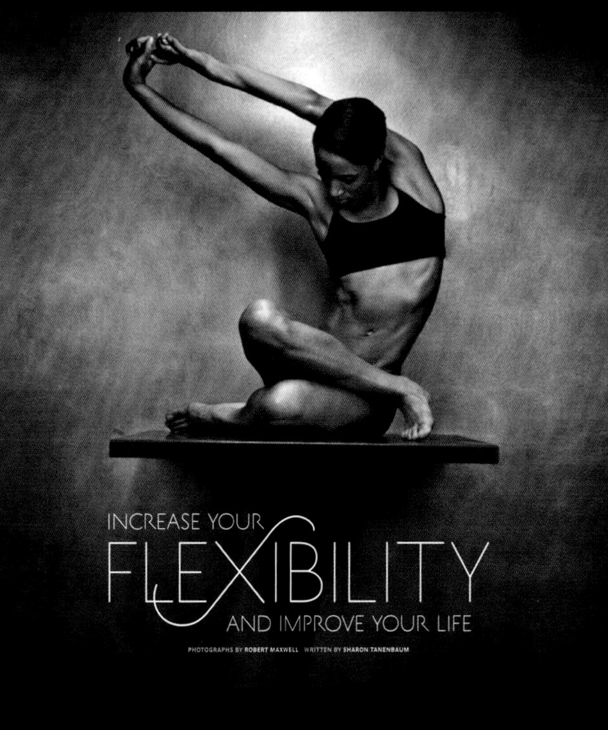
INCREASE YOUR
FLEXIBILITY
AND IMPROVE YOUR LIFE
PHOTOGRAPHS BY ROBERT MAXWELL WRITTEN BY SHARON TANENBAUM

THE SIMPLE ACT OF STRETCHING DOES A LOT MORE THAN MAKE YOU LIMBER. IT MAY HELP PREVENT INJURIES OR EVEN ILLNESS. AND YOU DON'T HAVE TO BE PRETZELIAN TO REAP THE BENEFITS—ALL IT TAKES IS 10 EASY MINUTES A DAY.
AUGUST 2010 | REALSIMPLE.COM 153

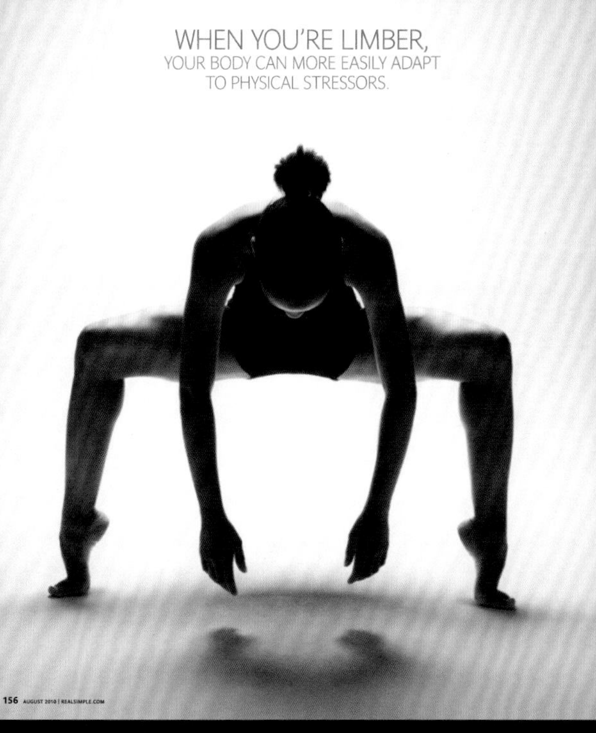
WHEN YOU'RE LIMBER,
YOUR BODY CAN MORE EASILY ADAPT TO PHYSICAL STRESSORS.
156 AUGUST 2010 | REALSIMPLE.COM

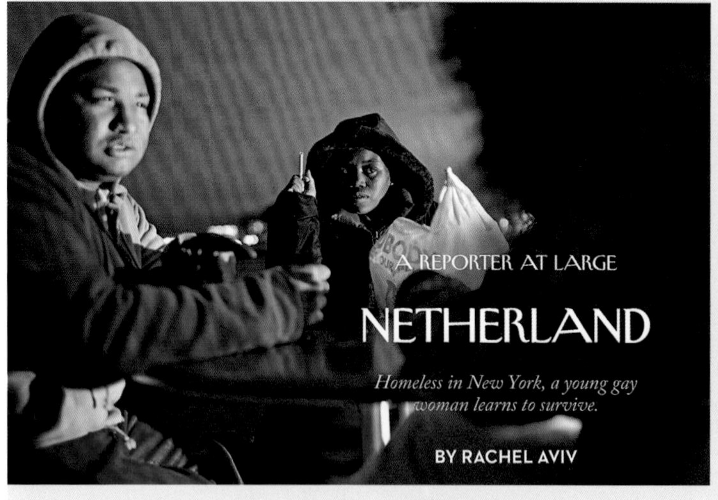

斯科特·达迪奇（Scott Dadich）

斯科特·达迪奇（Scott Dadich）来自美国得克萨斯州，曾在《得州月刊》（*Texas Monthly*）这家精美的杂志任职2006年加入*WIRED*杂志。并凭借创新杂志的版式以及探索与读者互动的新方式，获得多项大奖。传统纸质版杂志向苹果平板电脑版转型之际，达迪奇也成为优秀数字杂志设计的先锋之一。他是*WIRED*杂志App概念发起人，并率领团队着手研发了相应程序。*Wired*是第一个采用桌面出版系统制作App的杂志。他也为《纽约客》杂志设计了另一款App，这个App是2012年同类应用中的销售冠军。在知名出版公司康泰纳仕的支持下，达迪奇的团队在2012年推出《纽约客》苹果手机版。

达迪奇曾研究过如何从纸质杂志页面以及早期App的线性阅读（也就是逐页翻阅的PDF格式文档），转型成真正的数字阅读页面。他把这称为“堆叠（Stack）”系统，让页面像挂在晾衣绳上一样，可以横向快速切换、浏览。在康泰纳仕集团的投资下，达迪奇的设计与开发团队在向数字化转型的早期，就很有先见之明地进行了这样的探索。

达迪奇对于创新的渴望，让他升任康泰纳仕编辑平台与设计副总裁。因为对这种媒介设计创新抱有极大热忱并积极推广，他在设计界面临自印刷机发明后的最大转变时期中，作出了杰出贡献。

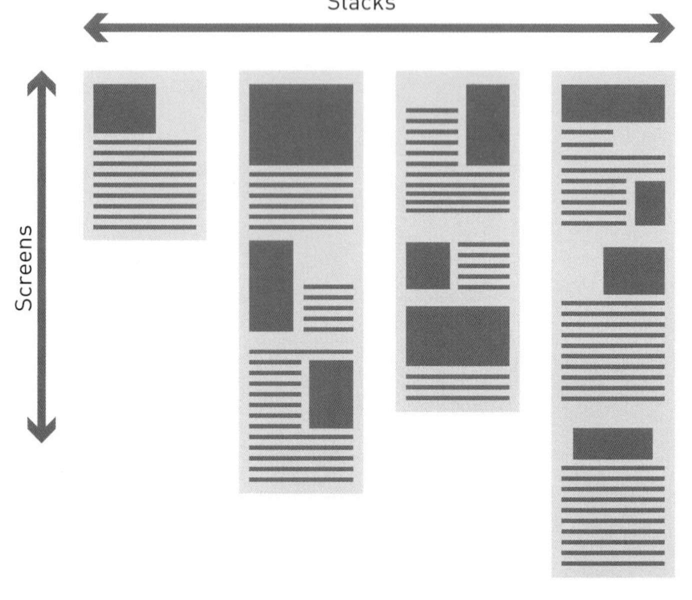

ELLEUK.COM

Rosie Huntington-Whiteley: supermodel superpower

赛克斯的设计大胆延续母品牌刊物的风格，并将其拓展到增强功能的App版本。封面上的模特罗西・亨廷顿（Rosie Huntington）也出现在Elle TV的视频中，让读者有不同的探索与分享内容的方式。

苏珊娜・赛克斯（Suzanne Sykes）

苏珊娜・赛克斯是英国籍的艺术总监，因为重塑当时时尚杂志的设计而闻名。1989年，她与编辑格伦达・贝利（Glenda Bailey）合作推出*Marie Claire*英国版。赛克斯在时尚周刊*Grazia*中，将精美时尚月刊设计手法融入出刊频率更高、节奏更快的周刊中。这对当时的畅销时尚杂志来说，是一种革命性的突破，也多次获得设计与艺术总监协会大奖项。*Grazia*的外观很有戏剧性，读者的评价也有好有坏，但是多层次的页面与配色，确实全面改变了时尚杂志的风貌。

在2007年到2012年，赛克斯曾担任*Marie Claire*美国版的艺术总监，之后又为英国*Elle*杂志大幅改版。英国*Elle*杂志在20世纪80年代曾风光一时，曾经热衷于使用超级名模的照片。然而赛克斯明白杂志风格需要跟上读者需求变化的脚步，因为现代读者不再持续购买印刷精美的杂志，反而喜欢使用社交媒体获得更多时尚信息。于是她和艺术总监马克・利兹（Mark Leeds）合作，设计出完全不同却相辅相成的线上内容和纸质版内容。而在当时，大部分杂志还是直接把纸质版本的内容转换成网站版本和数字版本。

在2013年3月，谈到纸质版杂志和数字版杂志的关系时，赛克斯表示：“纸质杂志依然是主角。我们运用纸质版杂志的优势，引导读者阅读数字内容。”2013年，*Elle*杂志纸质版每月的销售量为177094本，网站读者为45万人，也是英国最大的女性杂志，Facebook的粉丝超过500万人。赛克斯说，她与编辑洛林・坎迪（Lorraine Candy）的目标是“让读者可直接在页面上购物”。

展望未来

我们虽然几乎不可能预测出未来的设计趋势，但是思考科技与种种变迁会如何影响设计实践，依然有其价值。前面提到的所有大师都曾经受益于当时的新科技，例如彩色印刷普及，或弹性更大的排版系统问世等。如今，编辑设计师桌上都有苹果电脑，设计师们也变得身兼数职、多才多艺，要懂印刷知识，要能够进行艺术设计、制作图片、写代码、当编辑、写博客，还要懂得如何营销……这些还只是一部分。设计师还必须了解不同媒介的经济、文化、科技演变，并且能够与时俱进，适应新技术的发展。

在本章中，将会探讨在不久的未来，行业中可能出现的种种挑战及业界专家所提供的宝贵见解。

技术转变

数字杂志

如今编辑设计业最大的挑战在于，杂志的发行渠道多样化了，需要同时出纸质版、平板电脑等数字版以及社交媒体内容等版本。2010年苹果平板电脑上市之后，各家杂志竞相推出自己的App，希望能够通过数字版获利。Adobe、Woodwing、Mag+等软件很快推出专供InDesign使用的插件，让习惯纸质杂志的设计师能快速把纸质设计版面转换成平板电脑适用的版本。于是一夜之间，报刊编辑的工作流程中引入了一个全新的环节。纸质版完成后，设计团队还得把设计页面转成适用于平板电脑屏幕的小版面。这就需要重新调整所有版式，并且增加动画、视频、音效等互动元素。虽然这些App会不会成功尚未可知，但是设计师却意识到学习如何进行交互设计，已经势在必行了。

UM JORNAL A SERVIÇO DO BRASIL ★ ★ ★ WWW.FOLHA.COM.BR

FOLHA DE S.PAULO

copa 2006

Ou vai... ...ou racha

Alemanha e Itália vão fazer a 1ª semifinal

Portugal enfrenta Inglaterra

Nos intervalos da Copa, cuidamos do resto da realidade

Noivos mantêm casamento na hora do jogo da seleção

Secretário-executivo será o novo ministro da Agricultura

Premiê do Hamas acusa Israel de tentar derrubá-lo

Procurador quer inquérito contra 42 parlamentares

PT e PSDB são os partidos com mais candidatos nos Estados

Presidente sofre derrotas seguidas na Justiça Eleitoral

folhinha

Edição especial traz roteiro de férias

congresso

FOLHA DE S.PAULO

eleições 2006

O julgamento das urnas

eleições 2006

FOLHA DE S.PAULO

50%

38%

Lula cai, e 2º turno contra Alckmin está indefinido

Bombardeado pela crise do dossiê, presidente perde três pontos no Datafolha e tem 50% dos votos válidos, mesma marca da soma de seus adversários; resultado da eleição é imprevisível e tucano diz que "sorte está lançada"

9%

1.637 CARGOS ESTÃO EM DISPUTA HOJE

响应式设计

响应式设计最初的目的，是为了实现同一页面在电脑、平板电脑、手机等大小不一的设备上能够自动调整成适配的版面。使用者在浏览网站时，浏览器的窗口宽度会触动算法来调整版面，以适应不同的设备。原本为电脑的宽屏幕设计的多栏位版面，在智能手机上浏览时就可以自动变成单栏位设计。

为了帮助设计者应对上面所说的变化，避免设计师重复设计不同版本，有人已经开发出基于算法的响应式设计，不仅能够自动将纸质版设计的版式转换成数字版本，而且不同的平板电脑屏幕尺寸和格式，都能够自动调整匹配。

自动聚合内容

杂志可以利用RSS内容聚合服务自动生产内容。只要利用算法，就可以找出读者想要的内容，并通过RSS聚合订阅软件，将内容发送到不同的设备上。2012年，最受欢迎的聚合内容提供者是两年前由美国人迈克・麦克库伊（Mike McCue）和埃文・多尔（Evan Doll）成立的Flipboard。Flipboard是数字社交杂志应用，它通过聚合社交媒体上的内容，如Facebook与推特等，并在苹果平板电脑或安卓客户端上以杂志的形式呈现。使用者可以自由阅读所有与Flipboard合作网站的内容摘要，如果已经订阅了具体刊物如《纽约时报》，也可以直接通过Flipboard来阅读杂志内容。*magCulture*的创总监杰里米・莱斯利表示：

> “这种形式下，算法取代了艺术总监的角色Flipboard是第一个有缩放与空间概念的App。我很喜欢用手机阅读社交网站聚合的内容Flipboard的翻页方式也很精巧。以往，往往为了功能而牺牲移动设备上页面的美观性，看起来不太舒服的Flipboard很巧妙地运用杂志跨页的感觉，分为左右两个版面在设备上来呈现内容。”

融合纸质出版与数字出版

出版领域里一个非常有趣的主题，也是另一项基于算法的新项目，就是将传统印刷技术与数字技术相结合。《新闻俱乐部》的创办人表示，这项技术的宗旨是可以让即使不懂设计的使用者也能自行制作自己的纸质刊物。他们开发了一套定制化软件“Arthr”，并和印刷厂谈好了价格，利用对方机器不忙的时间来印刷。任何人都可以上传自己的文本，软件根据内容自行排版后，可以下单印制出小批量的报纸（懂设计的使用者可以自行上传PDF文档）。

伦敦伯格设计公司研发的“小小打印机（Little-Printer）”可以提供更加个性化的纸质与数字技术的交互体验。这台印刷机颠覆了过去纸质刊物印刷、发行、营销的传统路径，让使用者从网络仪表板（dashboard）选择内容。这些内容会通过WiFi，用“小小打印机”打印出来（参见232页）。小小打印机是伯格设计公司典型之作，有人认为它和Flipboard有些类似。然而公司的主管杰克・舒尔茨解释道：

> “小小打印机和Flipboard很不同，有自己的特点，其中最大的差异就是，小小打印机会将全部内容储存在云端，所有历史记录都可存留且查询。另外，小小打印机只在排定的时间送来指定的数据与资讯。而且简单易用，可以通过点击进行互动，产生一些个人使用数据。它可输出一些常见内容，例如猜谜、天气，或社交网站的内容，也有广播公司与大型出版社（例如《卫报》）提供的内容。重点是，小小打印机是个随时都能在身边的伙伴，无论内容是来自大型媒体还是小型媒体，都会以同样的格式呈现。小小打印机，一视同仁。”

对无限数字内容的有限呈现

通常说来，数字内容因其开放的特性，可以随时通过网络链接到其他内容，如果一直阅读下去，会有无穷无尽的内容在等着读者。于是设计师开始为读者阅读的数字内容设定边界，就像是纸质杂志有页数限制一样。这些网站或者App不再给读者提供数不清的网站内容，而是有定量限制。The Maga-

通过Flipboard自动生成的排版有杂志般的质感，鲜明的黑底白字或白底黑字，简单搭配图片与超链接，在苹果和安卓设备上都可用，就像上图所示。

zine是一款苹果手机端的App杂志，就是在进行这种尝试。这款应用由Instapaper（一款阅读应用）的创始人马可·阿蒙德推出，仅在苹果设备上可用。每两周只送四篇与技术相关的文章，没有纸质版本，也没有特效或装饰，只有少量文字供阅读，并有清楚的结尾，为读者设定了阅读的“边界”。

另一家初创公司Aeon Magazine，每个工作日都会发表一篇原创的长文，同样专注内容的质量而不是数量。这些材料具有高度的进步性，发人深省，由专家撰写，目的是让读者专注于每天提供的有价值的内容，而不是沉浸在没完没了的“必须阅读”中。Aeon的文章每天都会在网络上发布，也可以通过订阅电子邮件，每天或每周获得，并且可以适配台式电脑、平板电脑和智能手机的屏幕尺寸变化。

数据可得性：如今获取数据变得越来越容易，像《泰晤士报》这类公司往往可以利用积累的数据增加自身的优势。《泰晤士报》的设计编辑希乔恩·希尔（Jon Hill）表示：

> 随着数据获取越来越方便，我们可以利用数据做更多事情，GPS只是其中一种。另一项值得期待的方式是，我们可以利用累积了225年的档案库，以更有趣的方式讲述故事，进行图片展示，为报道提供脉络。这对《泰晤士报》这样的机构来说有很大的发挥空间……如今我们可以获得很多数据反馈，了解读者的阅读行为，在阅读上花费的时间等。我们还能通过微数据（microdata）来分析读者会花多少时间看我们设计的资料图表，这些详尽的数据反馈能告诉我们哪些内容是成功的。

改变商业模式

以前人们是彼此之间通过一对一的方式分享想法和观念，现在可以通过推特或Tumblr这些社交网站进行一对多的分享。这些工具有助于把读者带回纸质刊物，也是出版社用来分享内容的渠道与支持系统。新闻集团（News Corp）等新闻媒体公司也在研究如何通过各种渠道的产品与服务，来维持庞大媒体集团的运作。其中有些商业模式能够获利，但有些较比较老旧，无法获得高额利润。正如乔恩·希尔所说：

> “付费墙和数字版本的内容出现都是大事。我们必须重新审视《泰晤士报》的工作流程，看看如何更有效率地工作，确保设计团队能获得正确的讯息。我们非常努力，设法在全天候维持不同平台运作。然而我们无法帮每种新的平板电脑或移动设备都制作专属版本，因此要为设计工作排出优先顺序。我们希望设计出的每种版本都有不同之处。”

那么，杂志能光靠纸质版本生存下来吗？*Eye*杂志联合创办人西蒙·埃斯特森（Simon Esterson）回答：

> “很难一概而论。随着超市成为刊物的一大分销渠道，小型报刊经销渠道竞争力不在，小型出版商也跟着遭到淘汰。一方面市场被出版纸质版刊物的大出版公司垄断，另一方面，自助出版蓬勃发展。互联网带来了变化，读者可买到过期刊物，也可以从海外订阅，杂志直接送到家中。我们不用去找什么专有渠道或者千方百计去寻找特殊刊物的收集者，只要上网就买得到。像Eye杂志社这种小型出版社，只要读者能找到我们，我们就能和他们建立联系，让他们知道我们在做什么，推送新刊上市的资讯。对小型出版社来说，他们的图书可能无法出现在商店，但因为有网络渠道，所以这不再是问题。”

虽然苹果平板电脑深受众多用户的喜爱，但如果坚持认为苹果平板电脑或其他价格更低的安卓平板电脑，或是任何其他新设备会导致纸质刊物消失，恐怕言之过早。

杰克·舒尔茨（Jack Schulze）

杰克·舒尔茨（Jack Schulze）热爱文字编排，尊重好的想法，在涉足前卫设计之际，依然对设计充满美感的作品兴趣浓厚。他为*Mag+*杂志设计App版，可以让读者以一种简洁方便的形式来浏览不同的内容。2005年，舒尔茨和马特·韦伯（Matt Webb）在伦敦创立了伯格设计顾问公司（BERG），后来马特·琼斯也加入了这个团队，他们不断尝试开发新软件，探索不同内容的呈现方式。2012年2月，Fast Company将伯格设计顾问公司评为“世界50家最具创新性的公司”之一。Fast Company年度专刊专门列出以创新之举影响科技与文化的公司，它称赞伯格设计顾问公司“以天马行空的想象力融合纸质刊物与数字刊物”。公司以充满与疯狂的想象力，帮助许多企业利用数字技术开拓更多潜力，而伯格的创新产品（例如小小打印机Little Printer），也在常规的出版思维之外，提供了另一种选择。

在接下来的访谈中，舒尔茨就大型媒体公司的角色以及未来发展提出了自己的看法。

大型媒体公司如何保持灵活，并能够满足消费者不断变化的需求？

我认为大公司必须转型成为软件公司。史蒂夫·乔布斯与亚马逊（Amazon）已经改变了和消费者的关系规则。而要成功有两种办法：一是靠经验，二是靠数据。媒体公司必须开发自己的软件，仅仅是生产内容已经不够了。

大型媒体公司将何去何从？

大型媒体公司仍然有自己独特的价值。比方说，《卫报》编辑艾伦·拉斯布里杰（Alan Rusbridger）就很聪明，懂得如何转型。不过，大型媒体公司确实陷入了困境。以过去经验来说，媒体企业一定要做大规模，才会成功，要拥有许多资产，保持大量内容产出，让你的内容涉足电视领域，甚至在酒店的房间都能看到。但是在软件行业，三个人即使待在地下室也可以写出非常厉害的程序。

这些大型媒体企业依赖的是基础设施的价值，广电公司会有许多大楼，并到山上架十二米高的天线，接收电力，发出信号，播放电视节目和广播节目。在观众没有其他娱乐选择的时代，公司这样做就够了，可是现在大家能靠Facebook知天下事，广电公司空有一大堆硬件和业务，却只做出平平淡淡的内容。如果你有一家新闻公司，你需要有办公地点和印刷机印刷报纸，还得有卡车把印刷好的报纸送到各个渠道上去，而这一切只为了有人可能会花几块钱把报纸买回家，从头读到尾。这样做的代价太昂贵了。当没有选择的时候，这样做还可以。现在读者选择众多，这样看起来就很荒谬，浪费精力和金钱。

“小小印刷机”的工作原理是什么？

当我需要使用它时，先上网订阅想看的内容组，这些内容组每天会向读者发送一次内容，读者可以打印出刊。如果使用者想要每天早上收到发送的内容，也可以自行设置，这台印刷机上方有个按钮，有东西传来，灯就会亮，有点像电话答录机。如果读者选择不打印出来，而下一份刊物传内容来时，使用者会只拿到当前这一份刊物，而不是两份。如果使用者出门几天，回家时不会堆了一大堆刊物，只会拿到最新的一份，其他的都存在云端。

举例来说，我订阅了“现在太空上有多少人” 这个主题的资讯。如果网上的新闻资讯显示有六个人，只有等到人数出现变化时才会再向我发出新的新闻消息，因为之前的新闻我已经了解了，只需要在人数变成七人时知道就行了。日报必须每日出刊，得靠专栏作家拼命填满报上的空间，即使今天没什么新闻发生，也得填满，只能期盼读者有读报的习惯。但是小小打印机只在有值得阅读的消息时，才会向使用者传送内容。何不在只有新的事发生时再发消息呢?为什么要每天出版内容呢?不过传统媒体公司可不是这样运作的。

请您预测一下未来会出现哪些变化?

不出几年，中国制造的低价安卓平板电脑会到处都是，就像USB或CD一样普遍。之后还可能会搭配电影或书本赠送，而且就放在家里的水果盘上，我们想在看电子邮件时，不会再用电脑，而通过平板电脑就够了。

平板电脑将会是很便宜的移动设备，Google的安卓系统将成为一个开源系统，这样三星、诺基亚甚至苹果设备都可在硬件上使用安卓系统。苹果产品很重要，因为人们喜欢苹果的产品。

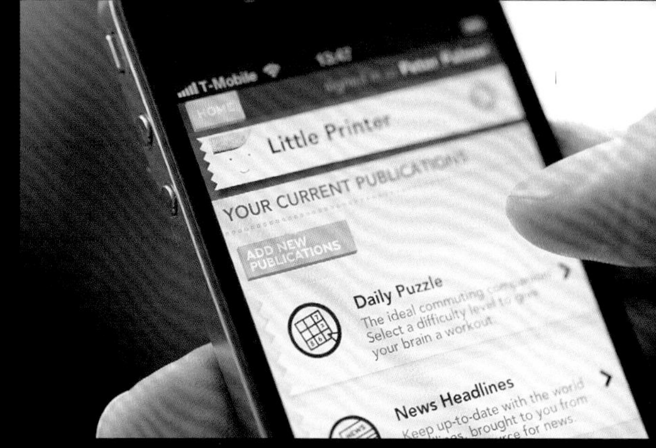

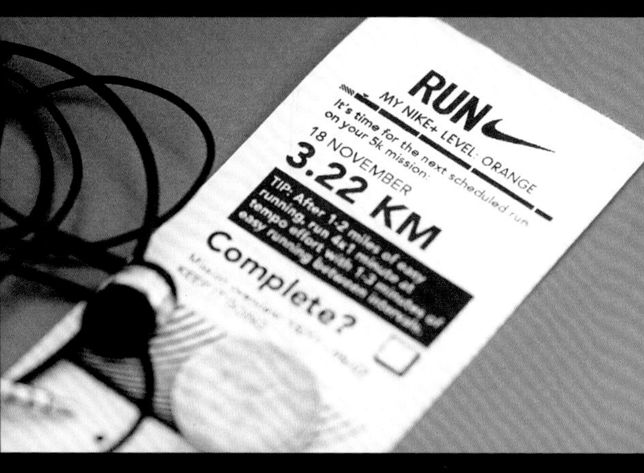

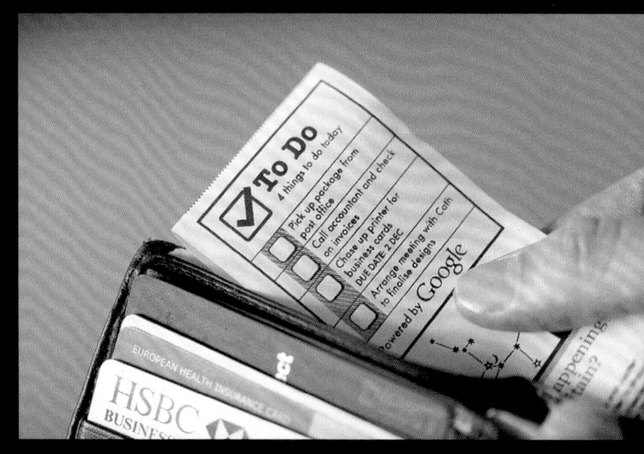

这份热感式终端打印机能把订阅的各种网站内容，全用这台小小打印机打印出来。这是用户的个人数据和内容，使用方便，通过家中的宽带就可以连接网络。

文化与行为转变

市场营销、出版、广播和广告之间的界限已变得模糊。现如今人们在日常生活中接触使用的媒体越来越多，经常一边看电视一边翻手机，或者一边上网一边听音乐。尼尔森图书调查公司的市场营销数据分析指出，从全球网络流量统计数字来看，人们比以往更希望获得更多资讯。

记者可在社交媒体上分享想法，编辑不再像以前一样能够掌控报刊读者的线性阅读体验了。渐渐地，编辑需越来越多元的方式来宣传推广出版物内容。纸质刊物与数字刊物的界限会更加模糊，因此杰里米・莱斯利表示：

> “现在面对纸质出版和数字出版，我们需要了解其间的差异是什么?编辑设计师需要了解制作小册子、发推特、开博客以及制作杂志之间的种种区别。问问自己，我们的声音在不同渠道会如何消失?随着互联网和数字技术的发展，出版和发表见解现如今已经不再是媒体人士的专利。”

旅游文化杂志*Monocle*就是一个例子。这份杂志一直以采用创新的方法，模糊广告与刊物内容的界限而闻名。2011年，编辑泰勒・布吕莱（Tyler Brülé）在彭博电视台及网络广播电台推出电视节目*Monocle 24*，在两个频道里都致力于为读者提供更亲切的体验，这也反映出泰勒・布吕莱对读者群的深刻了解，每个月下载量达到250万次。广播节目使用了类似BBC国际频道的广播风格，让品牌离读者更近，更有怀旧色彩。

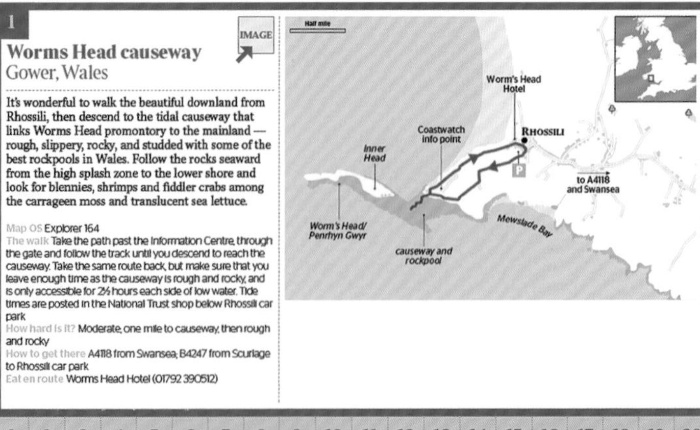

1

Worms Head causeway
Gower, Wales

It's wonderful to walk the beautiful downland from Rhossili, then descend to the tidal causeway that links Worms Head promontory to the mainland — rough, slippery, rocky, and studded with some of the best rockpools in Wales. Follow the rocks seaward from the high splash zone to the lower shore and look for blennies, shrimps and fiddler crabs among the carrageen moss and translucent sea lettuce.

Map OS Explorer 164
The walk Take the path past the Information Centre, through the gate and follow the track until you descend to reach the causeway. Take the same route back, but make sure that you leave enough time as the causeway is rough and rocky, and is only accessible for 2½ hours each side of low water. Tide times are posted in the National Trust shop below Rhossili car park
How hard is it? Moderate, one mile to causeway, then rough and rocky
How to get there A4118 from Swansea; B4247 from Scurlage to Rhossili car park
Eat en route Worms Head Hotel (01792 390512)

经济改变与全球新兴市场

虽然“金砖四国（BRIC，指巴西、俄罗斯、印度和中国）”的市场对大型媒体业者来说是新市场，但这些地区的新闻自由和英国、美国不太一样。这会影响《纽约时报》和BBC之类的新闻机构，他们很难在这些地方赚到钱。正如伯格公司的杰克・舒尔茨指出：

> “美国大型媒体公司对中国市场很感兴趣，因为规模差异。在中国，如果大型媒体能拥有百分之五的市场占有率就相当可观，足以支撑整个商业模式。这就是媒体集团想在中国市场推出媒体平台的最大原因。同样地，《卫报》也想进军美国市场，因为可以延续英国的商业模式。在非洲，

短信很重要，因为只要能建立连接（无论是社交或任何连接），都能在社会或经济成长时分一杯羹。斯特夫·马格达林斯基（Stefan Magdalinski）就在非洲做生意，通过短信来经营通讯黄页服务（Yellow Pages），非常受欢迎。他希望利用短信给人们带来信息和数据。”

在缺乏电力设施的地方，要解决电脑或移动设备充电的问题不难，只要用简单的太阳能板就可以。问题在于，偏远地区的网络覆盖率低，没办法下载数字刊物，所以这个问题要等到网络普及后才能突破瓶颈。但是网络公司要先保证基础建设的投资能有回报，才会在偏远地区搭建网络。

使用社交媒体的好处

下一代的出版人依然需要讲述那些有价值的故事，但他们会很自然地选择把社交媒体作为输出工具之一。不久的将来会不会有下一项突破技术或者产品来取代纸质印刷产品，尚未可知。不过可以确定的是，需要编辑与设计师在读者喜欢的媒体上讲述故事，传播内容。*RealSimple*的创意总监珍妮特·佛罗里克（Janet Froelich）举了下面这个例子：

“*Real Simple*杂志是第一本在Pinterest上获得超过10万粉丝关注的杂志，我们的大部分照片都会被读者复制到自己的博客里，发布或者在收藏夹里收藏。我们的照片与艺术团队会拍摄能产生视觉刺激、吸引读者的照片，并且发布到Instagram上，通常每张照片会得到成百上千的“赞”。这些照片帮助我们扩大了读者群，让我们能与读者交流，并展示了我们的愿景与才华。我们一直致力于实现纸质版、数字版与社交媒体的无缝整合。”

读者不再是安安静静地阅读内容，而是会移动阅读或者一心多用，在《泰晤士报》的“野外夏日散步（Wild Summer walks）”专题报道中，App版本加入了互动元素。读者可将内容当作散步时的指引，例如沿着威尔斯的海岸线散步，读者可以根据自己的所在位置来和内容互动。

*Real Simple*杂志读者常在厨房中使用苹果平板电脑版本来查阅食谱。内容是互动的，读者可以跟着步骤逐步去做。读者还会同时使用其他媒体，例如电视、电子邮件、社交媒体等。人们阅读杂志的方式在不断进化。

附录 纸质印刷演变史

公元105年 中国发明造纸术。

公元704—751年 中国出现雕版印刷品。

公元868年 中国以雕版印刷而成的《金刚经》是有确切日期的印刷品。

15世纪初 除了抄经僧侣，开始有专业作家加入书籍撰写的行列。此时商业发展、教育普及，欧洲中上阶层有了阅读更多书籍的机会。巴黎的作家组成公会，出版业因此形成。

1450年 德国美因茨的金匠约翰内斯·古登堡（Johannes Gutenberg）发明活字（也称为“铸字”或“热式铸字”），5年后印刷了180本古登堡圣经。

1457年 德国纽伦堡印刷发行第一份报纸《公报》。约翰·福斯特（Johann Fust）与彼得·肖（Peter Schoffer）制作的《美因兹诗篇》被认为是史上最早的彩色印刷品。

1476年 威廉·卡克斯顿（William Caxton）从德国科隆返回英国，带回许多活字，并在伦敦威斯敏斯特装设第一台印刷机。在此之前，他已在布鲁日印制出第一本英文书《特洛伊历史故事集》。

1486年 第一本有彩色配图的英文书在英国圣奥尔本斯发行。

1494年 文字编排者、教师与编辑阿尔杜斯·马努提乌斯（Aldus Manutius）在意大利威尼斯成立阿尔丁出版社（Aldine）。

1500年 全球约有35000种书，共完成1000万册印刷。

1501年 弗朗切斯科·格里夫（Francesco Griffo）设计的科体“Italic”首次在《维吉尔》诗集中使用，由马努提乌斯的阿尔丁出版社出版。

1588年 英国人蒂莫西·布莱特（Timothy Bright） 发明速记法。

1605年 第一份定期发行的周报在斯特拉斯堡出现。

1622年 英国报纸之父纳撒尼尔·巴特（Nathaniel Butter）在伦敦出版*Weekly Newes*，是英国第一份周报。

1650年 全球第一份日报在德国莱比锡诞生。

1633年 德国出版全球第一份杂志*Erbauliche Monaths Unterredungen*。

1690年 美国第一份报纸《海内外公共事件报》在马萨诸塞州波士顿发行，但只短暂发行就被勒令停业。

1702年 英国第一份日报《每日新闻》发行。

1703年 俄国彼得大帝创办《圣彼得堡新闻》。

1709年 英国通过《著作权法》。伦敦推出第一份主流杂志*Tatter*。

1714年 伦敦的亨利·米尔（Henry Mill）获得打字机专利。

1719年 德国雕刻师恩格尔弗·雅各布·布（Jakob Le Le Blon）获英王乔治一世特许，复制全彩图画，这是现代四色铜版印刷的基础。

1731年 英国出版第一本现代杂志《绅士杂志》。

1741年 本杰明·富兰克林（Benjamin Franklin）打算出本第一本美国杂志*General Magazine*，却在出刊前三天被*American Magazine*抢先。

1764年 法国的皮埃尔·福涅尔（PierreFournier）发明点数系统，测量字号大小。之后经过弗朗索瓦·迪多（Francois Didot）改良，建立全球统一的字号测量系统。

1784年 美国第一份日报《宾州晚邮报》出刊。

1785年 约翰·沃尔特（John Walter）在伦敦创办《每日环球记录报》，三年后改名为《泰晤士报》。

1790年代 德国巴伐利亚的艾罗斯·塞纳菲尔德（Alois Senefelder）发明平版印刷（Lithography），印刷流程不再需要雕版，因而简化。

1791年 伯恩（W. S.Bourne ）推出英国第一份周日报《观察家报》。

1814年 《泰晤士报》开始采用早期的轮转印刷，每小时印刷1100份。公元1830年，理查德·马奇霍（Richard MarchHoe）改良轮转印刷术后，全球才普遍运用这种技术。他将轮转印刷机改良到每小时可印2500页，到1847年还让印刷机扩增到5个滚筒。

1828年 《女性杂志》创刊，成为第一份受欢迎的美国女性杂志。

1842年 赫伯特·英格拉姆（Herbert Ingram）与马克·莱蒙（Mark Lemon）在英国创办《伦敦新闻画报》，采用木版画与雕刻工艺，促成促进了插图出版物的发展。

1844年 泰国出版第一份报纸《曼谷记录者》。

1845年 *Scientific America*在美国创刊至今从未停刊，是美国最历史悠久的杂志。

同年，平装书作为报纸增刊被引入美国（比德国晚四年），之后已出版的书籍就会用平装制作开本较小的再版书。

1850年 海德堡第一台印刷机在德国西南部弗兰肯塔尔的帕拉廷市完工，由安德烈亚斯·哈姆（Andireas Hamm）制造。

1851年 《纽约时报》创刊，价格为一美分。

1854年 《费加罗报》于法国巴黎发行。

1856年 《新奥尔良克里奥尔日报》上市，是第一份非裔美国人报纸。

1867年 日本第一份杂志——《西洋杂志》出版。

1874年 伊利诺伊州的雷明公司（E.Remington and Sons）制造出第一台商用打字机，这是七年前由威斯康星州的记者克里斯托弗·莱瑟姆·肖尔斯（Christopher Latham Sholes）发明的，只有大写字母，但有QWERT键盘。第二年改良后包含了小写字母。

1875年 平版胶印技术问世（平滑表面而不是凸版，印到蚀刻的金属板上）。

1878年 美国发明家威廉·拉瓦莱特（William A. Lavalette）获得印刷机专利，可大幅提升印刷品质，尤其是可读性。苏格兰的弗雷德里克·威克斯（Federick Wicks）发明铸字机。

1886年 奥特马尔·梅根塔勒（Ottmar Mergenthaler）发明莱诺铸排机（Linotype），可整合键盘、字模箱与铸字机，让铸字工每小时可以排17000字，可以按键盘制作铅条，而原本结合起来的字模可重新复位再利用。

1900年 美国约有1800种杂志出版，报纸发行量每日超过1500万份。

1903年 美国的伊拉华盛顿·鲁贝尔（IraWashington Rubel）与德国的卡斯帕·赫尔曼（Casper Hermann），分别开始使用平板印刷机。通过引进由华盛顿·勒德洛和威廉里德在伊利诺伊州芝加哥市开发的勒德洛排字机，排字得到了进一步的改进。

1911年 伊利诺伊州芝加哥的华盛顿·勒德洛（Washington Ludlow）与威廉·里德（ William Reade）推出勒德洛排字机，进一步改良排字过程。

1912年 *Photoplay*杂志在美国发行，成为第一份电影影迷的杂志。

1917年 《纽约时报》首次出现社论页。

1923年 《时代》杂志在美国创刊。

1933年 *Esquire*创刊，为美国第一份男性杂志。

1934年 纽约《时尚芭莎》聘请阿列克谢·布罗多维奇制作版面，带来了赏心悦目、充满动感的视觉效果。

1936年 艾伦·莱恩（Allen Lane）的企鹅出版社（Penguin Press）在英国重新引入平装书。美国时代公司的亨利·卢斯（Henry Luce）创立新闻摄影杂志《生活》，称霸美国新闻市场达40年之久，每周销量超过1350万份。

1941—1944年 新闻纪实摄影开始揭露第二次世界大战的画面，杂志推出以纪实照片为主的报道。有些照片被禁，遭到质疑。

1945年 *Ebony*是第一份专为非裔美籍市场设计的杂志，由美国人约翰·强森（Jonn H. Johnson）创刊。

1953年4月3日 第一份《电视指南》在美国十个城市的书报摊上市，发行量为156万份。*Playboy*杂志创刊，封面人物为玛丽莲·梦露。

1955年 俄亥俄州哥伦布的巴特尔纪念研究所（Battelle Memorial Institut）开发出干涂布纸。

美国*Esquire*品牌感强烈，因此率先在封面上大量使用极少的封面文案。当时只有少数刊物能使用彩色。

1956年 IBM制作出第一个硬盘驱动器。

1958年 亨利·沃尔夫（Henry Wolf）成为《时尚芭莎》的艺术总监，并推动图像与文字编排的整合。

1962年 英国全国性报纸《周日泰晤士报》推出全彩杂志增刊，由迈克尔·兰德（Michael Rand）设计。

1964年 美国统计数据显示，81%的成年人有阅读日报的习惯。

1965年 德国大型出巨头施普林格（Springer）推出*Twen*青少年杂志，设计者为威利·弗莱克豪斯，堪称报刊设计的突破之作。英国《每日镜报》推出杂志*Nova*，主编为丹尼斯·哈克特（Dennis Hackett），设计师为戴维·希尔曼（David Hillman）。

1967年 英国启用国际标准书号（ISBN）。《滚石》杂志在美国上市，而在1968年，《纽约客》杂志紧随其后推出，催生出一批地区性杂志。

1969年 安迪·沃霍尔创办了《访谈》杂志。

1971年 全球报纸开始从热式打字的凸版印刷改成平版印刷。

1975年 *Nova*杂志停刊。

1977年 苹果推出Apple II微型电脑。

1980年 瑞士日内瓦欧洲核子研究中心（CERN）的蒂姆·伯纳斯-李（Tim Berners-Lee）初次尝试万维网，开发了一款名为"万事皆可查"的软件程序（源自为他童年时读到的维多利亚时代百科全）。英国的尼克·洛根（Nick Logan）推出《TheFace》杂志。

1981年 《滚石》杂志推出全裸的约翰·列侬和穿着衣服的小野洋子的封面。这张由安妮·莱博维茨（Annie Leibovitz）拍摄的标志性照片，集中体现了杂志在音乐行业的影响力。

1982年 《今日美国》日报发行。这份模仿电视的视觉引导的全彩报纸，以多图为特色，立刻受到欢迎。并利用新技术辅助发行，让最终版本可以在国内各个地点印刷。

1983年 苹果推出了Apple Lisa电脑，引入了新的图形用户界面，使电脑在家庭计算与出版上更简单、更便宜。

1984年 苹果推出 Mac电脑，图形用户界面首次大受欢迎，如今电脑多使用此界面。*Emigre*杂志在加利福尼亚发行，迅速成为数字字体与图像的创意展示中心。

1985年 保罗·布雷纳德（Paul Brainerd）与奥尔德斯开发出第一款桌面出版软件Aldus Pagemaker1.0，供麦金塔电脑使用。桌面出版软件带动新形态的出版出现，让每个人都能掌握设计与编辑的工具。

1987年 Quark X Press软件推出。虽然晚了Aldus Pagemaker软件两年，但很快成为主要的桌面出版软件。

1991年 万维网首次亮相，使用蒂姆·伯纳斯-李（Tim Berners-Lee）的HTML（超文本标记语言），任何人都可以建立网站，分享者从最初的几百人立即飙升到数千万人。

1994年 康泰纳仕在意大利出版了A5手提包大小的*Glamour*杂志 。美国网景公司（Netscape）推出Mosaic浏览器的测试版。俄勒冈州波特兰的沃德·坎宁安（Ward Cunningham）发明了Wiki概念，能够让用户以非线性的协作方式，创造并连接内容网页。

1995年 美国自由派的杂志*Salon*只在线上出版，挑战传统纸质媒体的商业模式。

1997年 《纽约时报》在新闻页面采用彩色照片。

2004年 英国《独立报》从大报版式改为小报版式，同年《泰晤士报》也推出小报版式日报。

2005年 《卫报》改成柏林版式，全彩设计。英国《第一邮报》线上新闻杂志推出。

2006年 Google以16.5亿美元股票收购影片分享网站YouTube。美国报纸协会（Newspaper Association）的报告显示，美国新闻网站吸引了5800万读者。

2007年 英国《金融时报》线上版的广告收益提高了30%。电子书阅读器Kindle上市。

2008年 一份调查显示，前一天阅读过报纸的美国成年人比例为30%。

2009年 全球金融危机之后，报纸销量大幅下滑。美国人山姆·阿普尔（Sam Apple）推出《快速时报》，探索"按需新闻"模式。《泰晤士报》网站每日读者达到75万人。

2010年 苹果平板电脑上市，30天内卖出超过300万台。亚马逊书店宣布，电子书的销售量首次超过纸质书。*Wired*杂志推出苹果平板电脑版本。《泰晤士报》推出有互动功能的平板电脑版本。维基解密发表来自匿名来源的文件，违背业界常规，惹恼新闻媒体。在大型新闻出版商面临存亡之际，网络新闻报道的可靠性因此遭受质疑。

2011年 《卫报》推出平板电脑版本。苹果平板电脑第二版上市，第一年卖出超过1500万台。《纽约时报》宣布，将对网站内容使用者收取费用。新闻国际公司（News International）为《泰晤士报》建立付费墙。

2012年 跨平台杂志*Little White Lies*与*Letter to Jane*已经开发出不同格式和版本，并不再印刷纸质版。

2013年 独立杂志可以利用网络来降低制作成本，于是开始蓬勃发展。在莱韦森调查小组（Leveson Inquiry）调查媒体的不当行为之后，有提议通过皇家宪章，强化英国媒体的自律性，但这项提议受到众多编辑的反对。在伦敦举行的现代杂志大会（Modern Magazine Conference）上使用了"黄金时代"一词，表示在苹果平板电脑上市后三年，出版行业已经恢复正常。大型出版商与小型独立出版社都能利用数字媒体给行业带来的自由。

作者致谢

特别感谢：
所有图片提供者以及接受采访的设计师，尤其是珍妮特·佛罗里克（Janet Froelich）、琼恩·希尔（Jon Hill）、马克·波特（Mark Porter）、莎拉·道格拉斯（Sarah Douglas）、杰瑞米·莱斯利（Jeremy Leslie）、理查德·特里（Richard Turley）与斯科特·达迪奇（Scott Dadich），感谢劳伦斯·金出版公司（Laurence King Publishing）所有成员高超的编辑能力，特别是彼德·琼斯（ Peter Jones）与苏珊·乔治（Susan George），以及玛丽·韦斯特（Mari West）的图片编辑。
感谢中央圣马丁学院平面设计系的所有学生，慷慨分享工作坊中提到的作品： 斯特凡·亚伯拉罕斯团队（Stefan Abrahams' team）、桑德拉·阿杜奎特（Sandra Adukuaite），奥利弗·巴隆（Oliver Ballon）、萨罗米·德赛（Salomi Desai）、伊利亚娜·杜杜耶娃（Iliana Dudueva ）、杰特迈尔·德沃拉尼（Jetmire Dvorani）、乔丹·哈里森-特维斯特（Jordan Harrison-Twist）、约瑟夫·马歇尔（Joseph Marshall）、本·西尔弗敦（Ben Silvertown）团队。
谢谢我的父亲埃迪·柯德威尔（Eddie Caldwell），正是受到身为排版员的父亲鼓励，我才如此热爱出版。也要感谢我的先生约翰·贝尔克纳普（John Belknap），感谢他给我很多建议，并为我送上一杯杯的热茶。还要感谢萨姆（Sam）、埃德（Ed）与黛西（Daisy）的耐心。谢谢凯伦·西姆斯（ Karen Sims）与海布里读书俱乐部（Highbury Book Club）的支持。也要感谢中央圣马丁学院毕业的尤兰达·齐帕特拉（Yolande Zeppaterra）。她在2007年纸质出版的年代，完成了内容优质的《Print, Web & App：编辑设计人该会的基本功一次到位》第一版。